INTRODUCTION TO
PHYSICAL
SYSTEM
DYNAMICS

McGraw-Hill Series in Mechanical Engineering

Jack P. Holman, *Southern Methodist University*
Consulting Editor

Anderson: *Modern Compressible Flow: With Historical Perspective*
Dieter: *Engineering Design: A Materials and Processing Approach*
Eckert and Drake: *Analysis of Heat and Mass Transfer*
Ham, Crane, and Rogers: *Mechanics of Machinery*
Hartenberg and Denavit: *Kinematic Synthesis of Linkages*
Hinze: *Turbulence*
Hutton: *Applied Mechanical Vibrations*
Jacobsen and Ayre: *Engineering Vibrations*
Juvinall: *Engineering Considerations of Stress, Strain, and Strength*
Kays and Crawford: *Convective Heat and Mass Transfer*
Lichty: *Combustion Engine Processes*
Martin: *Kinematics and Dynamics of Machines*
Phelan: *Dynamics of Machinery*
Phelan: *Fundamentals of Mechanical Design*
Pierce: *Acoustics: An Introduction to Its Physical Principles and Applications*
Raven: *Automatic Control Engineering*
Rosenberg and Karnopp: *Introduction to Physical System Dynamics*
Schenck: *Theories of Engineering Experimentation*
Schlichting: *Boundary-Layer Theory*
Shames: *Mechanics of Fluids*
Shigley: *Dynamic Analysis of Machines*
Shigley: *Kinematic Analysis of Mechanisms*
Shigley: *Mechanical Engineering Design*
Shigley: *Simulation of Mechanical Systems*
Shigley and Uicker: *Theory of Machines and Mechanisms*
Stoecker and Jones: *Refrigeration and Air Conditioning*

INTRODUCTION TO PHYSICAL SYSTEM DYNAMICS

Ronald C. Rosenberg

Professor of Mechanical Engineering
Michigan State University

Dean C. Karnopp

Professor of Mechanical Engineering
University of California, Davis

McGraw-Hill Book Company

New York   St. Louis   San Francisco   Auckland   Bogotá   Hamburg
Johannesburg   London   Madrid   Mexico   Montreal   New Delhi
Panama   Paris   São Paulo   Singapore   Sydney   Tokyo   Toronto

This book was set in Times Roman by University Graphics, Inc.
The editors were Rodger H. Klas and Susan Hazlett;
the production supervisor was Dennis J. Conroy.
The drawings were done by J & R Services, Inc.
R. R. Donnelley & Sons Company was printer and binder.

INTRODUCTION TO PHYSICAL SYSTEM DYNAMICS

1 2 3 4 5 6 7 8 9 0 DOCDOC 8 9 8 7 6 5 4 3

ISBN 0-07-053905-7

Library of Congress Cataloging in Publication Data

Rosenberg, Ronald C.
    Introduction to physical system dynamics.

    (McGraw-Hill series in mechanical engineering)
    Includes bibliographical references and index.
    1. Dynamics.   2. Automatic control.   3. Vibration.
I. Karnopp, Dean.   II. Title.   III. Series.
TA352.R67   1983   620′.0042   82-17233
ISBN 0-07-053905-7

CONTENTS

Part 3   Modeling of Engineering Systems

PREFACE

We believe that the underlying motivation of all good books is a love of the subject, and this effort is no exception. Both authors have been actively involved in the study and the development of system dynamics for a number of years. Between us we have taught system dynamics at various levels in university and industry, both here and abroad, have written and consulted about the theory and applications, and have developed computer programs to help in system design.

This book is motivated also by the growing acceptance by engineering professors of the importance of system dynamics as a bridge between the mathematics of differential equations and the physics of engineering systems. Furthermore, system dynamics is recognized widely as a sound basis for the important engineering subjects of automatic control and vibration analysis. Since the subject is so widely applicable and so rich, we have chosen to concentrate on physical-system dynamics, stemming from power and energy exchanges between components, as distinct from information systems, such as computer networks, economic and management systems, or social systems.

Although many books treat the dynamics of physical systems, this one is unique in the degree to which it is able to exploit the underlying similarities between seemingly diverse components, elements, and systems, based on their energy interactions. The result is a comprehensive approach to modeling using bond graphs that yields considerable insight and affords clarity of thought to those who study it successfully. We have taken care to develop close ties between the physical and mathematical bases of the material and standard methods by which they are analyzed and studied, since there are powerful and widely used techniques that one well-versed in system dynamics ought to know.

The number of people who might profit from study of this book is large and the subject matter is widespread and rather diverse, so we have organized the material to emphasize the coherence of the subject. We have also allowed a large measure of flexibility in the choice of topics to be covered in a course based on the book. The plan is the following. In Parts One and Two the subject is introduced (Chap. 1) and developed in application to low-order, fairly simple mechanical and electrical systems (Chaps. 2 and 3). Then the models are converted into a standard-equation form (Chap. 4) and the dynamic behavior is found by Laplace transforms (Chap. 5). The helpful roles to be played by block diagrams, transfer functions, and computer simulation of dynamic systems are discussed in Chaps. 6 and 7. This completes a cycle from the engineering problem statement to modeling, equation formulation, equation solution, and interpretation of results. Thus the entire process typically involved in engineering design is covered.

Part Three, consisting of six chapters, considers the modeling of a number of physical areas in some detail. Here one is free to pick and choose, emphasizing either the breadth of coverage or depth in selected areas. The important topics of energy and power transducers and amplifiers and instruments are treated in Chaps. 12 and 13, respectively.

Part Four returns to methods for predicting dynamic response of systems and puts the subject on a more general footing. Building upon matrix methods and introducing some nonlinear concepts, Chaps. 14 to 16 enable one to extend previous knowledge to a wide class of models. This part provides an excellent basis for further work in modern control, vibration analysis, and finite-element techniques.

Finally, Part Five shows how the subjects of automatic control and vibration analysis can be developed as natural extensions of the system-dynamics core built earlier. In addition, some examples are given in Chap. 17 to suggest how the basic modeling methods are extended to include more complex and realistic engineering devices and components.

The flexibility of the organization permits the design of a number of course variations that can make good use of this book. A few examples should suffice to give some idea of the possibilities. Several of these suggestions are based on the authors' experiences with the material at their own universities. A typical first course in system dynamics at the sophomore or junior level can be organized mainly from Parts One and Two. In a quarter course there is usually enough time to select some additional chapters, which could come from Part Three to develop the modeling base. In a semester course one could select among applications in Part Three and response methods in Part Four. For a senior or graduate course one would probably move quickly through Chaps. 2 to 4, perhaps skim Chaps. 5 to 7, include some portion of Part Three, and cover Part Four in detail. If the system-dynamics course is a prerequisite to a control or vibration course, a bridge is provided by Chaps. 18 and 19. Clearly, many variations are possible. It has been our experience that as the instructor's familiarity with bond-graph methods increases, the number of topics which can be covered—and understood by the students—increases dramatically.

While the material covered in the book does not depend upon the use of computers, any available computing capability can be put to good use. The discussions in Chaps. 7 and 16, in particular, would benefit from some use of computing in conjunction with their study. The very nature of system dynamics lends itself to the wonderful experience of making a design decision, testing it, observing the results, and trying again.

Several system-dynamics courses have included a computer laboratory often based on the use of ENPORT. Many of the systems in the problems have been used for simulation studies in computer laboratories. At every stage of the book we have taken pains to give discussions which assist in the computer study of dynamic systems or in interpreting the results of computer studies.

We would like to thank the many other people who are responsible for bringing this effort to fruition, among them our editors, Diane Heiberg and Rodger Klas, and our typist, Suzanne Schueler. We also thank Professor H. M. Paynter, the discoverer of bond graphs; Dr. J. U. Thoma, a peripatetic disseminator of the word; Dr. R. R. Allen, and Professors J. J. von Dixhoorn and D. L. Margolis, all of whom have made valuable contributions to the spread of bond-graph techniques. Our particular thanks go to our students, many of whom caught our enthusiasm, and all of whom at least listened politely. They proved the truth of the saying that teaching is its own reward. Finally, although we feel we are experts in bond-graph methods, we must acknowledge that our bonds to Karen K. and Judy R., which are undeniably physical and dynamic and even involve power and energy, are of a unique sort which do not fit even the generalized scheme of this book.

Ronald C. Rosenberg
Dean C. Karnopp

ONE

AN OVERVIEW OF SYSTEM DYNAMICS

In the single chapter in Part One we introduce the *concept* of a dynamic-system model. In addition, we present some of the various forms mathematical models take in engineering practice. We hope to arouse your interest in the remainder of the book, where we present modeling, analysis, and simulation methods in considerable detail.

INTRODUCTION TO DYNAMIC-SYSTEM MODELS

If you believe that the only constant thing in life is change, you should be interested in the science of dynamics, which concerns itself with changes over time. If you think we should understand the behavior of a complete system as distinct from the individual behavior of its pieces or components, you should be interested in system dynamics. If you are studying engineering or physical science, you certainly can learn something useful from this book, which treats the sorts of physical systems which can be designed by engineers to serve some purpose.

In introducing the major concepts and techniques necessary to model, analyze, and simulate physical systems we expect that you have been exposed to differential equations and the engineering sciences concerned with mechanics, electric circuits, fluid mechanics, and heat transfer, at least to some degree. While some of the methods we present have been applied to social, economic, or biological systems, we shall emphasize the similarities between physical systems by means of a unified treatment of energy and power. Thus we hope that your knowledge and intuition in one field can be used to strengthen your understanding of other, analogous fields.

As you read on, you may be impressed to see what a variety of physical systems and techniques for studying them are presented, but you may also be disappointed to find that some particular types of systems and analysis methods are not included. In our selection we have tried to be guided by one of America's first scientists, Benjamin Franklin, who wrote:

It would be well if students could be taught *every Thing* that is useful, and *every Thing* that is ornamental: But Art is long and their time is short. It is therefore propos'd that they learn those Things that are likely to be *most useful* and most ornamental, Regard being had to the several Professions for which they are intended.†

---

† Cited in *Franklin Inst. News*, vol. 45, no. 1, p. 2, January–February 1981. Franklin was paraphrasing Hippocrates (460–377 B.C.), *Ars longa, vita brevis,* or "Art is long, life is short," *Aphorisms*, sec. I,1. The idea that education should be useful was Franklin's own, however.

1-1 WHY IS DYNAMICS IMPORTANT?

Most engineering systems are designed on the basis of a steady, or static, state. For example, the sizes of pumps, boilers, and turbines for power plants can be chosen on the basis of the steady flows of material and power necessary to produce the maximum output. Similarly, the main dimensions of buildings and bridges are determined by the static loads which must be supported. However, one must not be deceived into thinking that only steady or static situations need to be considered. The power demanded of a plant normally varies with time in a partly predictable way, and the transient operations of start-up and shutdown must be considered. Similarly, buildings and bridges are subjected to dynamic loads from winds and earthquakes, and their response to these excitations can be highly undesirable if the system dynamics of the structures have not been properly taken into account.

Another reason for understanding the dynamics of a system is the ever-increasing use of automatic control to improve the performance of engineering systems. The revolution in electronics has allowed intelligent controllers to be applied to virtually every type of engineering system, and it is impossible to design a control system without knowing something about the dynamic response of the system to be controlled. A major impetus to the study of multidisciplinary system dynamics has come from the field of automatic control, since systems typically involve a mixture of electrical, hydraulic, pneumatic, and mechanical systems. Study of this book will give you competence in dealing with such systems.

1-2 WHAT IS A MODEL?

Models are simplifications of reality. A model airplane resembles a real airplane in shape and perhaps color but not in size or complexity. Physical models are often used in engineering when it is impractical to experiment with real systems. A model airplane might be used in a wind tunnel, for example, to predict drag loads on the actual aircraft. In this book, however, we deal with what are usually called mathematical models of systems, which resemble the real systems they represent more abstractly than physical models. The idea is that the time history of important variables, e.g., temperatures, pressures, voltages, forces, and displacements, in the real system should also be produced by the mathematical model even before the real system is constructed. The mathematical model may yield the variable time histories by analytical solution of its equations, but nowadays it is more likely that a digital computer will be used to simulate the behavior. Our main efforts will be applied to showing how to construct models useful for predicting dynamic-system response and to methods for obtaining insight into their behavior.

Just as physical models are used for experiments so that a prototype system can be constructed with well-grounded expectations of satisfactory behavior, so too are mathematical models used. It is usually much easier to evaluate changes in a dynamic-system design by changing a mathematical model than by modifying a prototype system. To some extent, computer-simulated experimentation has supplemented or even replaced hardware experimentation in almost every field. When accurate mathematical models can be constructed with confidence, systems can be designed for optimum performance with a greatly reduced need for cut-and-try experimentation, and this pattern, in turn, greatly improves the productivity of the engineer.

1-3 WHAT FORMS DO MODELS TAKE?

Although you may imagine that mathematical models always involve equations directly, this is not quite true. As you will see, mathematical models must involve precise relationships, which may be implied by standard diagrams or incorporated into computer programs instead of being written out in direct equation form. In engineering practice the equations themselves may be so complex that direct study yields little or no insight into the dynamic behavior, especially for design purposes. Since we are interested in the behavior implied by the equations and in the results of computer simulation, if we can avoid the drudgery of writing out the equations themselves, so much the better. Nevertheless, equations are something that some people like to contemplate, and to help you understand models you will be asked to write equations for fairly simple models. This way you will be able to understand how more complex models work, even when the equations would be too complicated for human analysis.

In this introductory chapter we describe different forms in which a mathematical model may appear. The examples to be studied are shown in Fig. 1-1 and are kept simple enough for you to follow the derivations without knowing the techniques involved. The methods will be explained at length later so that you can apply the techniques to more complicated and realistic cases.

Schematic Diagrams

You probably recognize the diagrams shown in Fig. 1-1 although they are actually quite sophisticated mathematical models. The symbols for resistance and capacitance in the circuit (Fig. 1-1 a) do stem from the original ways in which resistors and capacitors were made with coils of high-resistance wire and parallel-plate capacitors,

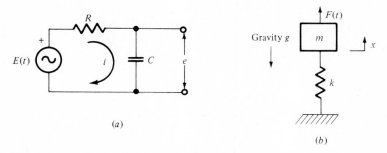

(a)

(b)

Figure 1-1 Schematic diagrams of two elementary system models: (a) electric circuit, (b) mechanical oscillator.

but many modern resistors and capacitors do not resemble these symbols. In addition, a real filter made from elements in the diagram of Fig. 1-1*a* would perform as the idealized mathematical model predicts only for a limited range of frequencies and amplitudes of voltage and current. The real system would have some inductance, for example, which is missing in the model. Also, if $E(t)$, the forcing voltage, were too high, the circuit would burn out. This effect is also omitted from the model. Such approximations are typical. Any model has its limits, but within those limits the mathematical model can predict what a real system will do.

Figure 1-1*b* seems to represent a lump of mass supported on a spring in a gravity field and subjected to a force $F(t)$. Think about it. If you put a lump of lead on top of a spring it might easily fall over. You could provide some guides to hold the mass in place, but that would add some friction to the system and Fig. 1-1*b* includes no friction. The schematic diagram may be intended to represent a machine tool on its vibration mounts or even an airplane on its landing gear. A skilled engineer may well decide on the stiffness of a spring in the vibration mount or in a landing gear on the basis of the model shown schematically in Fig. 1-1*b*, but no one would claim that this model represents the system in question beyond approximating part of its behavior.

A model is only a model, and its strength lies in simplicity; but therein lies its weakness also. You will see, after reading this book, that a whole sequence of more and more sophisticated models can be constructed. The appropriate model for a given situation is just complicated enough to answer your questions. An overcomplicated model is almost as bad as an oversimplistic one, in that it wastes time and effort and may introduce errors. Our approach will show you how to add or subtract certain effects in your model in an organized way to achieve the desired effects.

Equations of Motion

Mathematicians and many engineers or applied scientists feel most comfortable if their models can be expressed in equation form. For dynamic systems of the type we shall study these turn out to be ordinary differential equations.

For anyone who has studied electric circuits or mechanics it is a straightforward task to turn the schematic diagrams of Fig. 1-1 into equations. For mechanical systems, these are properly equations of motion, but for others there is motion only in an abstract sense. Also, there are several forms for the equations. For example, if we see a single loop current in Fig. 1-1*a*, we could use Kirchhoff's voltage law to add the voltage drops around a closed path to zero. Since the source drop is $-E(t)$, the resistor drop is Ri, and that of the capacitor is

$$\frac{1}{C} \int^{t} i \, dt$$

we can write

$$-E(t) + Ri + \frac{1}{C} \int^{t} i \, dt = 0 \tag{1-1}$$

We promised an ordinary differential equation and seem to have an integral equation, but if we define the charge q on the capacitor to be

$$q \equiv \int^t i \, dt \tag{1-2}$$

then

$$i = \dot{q} \tag{1-3}$$

is the current, (where $dq/dt \equiv \dot{q}$) and Eq. (1-1) can be rewritten as

$$-E(t) + R\dot{q} + \frac{1}{C} q = 0 \tag{1-4a}$$

or

$$\dot{q} = \frac{-1}{RC} q + \frac{E(t)}{R} \tag{1-4b}$$

Equation (1-4b) is an equation for charge, but maybe we really want to study the voltage e across the capacitor. Then we say

$$e = \frac{q}{C} \tag{1-5}$$

and Eq. (1-4a) can be manipulated to read

$$\dot{q} = C\dot{e} = \frac{-1}{R} \frac{q}{C} + \frac{E(t)}{R}$$

or

$$\dot{e} = \frac{-1}{RC} e + \frac{E(t)}{RC} \tag{1-6}$$

which is a first-order ordinary differential equation for the output voltage e in terms of the voltage forcing function $E(t)$.

The standard way to write the equation of motion for the model of Fig. 1-1b is to write Newton's law for the mass

$$m\ddot{x} = \Sigma F \tag{1-7}$$

where ΣF signifies the sum of all forces on the mass in the x direction. In Fig. 1-1b, we have the force input $F(t)$, the force of gravity $-mg$, and the spring force kx. (This example is so simple we can get along without a free-body diagram but we should really draw one, too.) Then

$$m\ddot{x} = +F(t) - mg - kx$$

or

$$m\ddot{x} + kx = F(t) - mg \tag{1-8}$$

Although many people might regard Eq. (1-8) as the mathematical model, since anyone who understood the meaning of Fig. 1-1b could arrive at Eq. (1-8), Fig. 1-1b can be regarded as the model just as well. As we shall see, not all schematic diagrams are translated into equations as easily as those in Fig. 1-1. Further, a number of equivalent equations can represent the same model, as Eqs. (1-1), (1-4b), and (1-6) demonstrate.

Block Diagrams

A mathematical model is often usefully represented in a pictorial form known as a *block diagram*. In a block diagram signals representing system variables flow along lines connecting blocks, which perform operations on the signals to enforce the equations of motion. These diagrams are useful, particularly in programming analog computers, which represent all signals as voltages and perform many types of block functions, e.g., summation and multiplication, electronically.

A simple way to create a block diagram is to start with an equation of motion, such as Eq. (1-6). It states that de/dt is just a sum (with signs) of e multiplied by the constant $1/RC$ and $E(t)$ also multiplied by $1/RC$. For such linear systems we need blocks which add, multiply by a constant, and integrate in time.

Figure 1-2a is an almost obvious diagram of Eq. (1-6). How do we get $\dot{e}$? We add $E(t)$ to $-e$ and multiply by $1/RC$, a constant. That is what the symbols in the left-hand part of the figure mean. The $\int dt$ block means that we integrate $\dot{e}$ to get e and then send the e signal back around following the arrows to form e. Since the blocks containing $+$, $1/RC$, and $\int dt$ can be realized electronically, this block diagram is essentially an analog-computer wiring diagram. Later we shall find ways to proceed from a schematic diagram directly to a block diagram without writing equations.

The block diagram of Fig. 1-2b simply shows that $\ddot{x}$ can be found by multiplying $F(t) - mg - kx$ by $1/m$. It is completed by integrating $\ddot{x}$ twice to find x, thus completing the picture. If you do not believe this, write Eq. (1-8) as

$$\ddot{x} = \frac{1}{m}[F(t) - mg - kx]$$

and see if this does not agree with Fig. 1-2b.

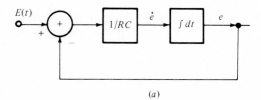

(a)

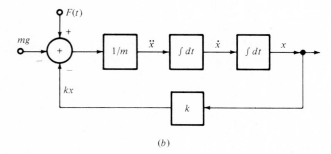

(b)

Figure 1-2 Block diagrams corresponding to the models of Fig. 1-1.

Block diagrams are very general, since it is possible to indicate all kinds of non-linear operations to be performed on the signals entering a block. In our examples only the simple linear operations of summation, multiplication by a constant, and integration in time have been illustrated. Although various professional engineering groups have proposed standard symbols for block diagrams, during your career you are certain to encounter various symbols for the same operation, and most practitioners develop a preference for their own set of partly unique symbols.

Signal-Flow Graphs

A simplified type of diagram is often used, particularly for linear systems. The idea is to use small circles to stand for the output signal and also to imply summation of the inputs. The direction of signal flow is indicated by an arrow, next to which the operation is indicated. If a signal is to be multiplied by a constant, just the constant is written; if integration is required, the symbol $1/s$ is placed next to the arrow. This notation comes from considering the Laplace transforms of the signals, discussed in Chap. 5. Don't worry about it now; just compare Figs. 1-2 and 1-3 and see whether you can perceive that exactly the same model is represented in two ways. It is probably a matter of taste whether you prefer the block diagrams of Fig. 1-2 or the signal-flow graphs of Fig. 1-3 for the simple linear systems of our examples.

Bond Graphs

Most of the mathematical models in this book will be expressed initially in terms of *bond graphs,* which have much in common with the various forms of models discussed above. Figure 1-4 shows bond graphs that are the equivalents of the schematic diagrams of Fig. 1-1. As you can see, bond graphs are more abstract than schematic physical diagrams, and at this point you are entitled to ask why you should learn such a modeling language when circuit diagrams and mechanical sketches, together with Kirchhoff's and Newton's laws, suffice for writing equations. There are several good answers to such a question.

First, many system models are less easily represented schematically than electric circuits and lumps of mass connected by springs. Bond graphs use the same (rather small) number of symbols or elements to represent all applicable types of systems. For simplicity, we restrict ourselves initially to electrical and mechanical systems,

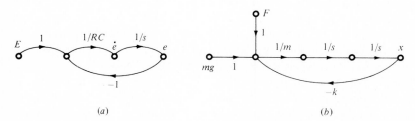

(a)   *(b)*

Figure 1-3 Signal-flow graphs corresponding to the models in Figs. 1-1 and 1-2.

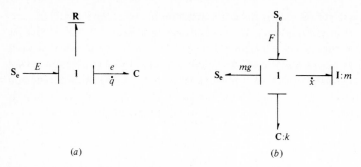

(a)   (b)

Figure 1-4 Bond graphs corresponding to the schematic diagrams in Fig. 1-1.

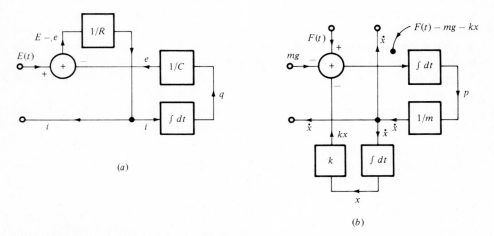

(a)

(b)

Figure 1-5 Block-diagram equivalents of the bond graphs in Fig. 1-4.

but when the basic ideas have been developed, we shall extend the bond-graph models to include hydraulic, thermal, and combination systems, for which alternative precise schematic diagrams to represent the model are hard to produce. A hydraulic piping diagram, for example, may resemble a circuit diagram but rarely makes it clear whether the analyst intends to include inertia, compliance, or friction effects in the pipe segments. A bond graph for the same system can be a precise statement of the model, however.

Furthermore, bond graphs can be processed in a standard way to produce block diagrams or equations. The marks you see in Fig. 1-4 indicate that the bond graph has been processed so that it can also be considered to be a minimal representation of a block diagram. Figure 1-5 shows the bond graphs of Fig. 1-4 expanded into block diagrams without passing through the equation-formulation phase.

If you compare the block diagrams of Figs. 1-5 and 1-2, you may be able to see that the same relationships are implied but the bond-graph versions are highly organized. Each line or bond in a bond graph implies the existence of a pair of signals whose flow is in opposite directions. These pairs are voltage and current for the electrical system and force and voltage for the mechanical one. The product of these signals is power. This power is flowing between various elements in the bond graph, and power, of course, is the rate of change of energy with time. The reason why bond graphs work for so many physical systems is that we can find signal pairs (their general names will be *effort* and *flow*) whose product is power for many types of systems. Bond-graph methods have a built-in analogy between all systems, which leads to economy of thought, and this repays some of the effort of learning a new modeling language.

The bond graphs of Fig. 1-4 also imply a set of equations. You will learn how to write equations from bond graphs in a prescribed way, but some initial insight can be gained just by comparing the equations below with those derived directly from the schematic diagrams. For the electric circuit the equation from the bond graph is in terms of the charge variable q of Eq. (1-2). It is, in fact, Eq. (1-4b) and will not be repeated here.

For the mechanical system the bond graph will produce two equations, one in the displacement x and one in the momentum p, which is just equal to $m\dot{x}$. The results are

$$\dot{x} = p/m \tag{1-9}$$

$$\dot{p} = F(t) - mg - kx \tag{1-10}$$

That these two first-order equations are equivalent to the single second-order equation (1-8) follows by finding p from Eq. (1-9), differentiating to find $\dot{p}$, and then substituting into Eq. (1-10).

It is no coincidence that Eqs. (1-9) and (1-10) are in the so-called *state-variable form* used in modern simulation, analysis, and automatic control. Bond-graph equations normally use as state variables such quantities as x and p, which together measure the energy state of a system. State equations derived from bond graphs always have physical meaning (and always deal with power and energy variables), whereas state equations derived in other ways may be mathematically significant but offer little or no physical insight.

Computer Models

Without computers, dynamic-system models have to be small and simple to be useful. Low-order, constant linear systems can be studied using pencil and paper, but a modest increase in size or the introduction of a few nonlinear functions is enough to make paper-and-pencil analysis unrealistic.

We tend to think of computers as numerically oriented equation solvers. For example, if we know the value of q at $t = 0$ in Eq. (1-4b), the differential equation can be rearranged into a difference equation that will give values of q at discrete sample points in time, $t = 0, T, 2T, 3T$, and so on. To do this, one might employ Euler's method. We introduce the approximation

$$\dot{q} \approx \frac{\Delta q}{\Delta t} \approx \frac{q(T) - q(0)}{T} \tag{1-11}$$

so that
$$q(T) \approx q(0) + \dot{q}T$$

We evaluate $\dot{q}$ at $t = 0$ using Eq. (1-4b); then

$$q(T) \approx q(0) + \left[\frac{-1}{RC} q(0) + \frac{E(0)}{R} \right] \qquad (1\text{-}12)$$

Those of you who like computer algorithms can probably see that if we can get $q(T)$ from $q(0)$ we can continue the process to get $q[(n + 1)T]$ from $q(nT)$ recursively

$$q[(n + 1)T] = q(nT) + \left[\frac{-1}{RC} q(nT) + \frac{E(nT)}{R} \right] \qquad (1\text{-}13)$$

A computer program incorporating Eq. (1-13) is a mathematical model that predicts $q(t)$ at discrete sample times. As an accurate equation solver, Euler's method has some defects. Since Eq. (1-11) is true only in the limit as $T \to 0$, if we pick T too large, the results of Eq. (1-13) will not resemble the solution to our original differential equation at all. The computer program may even become numerically unstable for large values of T. On the other hand, if T is too small, the number of steps required to cover a desired time span grows and the inevitable growth of roundoff error in the computation will make the solution inaccurate. Other methods than Euler's can achieve better compromises in accuracy, but the basic problems associated with large and small values of T remain.

There is a representational problem in using digital computers to predict the behavior of what we ordinarily consider to be analog (continuous) systems. We consider time and such variables as position, velocity, current, pressure, and temperature to be continuous in nature; i.e., we consider these variables to be analog variables. The digital computer can compute only at discrete time intervals and can carry only a finite number of decimal places. This is the fundamental reason why a digital computer is generally only an imperfect differential-equation solver. In fact, it is possible for a computer to produce a sequence of numbers using Euler's method even when the actual differential equation in question has no solution. Care and good judgment can reduce the representation problem to a minor one in most practical cases.

From another point of view, however, digital-computer programs can be considered mathematical models themselves, and they may be approximating the solution to differential equations only incidentally. Nature does not write a differential equation to decide how a mass should bob up and down on a spring; the oscillator equation is only a human model of the situation. Thus we can sometimes proceed directly from a physical system to a computer model without concerning ourselves with equations at all.

Among the several possibilities for such an approach is the Continuous System Modeling Program (CSMP), a well-known computer program developed at IBM, based on a block-diagram approach. It is typical of a variety of programs that basically make a digital computer simulate an analog computer. In addition to the simple linear blocks discussed in our examples, a CSMP can accept nonlinear blocks described by equations, tables of data, preprogrammed subroutines, or user-defined subroutines. Although CSMP contains the equivalent of differential equations, it is more than an equation solver, since the mathemathical model is really constructed directly using input-output blocks or subroutines.

While programs of the CSMP type require the system modeler to imagine input-output relations for system components, another class of computer models works directly with schematic diagrams. These programs organize the flow of signals themselves in order to create computable relationships which can be used to analyze or simulate the systems. A program called SPICE, for example, can accept a circuit graph and will produce the equivalent of differential equations as well as simulations of the system. When bond graphs are used, the ENPORT program can convert the bond graphs into equations or analyze or simulate the systems with no help from the analyst beyond the supply of necessary parameter and source data.

1-4 STRUCTURED MODELING

This book presents a structured approach to the modeling of many types of physical systems. Although we employ all the types of models described above at one time or another, we shall start from a unified base of description using bond graphs. You will see that bond graphs can be processed in an orderly way to produce state equations, block diagrams, transfer functions, and computer programs for analysis and simulation.

The advantages of bond graphs are many. First, a bond graph incorporates the physical assumptions you have made about the model in a clear and precise way. Second, after you decide how your system is forced, you can automatically produce a variety of other standard forms for your model, which may be useful for subsequent analysis or simulation. In many cases the bond-graph processing can be implemented in computer programs such as ENPORT so that you can concentrate on modeling decisions without being saddled with algebraic drudgery every time you think the model should be changed.

Still, for electric circuits alone, circuit diagrams are as good as bond graphs at representing models, and for many mechanical problems useful mechanical schematic diagrams can be made. Then the applicable physical laws can be used in each case to get some equations. Of course, this same approach can be applied to hydraulic and pneumatic systems, to thermal systems, and so on. Our aim, however, is to show you a unified approach to modeling, in which bond graphs are a great help since they apply to all parts of physical systems, even when several types of energy are involved.

Engineering systems are becoming increasingly complex as we search for more efficient ways to accomplish more demanding tasks. Mechanical typewriters have given way to electromechanical ones. Your wrist watch probably is a complex collection of electrical and mechanical parts, and your car contains mechanical, electrical, hydraulic, and thermal systems that interact in complex ways.

For complicated systems of this kind engineers have had to create mixed diagrams to try to represent the dynamics of the systems. A typical example is shown in Fig. 1-6, which combines elements of a mechanical schematic diagram, a block diagram, and a circuit graph. In such a mixed diagram a line sometimes represents

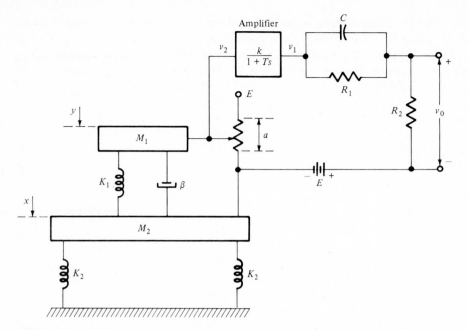

Figure 1-6 An example showing a combination of model types. *(From J. L. Melsa and D. G. Schultz, Linear Control Systems, McGraw-Hill, New York, 1969.)*

a mechanical connection, sometimes a conductor, sometimes a signal-flow path. Sometimes a voltage has a corresponding current, sometimes not. The bottom mass has two springs separately attached, so one might think it is meant to rotate as well as translate, but it turns out that it only moves up and down. Is gravity supposed to act on the masses? The figure does not tell us. What is the meaning of a on the slide-wire resistor? Is there a current flowing where the voltage v_0 is shown? The answers to these questions cannot be found from the diagram but must be stated elsewhere.

As you can imagine, no circuit-theory, vibration-theory, or block-diagram technique alone can be applied to this system. Someone has to add a number of assumptions to the model in Fig. 1-6 and then do quite a bit of creative combining of separate techniques to arrive at a single consistent model, expressible in direct equation form.

Although we strongly favor creativity, we think the organized approach to system dynamics based on bond graphs will allow you to use your creativity most effectively. After you have studied this book, you will be in a position to model many kinds of dynamic systems and to design them with confidence.

In preview, here is what you can expect. Part Two is a fairly complete exposition of modeling, analysis, and simulation of an important but restricted class of systems, namely, low-order mechanical and electrical systems. The models should be familiar enough for you to see the pattern of thought clearly. Part Three shows how a variety of other energy domains can be treated like electrical and mechanical systems. You may not need or want to study all the chapters of Part Three, but you should at least skim them to see the wide applicability of the modeling techniques. In Part Four we expand our techniques to handle more general cases not treated in Part Two. Finally, in Part Five we extend system dynamics to engineering applications, some of which, e.g., vibrations and controls, may turn up in subsequent courses and in practice. In any event, we trust you will recall the techniques learned here with respect, if not outright fondness.

PROBLEMS

1-1 The parallel LR circuit shown is the dual of the series RC circuit discussed in the text; i.e., what is true for voltages in one circuit is true for currents in the other. Note that there is a single voltage e and that the three currents, $I(t)$ from the source and i_R and i_L, must add up to zero algebraically at the node.

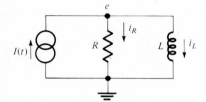

Figure P1-1

(a) Using the component relationships $I(t)$ = given current, $e = RI_R$, and $e = L\,di_L/dt$, reproduce the equivalent of Eq. (1-1) by expressing the current-summation condition in terms of e and $\int^t e\,dt$.

(b) Define $\lambda \equiv \int^t e\,dt$ or $\dot{\lambda} \equiv e$, where λ is the *flux-linkage variable*. Write an equation for λ like Eq. (1-4) or (1-4a), for q.

(c) Using $i_L = \lambda/L$, modify your equation into one for i_L similar to Eq. (1-6).

1-2 Add a damper to the mechanical oscillator, as shown. Assume that the damper force is b times the velocity with which it is extending and write an equation of motion by drawing a free-body diagram. Compare your result with Eq. (1-8).

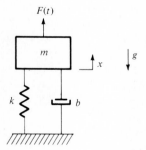

Figure P1-2

1-3 Modify the block diagram of Fig. 1-2 to include the damping introduced in Prob. 1-2.

1-4 Make a block diagram of the sort shown in Fig. 1-2a for Eq. (1-4b). Demonstrate that the block diagram derived from the bond graph in Fig. 1-5a is the same as your diagram if the blocks are moved around a little.

TWO

ELEMENTARY MECHANICAL AND ELECTRICAL SYSTEMS

In this part, simple mechanical and electrical systems are modeled using bond graphs. We choose these systems to start with because they are of the utmost practical importance. Furthermore, schematic diagrams and circuit graphs are often quite precise models in themselves, and thus the physical assumptions contained in the models can be transformed readily into bond-graph language. Part Two takes you through one entire cycle, from modeling to formulation, solution, and interpretation and introduces such important additional tools as block diagrams and computer simulation methods.

ELEMENTARY MECHANICAL SYSTEMS

2-1 INTRODUCTION

In this chapter simple mechanical systems are translated into the unified description of bond graphs. We start with simple translation and rotation because low-order mechanical systems are important in practice and because the physical assumptions contained in the models can readily be transformed into bond-graph language. We assume that you already understand what is meant by the terms mass, spring, and damper. If you have forgotten, we shall refresh your memory, briefly. Later we describe more complex and subtle system models with the same general description, thus building in a unified way on the base of this chapter.

2-2 THE DYNAMICS OF MECHANICAL TRANSLATION

We begin by considering mechanical systems of a very restricted sort. Masses will be allowed to move back and forth in straight lines, without rotating, subject to the actions of springs, dampers, and prescribed forces. Even for large, complex system models the number of basic elements required is quite small if only translation is involved. Thus, translational mechanical systems provide good examples for an introduction to a generalized approach to physical-system dynamics. Furthermore, despite the apparent simplicity of translatory systems, quite a few engineering design decisions are based on such models.

The basic connection between two mechanical subsystems is shown schematically in Fig. 2-1. The parts of the system labeled A and B are connected with a rigid, massless bar. This means that both parts A and B must have a common velocity

$V(t)$. Inside the bar a force exists to satisfy the common-velocity constraint. In Fig. 2-1b the rod is shown symbolically broken apart so that the force $F(t)$ is evident. We have shown F acting as a *compressive* force in the bar; i.e., if F is positive, it tends to push B to the right and A to the left. (Note that our convention is that a *single force F exists. F acts to the right on B and the left on A in an action-reaction pair.)

The ideal bar can be represented by a mechanical *bond,* shown in Fig. 2-1c and d as a line connecting the symbols A and B. Above the line in Fig. 2-1c we note the force F, and below the line we note the velocity V. We observe that the bond marks a power connection, in the sense that FV is the power that flows between A and B at any instant of time. If V is positive, as shown, and F is positive (compressive), a moment's reflection will show that F is doing work on B but absorbing work from A. This fact is indicated by the half arrow on the bond pointing from A to B. A point to remember is that the half arrow points in the direction of actual power flow *if FV > 0*; but since at any particular instant V or F could be negative, FV may not always be positive. The half arrow points in the direction of power flow when the product FV is positive, i.e., when F and V are both positive or both negative.

Figure 2-1d again shows the two parts of the system, A and B, bonded together, but the letters e and f, which stand for effort and flow have been substituted for F and V, respectively. *Effort* and *flow* are generalized bond-variable names that will be used in all energy domains. In mechanical translation effort stands for force and flow stands for velocity, and the product ef is power on a bond. For this reason, effort (force) and flow (velocity) are called *power variables.*

In bond graphs we need to recognize only four basic variables. They are effort, flow, and the time integrals of effort and flow, which we call momentum p and displacement q, respectively. These variables are shown in Table 2-1 both in general notation, which will be used for all energy domains, and for the particular case under study, mechanical translation. For reasons that will become clearer shortly, the inte-

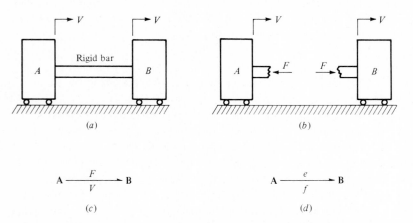

Figure 2-1 The mechanical bond: (a) rigid, massless bar connects A and B, (b) the force F in the bar, (c) a bond graph indicating force F and velocity V, and (d) a bond graph indicating effort e and flow f.

Table 2-1 Variables in mechanics: translation

| Variable | Notation | | SI unit |
	General	Translation	
Effort	$e(t)$	F, force	N
Flow	$f(t)$	V, velocity	m/s
Momentum or impulse	$p \equiv \int e \, dt$	P, momentum	N·s
Displacement	$q \equiv \int f \, dt$	X, distance	m
Power	$\mathcal{P}(t) \equiv e(t)f(t)$	$F(t)V(t)$	W
Energy	$E(p) = \int f \, dp$	$\int V \, dP$, kinetic	J
	$E(q) = \int e \, dq$	$\int F \, dX$, potential	J

grated variables p and q (or P and X) are called *energy variables* because stored energy can be expressed naturally in terms of them.

Table 2-1 also shows the units of mechanical variables in the International System (SI, from Système International).† Although some people feel more comfortable in other units, there are strong reasons for using SI units when a variety of physical systems is to be modeled. The main reason is that power and energy are measured in exactly the same units whether the system is mechanical, hydraulic, electrical, or thermal. This will greatly simplify our modeling of transducers, which convert power from one domain to another, e.g., electric motors and generators. Also, the SI rigorously distinguishes between mass (in kilograms, kg) and force (in newtons, N), and you can forget about slugs and poundals forever.

The Passive 1-Ports

You may have seen mechanical-system models containing mass elements, springs, and dampers. Here we consider these ideal elements, which represent power dissipation and two forms of energy storage, in a fundamental way. We call these idealized elements *passive* because they contain no sources of power. We call them *1-ports* because they can be imagined to exchange power (and hence energy) at a single location or *port*.

At a port an element may be connected or *bonded* to another element to form part of the system model. Elements bonded together in a *bond graph* to represent a system resemble chemical elements that can bond together and can be shown in a bond diagram. This analogy between chemical-bond diagrams and physical-system models was noticed in the late 1950s by H. M. Paynter, who coined the name bond graph and invented the basic notation.

---

† Table A-1 in Appendix A lists SI units used in this book and some equivalences; Table A-2 lists prefixes.

We denote the passive 1-port elements by the letters R, C, and I, which identify the type of element. Each has a single line attached, which indicates how the element is bonded to another element. At each port both an effort (force) and a flow (velocity) are implied. In the discussion to follow, the mechanical elements will be presented, but the 1-ports will also be shown in general notation. We can then use the same general elements for other forms of energy after we have decided what the effort and flow variables are in the other domains. Thus our efforts at generalization will repay us later.

The first element considered, called the *resistor* in general, is often represented schematically in mechanics by the damper or *dashpot* symbol shown in Fig. 2-2a. In reality, not many actual resistive devices look like dashpots—automotive shock absorbers are intended to be dashpots, roughly speaking—but the dashpot symbol is inserted into schematic diagrams to indicate frictional losses of all kinds. The bond-graph element R shown in Fig. 2-2b serves the same purpose. An R element always indicates that the effort (force) and the flow (velocity) on its bond are directly related, as shown by the so-called *constitutive law* graphed in Fig. 2-2c. Two constitutive laws are shown in the figure. The relationship

$$F = bV \qquad (2\text{-}1)$$

or, in generalized notation,

$$e = Rf \qquad (2\text{-}2)$$

is called *linear* because of its straight-line representation. In algebraic terms, the force is proportional to the velocity. The most common case is when b is constant, but sometimes b varies with temperature or some other variable.

The half arrow pointing toward R means that whenever the product FV (or ef) is positive, power is flowing into the R. We would expect FV always to be positive,

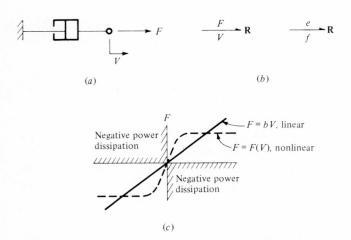

(a)

(b)

Negative power dissipation

$F = bV$, linear
$F = F(V)$, nonlinear

Negative power dissipation

(c)

Figure 2-2 The mechanical resistor: (*a*) schematic diagram, (*b*) bond-graph symbols, and (*c*) constitutive laws.

since friction always dissipates power. Using Eq. (2-1), we can compute the power
$\mathcal{P}$ as

$$\mathcal{P} = FV = bV^2 \tag{2-3}$$

which shows that $\mathcal{P}$ is positive for a linear dashpot when b is positive.

Consider the more general way to plot the constitutive law of a resistor shown
in Fig. 2-2c. The idea is that FV will be positive when both F and V are positive or
both are negative. Thus all possible constitutive laws that are purely dissipative lie
in the first and third quadrants of the F-V plane. For nonlinear resistors the effort-
flow relationship will plot as a curved line or one with segments, rather than as a
straight line. For the nonlinear damper we can indicate this as

$$F = F(V) \tag{2-4}$$

where the notation indicates that there is a functional relationship between F and V.
Equation (2-1) is the linear special case of Eq. (2-4).

Here is a useful procedure to remember. Always show the half arrow on the
bond pointed at the R. Then FV represents power into the resistor, and the constitutive law will always plot in the first and third quadrants. Further, the damper coefficient in Eq. (2-1) will be positive.

The next element to consider is the mechanical spring shown in Fig. 2-3a. The
bond-graph element that represents a spring is the *capacitor C.* (If the general term
capacitor sounds too electrical, you might consider the mechanical symbol C to stand
for compliance.)

A comparison of Figs. 2-2 and 2-3 makes it seem that R and C are analogous.
Physically, however, there is an important difference. The capacitor's constitutive
law is a relationship between force and *displacement,* rather than force and velocity.
This makes all the difference. In Fig. 2-3a, one end of the spring is fixed so that the

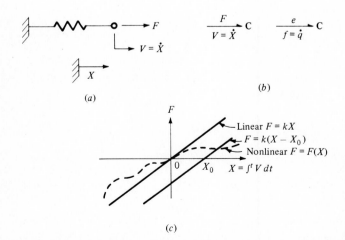

(a)

(b)

(c)

Figure 2-3 The mechanical capacitor: (a) schematic diagram, (b) bond-graph symbols, and (c) constitutive laws.

velocity V is the rate of change of X, the deflection or displacement of the spring. Thus, while we speak of effort and flow for the spring, the spring relates a force to the time integral of the flow (the deflection). To emphasize this crucial difference, we often write $V = \dot{X}$ or $f = \dot{q}$ near the bond of a C element, as in Fig. 2-3b.

Some possible constitutive laws are sketched in Fig. 2-3c. In the linear case, if we measure X from the position the spring end assumes when no force is applied, the law reads

$$F = kX \tag{2-5}$$

or

$$F = \left(\frac{1}{C}\right)X \tag{2-6}$$

where k is the *spring constant* and C is the *compliance*. These two quantities are reciprocals of each other. If we choose to measure X from some point such that the force vanishes when $X = X_0$, the same linear spring will have the constitutive law

$$F = k(X - X_0) \tag{2-7}$$

showing that capacitor constitutive laws may pass through any quadrants of the F-X plane.

If we point the bond half arrow toward C, positive FV represents power flow *to* the capacitor. For springs and other types of capacitors, the power flow represents the rate of storage of energy. Since power is the time rate of flow of energy, we can compute the energy by integrating the power in time

$$E(t) = \int^t \mathcal{P}\, dt = \int^t FV\, dt \tag{2-8}$$

When we recognize that $V\, dt$ is just dX, this integral can be converted into an expression for energy as a function of X alone, i.e.,

$$E(X) = \int^X F\, dX \tag{2-9}$$

The right-hand side is general, but only when $F = F(X)$ can one directly find an energy expression as indicated. Of course the mechanical spring is just such an element. For nonlinear springs, the energy is

$$E(X) = \int^X F(X)\, dX$$

and, for the linear case, using Eq. (2-5), the energy stored is

$$E(X) = \int_0^X kX\, dX = \tfrac{1}{2}kX^2 \tag{2-10}$$

This last expression specifically indicates that the energy stored in a linear spring is just $\tfrac{1}{2}kX^2$ if one starts from zero energy at the position $X = 0$ and if $F = 0$ when $X = 0$.

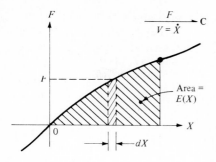

Figure 2-4 Energy storage in a mechanical capacitor. The shaded area represents energy stored as the spring is deflected from 0 to X.

Figure 2-4 shows the general case. The increment $F(X)\,dX$ in energy appears as in infinitesimal area, and thus $E(X)$ corresponds to the shaded area under the constitutive-law curve as shown. When dX is positive, an additional increment of energy is stored; the area increases. When dX is negative, an increment of stored energy leaves the spring; the area decreases. Thus if we were to extend the spring to X from zero and later to return to zero, the energy would vanish; so too would the net area. Energy would have been stored and subsequently released in exactly the same amount. For this reason, any 1-port mechanical spring represented by a C is *energy-conservative,* regardless of the details of its constitutive law.

In mechanics $E(X)$ is usually called *potential energy.* It has been so labeled in Table 2-1. Since potential energy is a function of displacement, we call X an *energy variable* to distinguish it from the *power variables F* and *V.* As Table 2-1 hints, the term "potential" does not seem so universally applicable when we deal with non-mechanical systems, even though arguments we have just made will hold for other types of elements that can be represented by capacitors. Therefore, we simply indicate that q, the displacement, is the energy variable corresponding to X and there is an energy expression corresponding to Eq. (2-9) yielding

$$E(q) = \int^{q} e(q)\,dq \qquad (2\text{-}11)$$

which will be appropriate to any C element.

The third and last passive 1-port to be considered is the mass element, represented by a bond-graph *inertia* element. The symbol is I; since it represents inertia as well as inductance in electrical systems, it should be easy to remember. The properties of this element are shown in Fig. 2-5.

The conventional description of a mass particle asserts that the absolute acceleration of a mass is proportional to the net force. If $\ddot{X}$ is measured in an inertial coordinate frame in Fig. 2-5, then V is an absolute velocity and $\ddot{X}$ is an absolute acceleration and we can write

$$F = m\ddot{X} \qquad (2\text{-}12)$$

for motion along a straight nonrotating line. (Since the time of Einstein we have been willing to concede that Newton's concepts of time and space are philosophically

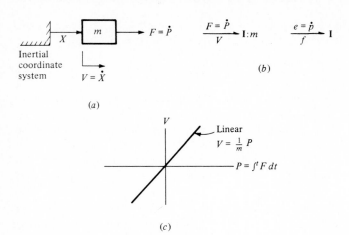

Figure 2-5 Mechanical inertia: (*a*) schematic diagram, (*b*) bond-graph symbols, and (*c*) constitutive laws.

untenable, but as practical engineers we normally operate within newtonian mechanical assumptions unless velocities approach the speed of light.)

In order to fit Eq. (2-12) into a general scheme of things, we integrate in time to get

$$P = mV \tag{2-13}$$

where we have used the definition

$$\dot{P} \equiv F \tag{2-14}$$

This corresponds to $P \equiv \int^t F\, dt$. Thus we regard $F = ma$ as the combination of a constitutive law, $P = mV$, and a law that states that force is the time rate of change of momentum P, Eq. (2-14). Although we tend to regard Eq. (2-12) as Newton's law now, historians of science often claim that Newton never wrote his law in a form like Eq. (2-12) (crediting Euler instead) and that Newton's verbal statement is better rendered by Eq. (2-14). Of course, Eqs. (2-13) and (2-14) do imply Eq. (2-12), but the structure of our mass element is better seen by introducing the P variable we call *momentum* if a mass is involved. We call P an *impulse* if we merely integrate an arbitrary force in time.

On a bond graph we can write the force or effort on an I element as the rate of change of momentum, as in Fig. 2-5*b*. This serves to remind us that the I element relates velocity and momentum (flow and momentum in general). The constitutive law is sketched in the V-P plane in Fig. 2-5*c*, which shows only the linear relation of Eq. (2-13) because in newtonian mechanics that is all there is. A nonlinear velocity-momentum law appears in special relativity, and in other energy domains the I element will frequently be nonlinear; but for now let us accept the linearity of Newton's law with gratitude.

The I element is energy-conservative, just like the C element. To see this, we start again to compute $E(t)$, the energy stored as the integral of the power flowing into I, as in Eq. (2-8). Note that the power-flow half arrow again indicates that whenever FV is positive, energy is being stored by the inertia element

$$E(t) = \int^t \mathcal{P}\, dt = \int^t VF\, dt \qquad (2\text{-}8)$$

Now, however, that we pair F with dt and use Eq. (2-14) to write $F\, dt = dP$

$$E(P) = \int^P V\, dP \qquad (2\text{-}15)$$

Believe it or not, Eq. (2-15) is just as valid an energy expression as the familiar Eq. (2-9). In fact, for a mass element, V is a function of P, so that Eq. (2-15) works out to be

$$E(P) = \int^P V(P)\, dP = \int^P \frac{P}{m}\, DP = \frac{P^2}{2m} \qquad (2\text{-}16)$$

This is the correct form for the *kinetic energy* in a mass. It is a function of the *momentum variable,* which explains why a p variable, like a q variable, is an energy variable.

You may argue that since $mV^2/2$ is how kinetic energy was expressed in your physics text, V could be an energy variable too, not just a power variable. A simple combination of Eqs. (2-13) and (2-16) does give $mV^2/2$ as an alternative energy expression, but our choice of the variables P and X (or p and q generally) as energy variables stems from the two ways to associate the dt in the energy integral of Eq. (2-8), that is, $dX = V\, dt$ or $dP = F\, dt$, together with the appropriate constitutive equation. One way we get potential energy and the other kinetic energy.

The general pattern of this derivation works for all I elements even when the term "kinetic" is not directly applicable. Therefore, in Table 2-1 we simply note that for any I element, stored energy is a function of p

$$E(p) = \int^p f(p)\, dp \qquad (2\text{-}17)$$

In the mechanical case we call this energy kinetic.

Figure 2-6 shows that kinetic energy is representable as an area under the I-element constitutive law plotted in the V-P plane. This comes from associating the increment of energy $V\, dP$ with an elementary area, much as was done for potential area in Fig. 2-4. We note that if a mass has its momentum increased from zero to

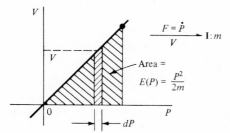

Figure 2-6 Energy storage in a mechanical inertia. The shaded area represents energy stored as the momentum is increased from 0 to P.

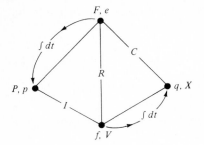

Figure 2-7 The tetrahedron of state, showing the variables related by the three passive 1-ports, based on notation originated by H. M. Paynter.

any value and then back to zero again, the net area vanishes and any energy stored is recovered. Thus any 1-port I element, whatever its constitutive law, is *conservative*.

This is a good time to look at Table 2-1 again, since we have discussed mechanical versions of the four variable types e, f, p, and q and have used expressions for power and energy. Figure 2-7 shows a mnemonic device to help you keep the three passive 1-ports clear. We imagine a tetrahedron with e, f, p, and q (or F, V, P, X) labeling the vertices. The five visible edges have special significance. For example, the edges connecting e to p and f to q have integration symbols, because p is the integral of e and q is the integral of f. The other edges, labeled I, R, C, stand for the three passive 1-ports; the edges show the variables which the three constitutive laws relate.

A Word about Notation

Consistency is a virtue in system dynamics, particularly when many different types of systems are to be studied. There are inevitable conflicts between the symbols used in various fields, as we have already seen. It is common to use F for force and we use f for flow. Also, we use p for generalized momentum, P for mechanical momentum (no problem so far), but also $\mathcal{P}$ for power, and eventually P for pressure. A hardworking symbol is q, since it is used for generalized displacement, electric charge, volume flow rate, and heat flow.

The problems disappear when we use only the generalized variables e, f, p, q, $\mathcal{P}$, and E, but during the modeling process it is helpful to use familiar symbols in the various energy domains. The following rules will be observed as rigorously as possible in this text and are recommended to keep the confusion level low:

1. Anything written next to a bond should be an effort or a flow. Thus we write V or $\dot{X}$ for a flow but not X. We also write either F or $\dot{P}$ for an effort, but not P, near a bond.
2. Write efforts above horizontal bonds and to the left of vertical bonds; write flows below and to the right. For example,

$$\frac{e}{f} \qquad e \bigg| f \xrightarrow{\quad F \quad} \qquad F \bigg| \dot{X} \xrightarrow{\quad \dot{P} \quad}$$

Some judgment may be required, of course, if bonds appear at an angle.

3. For C and I elements always plot constitutive laws with the p or q variables on the horizontal axis. Then the energy interpretation of area will always be below the constitutive law, as shown in Figs. 2-4 and 2-6.
4. For R elements plot the flow on the horizontal axis and the effort on the vertical axis.
5. Always distinguish the type of element by R, C, or I. If the element is linear, the slope of the constitutive law is related to a parameter such as mass m, stiffness k, or damping coefficient b. If you want to remember the parameter, you can write it next to the element using a colon

$$\longrightarrow I\!:\!m \qquad \longrightarrow C\!:\!k \qquad \longrightarrow R\!:\!b$$

Do not confuse an element type with its linear parameter; R, C, and I may denote nonlinear elements, which do not have constant m, b, or k parameters.

The Active 1-Ports, or Sources

When you step into an elevator and push the button for your floor, you expect the soles of your shoes to be subjected to a prescribed vertical velocity as a function of time. If integrated in time, this velocity will show the proper vertical displacement to move you to your floor. On high-speed elevators you may notice that your stomach responds dynamically, not remaining at a constant distance from your feet. To a first approximation, the elevator imposes a velocity on your feet independent of the dynamics of your body.

This kind of behavior is modeled by a *velocity* source acting on your feet. Although your feet react with a force on the elevator or source, the velocity source pays no attention. The source stands ready to supply or absorb any amount of power in order to impose its velocity on whatever it is connected to. For this reason, sources are called *active,* and it is also clear that a real elevator acts only as an approximately ideal source when the reaction forces are not too large.

Fig. 2-8a shows a velocity source in schematic form. Often the drive system is omitted, and one simply indicates the presence of the source by the arrow and symbol $V(t)$. The notation $V(t)$ is meant to imply that the velocity V is a *given* function of time, independent of the system it is acting upon.

The bond graph corresponding to the schematic diagram of Fig. 2-8a is shown in Fig. 2-8b. Note that while V is easy to denote in Fig. 2-8a, the force F is not. In the bond graph both F and V are denoted clearly. The symbol S means a source, and the subscript f means that a flow source is involved. In this case, the flow is a velocity and $V(t)$ helps to remind us that V is given.

Sources often have half arrows pointing away from the S symbol because one tends to think of the source as supplying power and of passive elements as absorbing power. With a velocity source the power at any instant may actually be supplied or absorbed, depending upon the instantaneous value of the reaction force.

In Fig. 2-8a $V(t)$ is also the compression velocity of the spring; if F is positive in compression, $FV(t)$ represents power coming out of the source and flowing into the spring. Figure 2-8b reflects this by showing the half arrow pointing from S_f toward C. A similar situation occurs if we imagine a force $F(t)$ applied to part of a

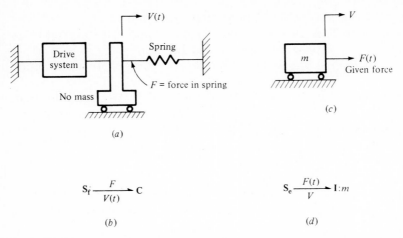

Figure 2-8 Velocity and force sources: (*a*) schematic diagram of velocity source acting on a spring, (*b*) bond graph for (*a*), (*c*) schematic diagram of force source acting on a mass element, and (*d*) bond graph for (*c*).

system such as a mass, as shown in Fig. 2-8*c*. The drive system that produces F is not shown, but the notation $F(t)$ is meant to indicate that F is a given function of time independent of the system response.

The bond graph of Fig. 2-8*d* denotes the force source as an effort source S_e. If both F and V are positive, power is being supplied by S_e and absorbed by the mass; hence the half arrow points toward I. Remember that it is quite possible for FV to be negative at some time. Then the power would be flowing in the direction opposite the arrow at that instant.

Although sources are common in every field, the effort and flow sources described are highly idealized in most cases; i.e., a device that will apply a prescribed force or velocity to any system will always be found to have limitations in magnitudes of force, velocity, and power, although ideal sources do not. Later we shall see that physical sources can be modeled as dynamic systems themselves, by using ideal sources and other elements to represent nonideal effects.

One case in mechanical systems in which an ideal source is a good model by itself involves the force of gravity. For many practical problems we can assume that a constant force of magnitude mg acts on a mass m in a gravity field with constant acceleration g. This effect can be represented by an ideal constant-force source.

The Ideal 3-Port Junctions

Figure 2-8 contains two bond graphs, each consisting of a 1-port source bonded to a passive 1-port element. If you think about it, this type of graph is all that can be created with 1-port elements alone. We shall study some important 2-port elements, but even with them only long chainlike system models can be made (see Fig. 2-9*a*). General system models become possible when 3-port elements are used. For example,

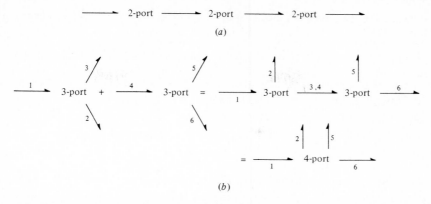

(a)

(b)

Figure 2-9 Combination of 2- and 3-port elements: (a) several 2-ports strung together still have only two external ports, (b) two 3-ports bonded together from a 4-port.

if two 3-port elements are bonded together, the result is a 4-port element. (Figure 2-9). A generalization of this idea is that topological models of arbitrary complexity become possible when 3-ports are used. We shall find that two special 3-ports are uniquely powerful in physical-system modeling. They are discussed next.

The 1-Junction

The 1-junction 3-port is shown in Fig. 2-10. Schematic diagrams of mechanical systems often include a symbol that looks like a little cart on rollers, as in Figs. 2-1 and 2-8. This symbol defines a single translational velocity for several connections. In Fig. 2-10a the cart is assumed massless; three elements are attached to it. This figure is really a free-body diagram, and the three forces F_1, F_2, and F_3 arise from the connected elements, such as springs, dampers, and mass elements, which are not shown.

The bond-graph symbol for this cart is a 1 with three bonds connected, corresponding to the three forces. In Fig. 2-10b the force and velocity associated with each bond are labeled, while in Fig. 2-10c the bonds themselves are numbered.

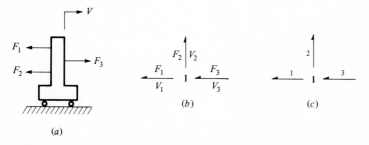

(a)

(b)

(c)

Figure 2-10 The 1-junction or common-velocity junction: (a) the massless-cart symbol, (b) bond graph with forces and velocities indicated, and (c) bond graph with numbered bonds.

The constitutive laws for this type of 3-port are simple but very useful. All velocities are identical, so part of the laws can be written as

$$V_1 = V_2 = V_3 = V \tag{2-18}$$

where V is a common velocity. The 1-junction can thus also be called a *common-velocity junction* or, in general terms, a *common-flow junction*.

The other part of the constitutive laws comes from balancing forces on the free-body diagram of Fig. 2-10a. The result is

$$F_3 - F_1 - F_2 = 0 \tag{2-19}$$

which is true because the cart is massless. The special nature of 1-junctions has to do with power conservation. Since all velocities are the same, if we multiply the terms in Eq. (2-19) by V, we derive a power law

$$\mathcal{P}_3 - \mathcal{P}_1 - \mathcal{P}_2 = 0 \tag{2-20}$$

The signs in the power law are reflected in the bond half arrows in Fig. 2-10b and c. If we count power *supplied* to the cart or 1-junction as positive and power *delivered* by the cart to whatever is attached to it as negative, Eq. (2-20) states that the net power supplied $(+\mathcal{P}_3, -\mathcal{P}_1 -\mathcal{P}_2)$ adds up to zero. The half arrows point toward the 1-junction for power supplied and away for power delivered.

The half arrows can be applied to the bonds from the free-body diagram as follows. Imagine that V is positive as shown; i.e., the cart is moving to the right. If F_3 is positive as shown, it is pulling the cart along and is doing work on it. This means that $F_3 V_3$ is positive and power is flowing *to* the cart. The half arrow on bond 3 therefore points *to* the 1-junction. On the other hand, if V is greater than 0 and both F_1 and F_2 are positive in the direction shown, those elements are being dragged along by the cart; power is flowing *from* the 1-junction to them. Thus bonds 1 and 2 have half arrows pointing away from the junction.

It will take you a while to become practiced at using 1-junctions. but for now remember these several properties:

1. There is a single flow variable (in our case, a single velocity) common to all bonds which are connected to a 1-junction.
2. An algebraic sum of all efforts (in our case, forces) on the bonds attached to a 1-junction is zero. The signs in this algebraic sum are determined by the assumed direction of the forces in a free-body diagram or the half-arrow directions in a bond graph.
3. The net power flowing to a 1-junction on all its bonds is zero at any instant of time. In other words, the total power coming into a 1-junction is balanced exactly by power flowing away. This is a direct consequence of properties 1 and 2.
4. We can combine two 3-port 1-junctions as in Fig. 2-9 to make 4-port 1-junctions. Then the velocity on all four bonds will be the same, and the four port forces will add to zero algebraically. Although the 3-port 1-junction is fundamental, we shall often find it convenient to use 1-junctions with various numbers of ports.

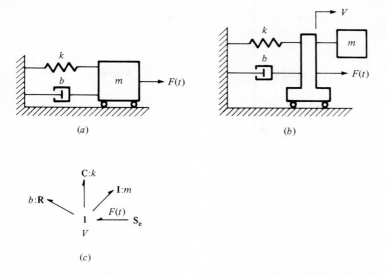

Figure 2-11 Example system using a 1-junction: (*a*) typical schematic diagram, (*b*) equivalent diagram showing massless cart, and (*c*) bond graph with 4-port 1-junction.

The reason for using the word "junction" for this element may be clear if you see its use in a system. Figure 2-11*a* shows an oscillator model as it is sketched conventionally. Figure 2-11*b* shows an equivalent system in which the mass is indicated separately from the massless cart. Now you see the mass force acting at the attachment point on the cart connector, and the net force acting on the mass can be seen clearly. The cart defines a common velocity for (1) the force source, (2) the end of the spring, (3) the end of the damper, and (4) the mass. The cart is in equilibrium under the action of the four forces (which we could show by redrawing the system as it is cut apart). The cart is represented in the bond graph of Fig. 2-11*c* by a 4-port 1-junction. Attached to it are the three passive 1-ports R, C, and I. We have indicated the existence of three linear constitutive laws with parameters b, k, and m. The effort source S_e has $F(t)$ indicated as the given force.

Since $F(t)V$ represents power flowing from the source to the cart, the half arrow points from S_e to 1. The other half arrow points toward the passive elements, as usual, and the parameters b, k, and m will be positive. Our choices set up sign conventions for the damper, spring, and mass forces. Calling these forces F_d, F_s, and F_m and following the signs shown on the 1-junction, we have

$$F(t) - F_d - F_s - F_m = 0 \qquad (2\text{-}21)$$

or, in terms of power flowing into the 1-junction,

$$F(t)V - F_d V - F_s V - F_m V = 0 \qquad (2\text{-}22)$$

Alternatively, you could start with Eq. (2-22), which essentially states that the power flowing into the 1-junction is instantaneously balanced by the power flowing out, and by dividing out the common velocity arrive at Eq. (2-21). Equations (2-21) and (2-

22) allow us to sketch a free-body diagram of the cart to see in which direction the forces F_d, F_s, and F_m point. Try it now for yourself.

Perhaps you can see now that the 1-junction joins up the four 1-port elements of this system, and this is the origin of the name. Surprisingly, there is only one other fundamental junction needed to model many physical systems.

The 0-Junction

The 0-junction is just the bond dual of the 1-junction, in the sense that the roles of effort and flow are reversed. A 0-junction has a *common effort* which appears on all bonds, and an algebraic sum of the bond flows vanishes. (You can remember these properties by recognizing that 0 and 1 are bond duals of each other.) In mechanical systems, a 0-junction is a *common-force* junction, and a sum of port velocities is zero. As you might imagine, the 0-junction, like the 1 junction, conserves power.

The 0-junctions do not appear in mechanical schematic diagrams as obviously as 1-junctions, but they are there all the same. It is easier to show how an 0-junction is used than to show it in isolation. Thus in Fig. 2-12 we study a damper connected between two carts. (The carts can be modeled by 1-junctions, but for now we use them to define some velocities and forces.)

The two ends of the damper move with velocities V_1 and V_2, but the damper force responds only to the rate of extension or compression. We define V_3 to be the *relative velocity* $V_1 - V_2$, which is positive if the damper is *compressing*. Figure 2-12b shows free-body diagrams for the carts and the damper. Corresponding to the compression velocity V_3, we have a compressive force F_3, as shown. In fact, if the damper is linear with coefficient b, its constitutive law is

$$F_3 = bV_3 \tag{2-23}$$

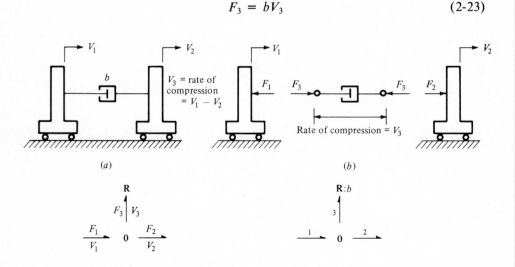

Figure 2-12 The mechanical 0-junction: (*a*) schematic diagram, (*b*) free-body diagrams, (*c*) bond graph with all forces and velocities indicated, and (*d*) bond graph with numbered bonds.

Corresponding to the velocity V_1 on the left-hand cart we show a force F_1, and corresponding to V_2 on the right we show a force F_2. Since ideal dampers are massless, you can see that all the forces are really aspects of a single, *common force* in all its action-reaction guises. Thus

$$F_1 = F_2 = F_3 = F \qquad (2\text{-}24)$$

The other part of the constitutive law is the definition of relative velocity

$$V_3 = V_1 - V_2 \qquad (2\text{-}25)$$

Multiplying Eq. (2-25) by the common force F yields a power relation,

$$\mathcal{P}_3 = \mathcal{P}_1 - \mathcal{P}_2 \qquad (2\text{-}26)$$

which states that the power lost in the damper is the power supplied by the left-hand cart minus the power absorbed by the right-hand cart. [Of course, we could rearrange Eqs. (2-25) and (2-26) to express the facts that an algebraic sum of velocities vanishes and a sum of powers also vanishes.]

The bond graphs of Fig. 2-12c and d show the 1-port R for the damper connected to our 0-junction. Equations (2-24) and (2-25) are constitutive laws for the 3-port 0-junction. You can compare them with Eqs. (2-18) and (2-19) for the 1-junction to see the duality in efforts and flows.

The bond half arrows follow from a careful study of the free-body diagrams. We draw the diagrams in such a way that all the forces are shown as aspects of a single force; i.e., no plus or minus signs will appear in Eq. (2-24). Then we ask ourselves: If the force and velocities are positive as indicated, which way is the power flowing? We then point the half arrow on each bond in the positive power-flow direction.

You will find that with experience it is easier to put half arrows on the bond graph than to indicate in detail the positive senses for each force and velocity in the system. Nonetheless it is wise to understand the sign conventions in Fig. 2-12 in some detail. Can you see that if V_3 is positive, the damper is compressing and F_3 is positive? Can you see that positive $F_3 V_3$ represents power flowing to the resistor and away from the 0-junction? Can you see that $F_1 V_1$ represents power *supplied* by the cart to the 0-junction?

Some System Examples

Let us consider some examples of the use of 0- and 1-junctions in system modeling. As a start, we combine the ideas contained in Figs. 2-11 and 2-12.

Example 2-1 In Fig. 2-13 two masses are acted upon by springs, a damper, and a force source. The bond graph of Fig. 2-13b is constructed as follows. Each of the two masses is represented by an I element attached to a 1-junction. The 1-junctions represent the ideal connection behavior, as in Fig. 2-11b. Any other bond connected to a 1-junction will have as its flow the common velocity, either V_1 or V_2. For example, the force source S_e is bonded to the V_2 1-junction and the left-hand spring is bonded to the V_1 1-junction because V_1 is the rate of extension of this spring. The spring and the damper which fit between the masses can be assembled in the bond graph by the technique of Fig. 2-11. Each of these elements uses a 0-junction for two reasons: (1) $V_3 = V_1 - V_2$ is the rate of compression of both the spring and the damper. A 0-junction with the sign arrows as shown enforces the proper relation between the three velocities V_1, V_2, and V_3. (2) Each 0-junction has a common force. The force in the spring acts (back) on m_1 and (ahead) on m_2. Similarly, the damper force acts on the two masses. Of course, the spring force and the damper force are different, in general.

To help set up the bond graph we have labeled the two 1-junctions with V_1 and V_2 to remind us that every bond incident upon one of these junctions will have the appropriate velocity as its flow variable. In Fig. 2-13b the graph implies that both the coupling spring and the damper react to the same relative velocity, V_3. This leads to the slightly different but equivalent bond graph in Fig. 2-13c. In this case, we start by recognizing three important velocities, V_1, V_2, and $V_3 = V_1 - V_2$. For each velocity there is a 1-junction, and the 1-junctions are connected by a single 0-junction, which enforces the relationship between the velocities. Now we bond each element to the 1-junction which has the appropriate velocity. In this new bond graph V_3 is constrained by the 0-junction but is common to all the bonds of the upper 1-junction.

If you followed this example, you should be proud of yourself, since we have come quite a way from the "little cart" ideas. Although you could show carts for the V_1 and V_2 1-junctions, it would be harder to show one for V_3. Try it! Also, making detailed free-body diagrams for every part of the system so that *every* force could be shown would be tedious. Fortunately, this is not necessary because the bond graph takes care of things like power conservation when it is appropriate. For example, the

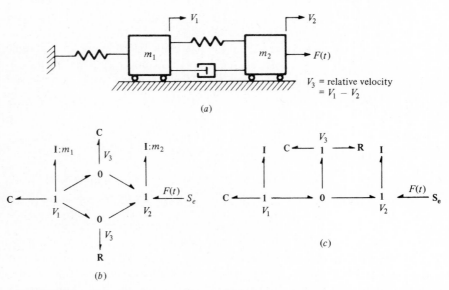

(a)

(b)

(c)

Figure 2-13 System for Example 2-1: (*a*) schematic diagram, (*b*) a bond graph based on Fig. 2-12, and (*c*) equivalent bond graph.

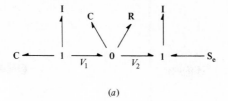

(a)

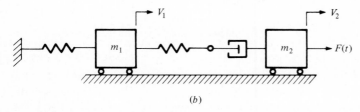

(b)

Figure 2-14 System for Example 2-2: (a) a bond graph and (b) schematic diagram.

0-junction in Fig. 2-13c says that a common force acts (back) on m_1 and (forward) on m_2, but what is this force? From the V_3 1-junction we see that it is the *sum* of the forces in the C and R elements. In Fig. 2-13b the 0-junctions separately apply the spring and damper forces to the masses, while in Fig. 2-13c the single 0-junction applies the two forces in combination. Since we arrived at a correct representation of V_3 in constructing the bond graph of Fig. 2-13c, the forces automatically worked out correctly also.

Example 2-2 Now let us try a variation in procedure. A bond graph similar to that of Fig. 2-13c is shown in Fig. 2-14a. Note that now both the C and R are joined to a 0-junction directly. Velocity variables V_1 and V_2 are indicated. What does the corresponding mechanical system look like?

Since the two 1-junctions and their adjoined elements are identical to the previous model, we can start with them. The spring C and damper R must share a common force, due to the 0-junction. That force acts on both mass 1 and mass 2. Hence we arrive at the mechanical schematic of Fig. 2-14b by inexorable logic. To be absolutely sure consider the velocity relationships. In Fig. 2-14a the 0-junction has four ports and implies that an algebraic sum of V_1 and V_2, the spring relative velocity and the damper relative velocity, must be zero. In checking Fig. 2-14b we see that this is the case. Although more than one bond graph may represent a system, you should check the relations implied by the graph because apparently minor changes in a bond graph may imply very different system models.

The Ideal 2-Port Elements

Many system models require only the basic active and passive 1-ports joined by 0- and 1-junctions, but we sometimes also need 2-ports. We now consider two ideal 2-

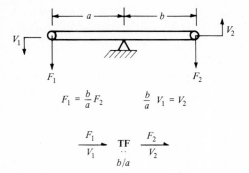

$$F_1 = \frac{b}{a} F_2 \qquad \frac{b}{a} V_1 = V_2$$

$$\xrightarrow{\frac{F_1}{V_1}} \quad \text{TF} \quad \xrightarrow{\frac{F_2}{V_2}}$$
$$b/a$$

Figure 2-15 A lever represented as a transformer.

port elements, both of which are ideal in the sense of power conservation, just as the junctions are. One of the ideal 2-ports is called a *transformer* (TF); in fact, the ideal electric transformer is modeled by a bond-graph transformer. In mechanical systems a variety of devices are represented in idealized form as transformers, including the rigid, massless lever shown in Fig. 2-15. The transformer has two ports, and the efforts (forces) at the two ports are proportional to each other, as are the flows (velocities). For the lever, if F_1 and F_2 are vertical forces as shown and the pivot is frictionless, by taking moments about the pivot we find the equilibrium relation

$$aF_1 = bF_2$$

or
$$F_1 = \frac{b}{a} F_2 \qquad (2\text{-}27)$$

Another relation is found for the vertical velocities V_1 and V_2 (assuming small angular rotation) by computing the angular velocity of the rod as

$$\frac{V_1}{a} = \omega = \frac{V_2}{b}$$

or
$$\frac{b}{a} V_1 = V_2 \qquad (2\text{-}28)$$

The ratio b/a is called a *modulus* of the transformer. If Eqs. (2-27) and (2-28) are multiplied together appropriately and the common factor b/a is canceled out, we get the power relation

$$F_1 V_1 = \mathcal{P}_1 = \mathcal{P}_2 = F_2 V_2 \qquad (2\text{-}29)$$

which indicates that power flowing into port 1 is always equal to power flowing out of port 2. This power relation is embodied in the bond-graph symbol in Fig. 2-15, which shows the sign convention of half arrows pointing *through* the TF symbol.

Since both forces are shown pointing down in Fig. 2-15, Eq. (2-27) must be written with a positive proportionality factor b/a. Then with V_1 down and V_2 up Eq. (2-28) is written with the same positive proportionality factor b/a. The result is that $F_1 V_1$ is power being supplied *to* the left-hand end of the lever and $F_2 V_2$ is power being expended *by* the right-hand end of the lever on whatever is connected to it. You should study the schematic diagram and the bond graph in Fig. 2-15 to see that the sign conventions are equivalent.

Again we remind you that the variable and power directions are for reference purposes. They do not assert that the actual powers, as computed or measured, always flow in the directions shown. As an engineer you will often be thinking about signs, so it is not surprising that signs take time to puzzle out at first. They do require attention.

To indicate that transformers appear in guises other than levers, consider the pulley system of Fig. 2-16a. The parts of the bond graph connected with the two 1-junctions in Fig. 2-16b follow essentially the same development as in Fig. 2-11. The velocity V_1 is common to the mass M, the friction element, and the cable leading to the moving pulley. The vertical velocity V_2 is common to mass m, the gravity force, and the attached cable. Hence, we need two 1-junctions. We assume that the fixed pulley around which the cable is bent changes the direction of the cable velocity but not its magnitude. Thus, the top of the moving pulley has velocity V_2. The bottom of the moving pulley has zero velocity because the lower run of the cable is fixed to ground. What about the center of the moving pulley to which M is attached? If you think the center's velocity is $V_2/2$, you are right. This is a problem in kinematics which requires some thought.

Below the completed bond graph in Fig. 2-16b we have listed the correct velocity relationship. As you can see, if the TF representation is correct, the two forces F_1 and F_2 should be related by the ratio of 2:1. Is this correct? F_1 is the tension force in the cable attached to M, and F_2 is the tension force in the cable attached to m. Make a free-body diagram for the moving pulley to see that $F_1 = 2F_2$, as the transformer requires. The bond half arrows indicate that when $F_1 V_1 > 0$ and $F_2 V_2 > 0$ (forces in tension), $F_2 V_2$ is the power expended on the cable system and $F_1 V_1$ is the power supplied to M.

There is one other ideal 2-port to be introduced, and although simple mechanical systems can usually be modeled without using it, we can show how one could arise in a translational system. Its name *gyrator* (GY) is derived from the gyrational cou-

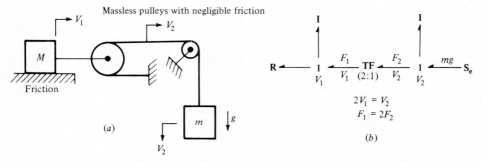

Figure 2-16 Example of a pulley system: (*a*) schematic diagram and (*b*) bond-graph representation using a transformer.

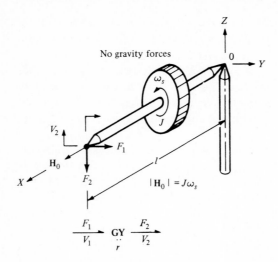

No gravity forces

Figure 2-17 A mechanical gyrator.

pling terms that occur frequently in nonlinear mechanical systems, but the simplest gyrator is a linear element no more complicated than the transformer.

Figure 2-17 shows the gyrator bond-graph element GY and a mechanical system that could be modeled as a gyrator. The parameter r, or *modulus*, enters the gyrator constitutive laws as follows:

$$F_1 = rV_2 \tag{2-30}$$

$$rV_1 = F_2 \tag{2-31}$$

As in the case of the transformer, power is conserved, as can be seen by multiplying Eqs. (2-30) and (2-31) appropriately

$$F_1V_1 = \mathcal{P}_1 = \mathcal{P}_2 = F_2V_2 \tag{2-32}$$

The difference between transformers and gyrators is that transformers relate efforts to efforts and flows to flows, while gyrators relate the effort at one port to the flow at the other. The parameter r in the constitutive law has the dimensions of effort divided by flow. This is like a resistance coefficient, which is why we use the letter r. In the case at hand, r resembles a dashpot coefficient as used in Eq. (2-1) in terms of its units. Notice that a transformer modulus may be a dimensionless quantity, e.g., ratio of lengths, while the gyrator parameter always has units.

Now we compute the modulus r for the system of Fig. 2-17. Imagine that the disk is spinning with large angular velocity ω_s around the rod of length l presently pointed along the X axis. One end of the rod is fixed on a pivot at point 0, and we neglect gravity. We could also assume that the point 0 is the center of mass of the disk and that a clever gimbal mount takes the place of the simple pivot. Forces F_1 and F_2 now can be applied to the other end of the rod as shown. The forces cause moments about 0, and for a rigid body the moment about a fixed point must equal the time rate of change of the angular momentum about 0. Gyroscopes may act peculiarly because applied moments often cause changes in *direction* of the angular-momentum vector rather than changes in its *magnitude*. This is also true of our gyrator. The angular-momentum vector $\mathbf{H}_0$ is mainly directed along X and has magnitude $J\omega_s$, where J is the moment of inertia around the X axis.

The force F_1 causes a moment $F_1 l$ around the Z axis, so $\dot{\mathbf{H}}_0$ must also be in the Z-axis direction. This is accomplished by having $\mathbf{H}_0$ swivel so that the rod tip moves upward with velocity V_2. The magnitude of $\dot{\mathbf{H}}_0$ is then the length of $\mathbf{H}_0$ times the angular rate of swiveling V_2/l around the negative Y axis. In component form the relation is

$$F_1 l = M_{0Z} = |\dot{\mathbf{H}}_0| = J\omega_s \frac{V_2}{l}$$

or
$$F_1 = \frac{J\omega_s}{l^2} V_2 \tag{2-33}$$

A similar analysis for the effect of F_2 yields

$$F_2 = \frac{J\omega_s}{l^2} V_1 \tag{2-34}$$

so that
$$r = \frac{J\omega_s}{l^2} \tag{2-35}$$

Of course, the system we have just studied is only approximately a gyrator since large enough forces would cause the gyroscope to move through large angles or even to attain significant angular momentum around axes other than the spin axis, but you see that the gyrational coupling effect does exist. In many cases gyrators are indispensible although they rarely appear in simple translational mechanical systems.

We have now introduced the basic forms of all the elements needed to represent all types of physical-system models. There are nine in all: C, I, R, S_e, S_f, 0, 1, TF, and GY. It is time for you to practice using these elements to model translational mechanical systems. To help you we present a procedure for representing mechanical systems in bond-graph form.

A Procedure for Modeling Translatory Systems

Most people find it easier to visualize velocities and to consider how they are related in a mechanical system than to consider forces in all parts of the system. For this reason, the following procedure is expressed in terms of velocities. However, the forces will be properly constrained by the bond-graph model also. The procedure has five steps.

1. For each velocity you wish to use, establish a 1-junction. For inertial elements, you should use absolute velocities, but you may also want to establish relative-velocity 1-junctions.

2. Bond inertia elements to their absolute-velocity 1-junctions. Connect C and R elements between 1-junction pairs using 0-junctions, unless you already have established relative-velocity 1-junctions for such elements. Connect force and velocity sources appropriately.
3. When velocities on the 1-junctions are related directly, use 0-junctions and transformers to enforce the constraints. Check the correctness of your bond graph by seeing that the implied force relations are correct.
4. Add sign-convention half arrows on any bonds not already oriented.
5. Simplify the graph where possible. In particular, if there is ground velocity on a 1-junction with zero velocity, you can eliminate all its bonds. Also, 2-port versions of 0- and 1-junctions with in-out bond arrows can be replaced by simple bonds.

The last step deserves further discussion. Note that

$$\rightarrow 1 \rightarrow \; = \; \rightarrow$$

$$\rightarrow 0 \rightarrow \; = \; \rightarrow$$

but other 2-port 1- and 0-junctions should be left alone. The reasoning is simple; e.g.,

$$\begin{array}{cc} F_1 & F_2 \\ \rightarrow 1 & \rightarrow \\ V_1 & V_2 \end{array}$$

means that $V_1 = V_2$ (common velocity on a 1-junction) and $F_1 - F_2 = 0$ or $F_1 = F_2$. A single bond can represent the same information, but

$$\begin{array}{cc} F_1 & F_2 \\ \rightarrow 1 & \leftarrow \\ V_1 & V_2 \end{array}$$

means $V_1 = V_2$, as always, together with $F_1 + F_2 = 0$ or $F_1 = -F_2$, so this element changes the sign of the force. A single bond cannot represent these relations. Similar results hold for the other cases. Figures 2-18 to 2-20 show three examples of the use of the modeling procedure. In each case, a schematic diagram is converted to a bond-graph model in several logical steps.

Example 2-3 In Fig. 2-18 we model a system like that of Fig. 2-13 using the procedure. First, 1-junctions are established for the velocities V_a, V_b, $V_a - V_b$ and the ground velocity V_g. Next, we begin bonding elements to the appropriate junctions. The mass elements are attached to the V_a and V_b junctions; one spring and the damper are attached to the relative-velocity junction; the other spring is attached to a 0-junction, which indicates that its velocity of compression is $V_b - V_g$; and the force source is attached directly to the V_a junction. If we mean the ground velocity V_g to be zero, we can simply remove the junction and its bond and simplify the 2-port 0-junction going to C. (We could have eliminated V_g from the beginning by recognizing that if $V_g = 0$, the spring's compressional velocity is V_b and bonding the C directly to the V_b junction.)

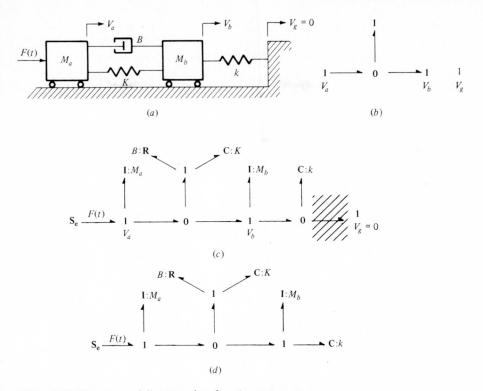

Figure 2-18 Five-step modeling procedure for a two-mass system.

Because we decided at the beginning that the relative velocity $V_a - V_b$ would be necessary and established a 1-junction for it, we came immediately to a form like Fig. 2-13c. If we had simply strung the damper and spring between the V_a and the V_b junctions using separate 0-junctions, we would have achieved the graph equivalent to Fig. 2-13b.

Example 2-4 Figure 2-19 shows a system which can slide down an inclined plane and oscillate. First, we establish 1-junctions for velocities in the direction of motion, as well as a junction for the relative velocity common to a spring and damper. One spring is attached on a 0-junction between two absolute-velocity junctions, while the spring-and-damper combination is bonded to the relative-velocity junction. The inertia elements are bonded to the appropriate absolute-velocity junctions. The masses are acted upon by components of the weight forces in the direction of motion, and these forces are supplied by directly bonded force sources. The nonlinear friction R elements relate forces on two of the masses to their absolute sliding velocities V_a and V_b, so R elements are directly bonded to the appropriate junctions.

Although there are forces to the inclined plane, since there is no motion normal to the plane, no power is involved and such forces do not appear directly in the bond graph. The normal-force components $M_a g \cos \alpha$ and $M_b g \cos \alpha$ would influence the constitutive laws of the nonlinear friction R elements, however.

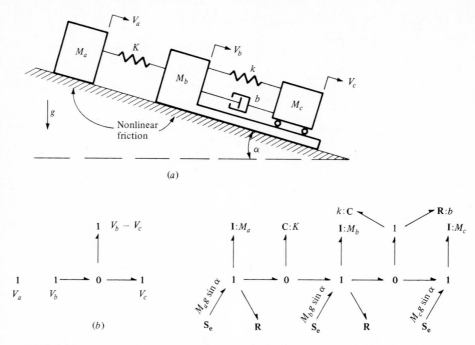

Figure 2-19 Five-step modeling procedure for a three-mass system.

Example 2-5 A third system is modeled in Fig. 2-20. Several useful velocities have been indicated on the schematic diagram. V_2 appears twice since we have decided that the upward movement of the lever is the same as the movement of one end of the spring because of the cable-and-pulley arrangement. Next, four 1-junctions are shown, labeled by their common velocities. Then a flow source has been attached, the spring between V_1 and V_2 has been inserted on a 0-junction, V_3 and V_4 are related to V_2 using TFs, and V_3 and V_4 are recognized as the velocities of a spring and a damper. Finally, with the sign convention shown, three 2-port 1-junctions can be replaced by single bonds.

The problems at the end of the chapter provide an excellent opportunity for you to learn the five-step procedure for modeling mechanical translatory systems. Additional examples appear throughout the book.

2-3 FIXED-AXIS ROTATION

There is a strong analogy between the mechanics of bodies which translate and those which rotate around a fixed axis. Table 2-2 shows how we can define effort as torque and flow as angular velocity and complete a table of power and energy quantities just as we did for translation (Table 2-1). Since the nine basic bond-graph elements have been defined in terms of generalized variables as well as by variables appropriate to translational mechanics, we should be able to model schematic diagrams of

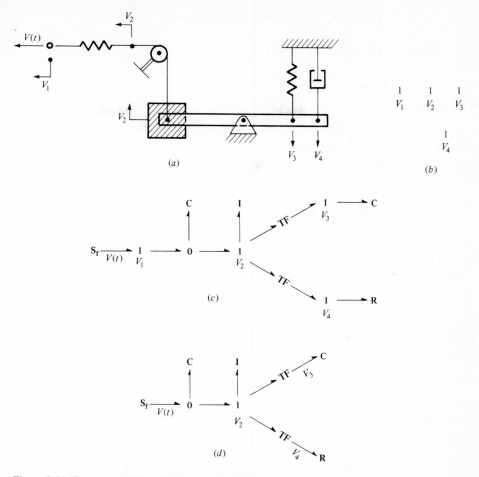

Figure 2-20 Five-step modeling procedure for a lever system.

Table 2-2 Variables in mechanics: rotation

Variable	Notation		
	General	Rotation	SI unit
Effort	$e(t)$	τ, torque	N·m
Flow	$f(t)$	ω, angular velocity	rad/s
Momentum or impulse	$p \equiv \int^{t} e\,dt$	H, angular momentum	N·m·s
Displacement	$q \equiv \int^{t} f\,dt$	θ, angle	rad
Power	$\mathcal{P}(t) \equiv e(t)f(t)$	$\tau(t)\omega(t)$	W
Energy	$E(p) = \int F\,dp$	$\int \omega\,dH$, kinetic	J
	$E(q) = \int e\,dq$	$\int t\,d\theta$, potential	J

rotational systems directly into bond graphs using Table 2-2. A little practice is all that is required.

The Passive 1-Ports

As before, we find R, C, and I elements, as can be seen in Fig. 2-21. The elements will not be discussed in as much detail as their analogous translational elements, since the arguments are merely repeated with τ, ω, H, and θ replacing F, V, P, and X, respectively. For example, the rotary damper shown in Fig. 2-21a relates torque to angular velocity and dissipates power, much as a translational damper relates F to V. A linear damper has the law

$$\tau = b\omega \tag{2-36}$$

where the coefficient b now has units $\text{N}\cdot\text{m}/(\text{rad/s})$, instead of $\text{N}/(\text{m/s})$. With the standard R sign convention, the power dissipated is

$$\mathcal{P} = \tau\omega = b\omega^2 \geq 0 \tag{2-37}$$

The C element of Fig. 2-21b relates τ to θ in a manner analogous to the case shown in Fig. 2-3. The C element represents a torsional spring that is energy-conserving. A linear version of such a spring is

$$\tau = k\theta \tag{2-38}$$

where k in this case has the units $\text{N}\cdot\text{m}/\text{rad}$. The I element represents a flywheel or other rotary inertia. You may be familiar with the law

$$\tau = I_0\alpha \tag{2-39}$$

where I_0 is the moment of inertia about a fixed axis and $\alpha = \dot{\omega}$. We prefer to integrate this relation once to obtain

$$H_0 = \int^t \tau \, dt = I_0\omega \tag{2-40}$$

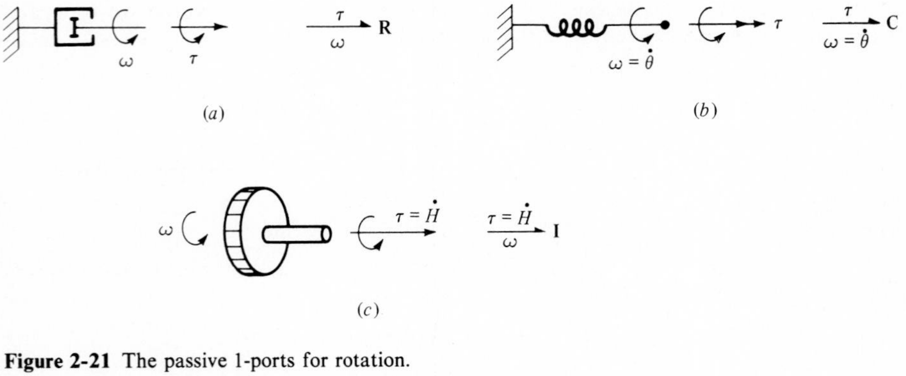

(a)

(b)

(c)

Figure 2-21 The passive 1-ports for rotation.

where H_0 is the angular momentum. (The 0 subscripts are to remind us that H and I refer to fixed axis 0.) This development parallels that leading to Eq. (2-13). The rotary version of kinetic energy is expressed in terms of H, rather than ω, as indicated in Table 2-2

$$E(H_0) = \int^t \omega \, dt = \int_0^{H_0} \omega \, dH_0 = \int_0^{H_0} dH_0 = \frac{H_0^2}{2I_0} \qquad (2\text{-}41)$$

Of course, the more familiar version of kinetic energy $\frac{1}{2}I_0\omega^2$ follows from Eq. (2-41) upon substituting Eq. (2-40).

You may find it worthwhile to convert Figs. 2-2 to 2-8 into their rotational equivalents using Table 2-2. It is not difficult and may provide some valuable insights. For example, it is easy to describe torque sources and angular-velocity sources in strict analogy to force and velocity sources.

The 3-Port Junctions

The common-effort 0-junction for rotary systems has a *common torque* on its bonds, and an algebraic sum of the angular velocities vanishes. The common-flow 1-junction has a *common angular velocity* on all bonds, and a sum of the torques adds to zero. Since these elements are so similar to their counterparts in translational systems, we simply show how they are used in practice.

Example 2-6 Figure 2-22 shows a fixed-axis system in which the 0- and 1-junctions are quite prominent. If we call the angular velocities of the flywheels ω_1 and ω_2, we can establish two 1-junctions that have common angular velocities ω_1 and ω_2 and can bond an I element to each. The parameters of the I elements are the moments of inertia J_1 and J_2. The springs react to the amount they are twisted. The time rate of twist of the right-hand spring is $\omega_3 = \omega_1 - \omega_2$. A 0-junction represents this relation, but it also enforces the constraint that three equal torques are involved; the torque on flywheel 1, the torque on flywheel 2, and the torque in the twisted torsional spring are all the same. Thus, we connect the flywheels with a C element bonded to a 0-junction and show ω_3 as its angular velocity of deflection.

On the input side there is a driving angular velocity $\omega_0(t)$, which we represent as a S_f element (flow source). The input-side spring is deflected at the rate $\omega_0 - \omega_1$, as shown by the left-hand 0-junction. A load torque $\tau(t)$ is denoted by the S_e (effort-source) element, to complete the model. The translational system of Fig. 2-13 and the rotational system of Fig. 2-22 have a lot in common, although there are some significant differences. As you can see, the two bond graphs are similar and if $\omega_0(t)$ were set to zero and the dashpot removed, virtually identical bond graphs would result.

To further test your understanding, imagine that the bearings supporting the flywheels in Fig. 2-22 have some friction you want to include in your model. Bearing friction implies torques related to the angular speeds ω_1 and ω_2, so friction in the torsional system should be modeled by R elements bonded directly to the two 1-

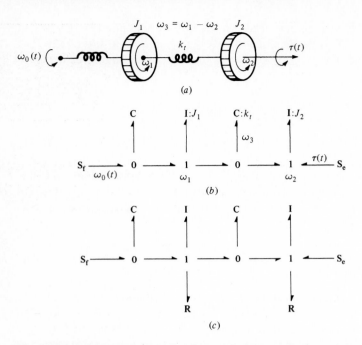

Figure 2-22 Torsional system illustrating typical use of 0- and 1-junctions

junctions. The resulting bond graph is shown in Fig. 2-22c; is it just what you expected?

Rotational and Rotational-Translational Transformers

Of the two basic types of 2-ports, GY and TF, the transformer is much more common in mechanical translation and fixed-axis rotation systems. For example, the simplest useful model of a rotational power transmission is a transformer. Consider the gear pair shown in Fig. 2-23a. With ω_1 and ω_2 defined in the figure, one property of the gears is to enforce a proportional relationship between the two angular velocities. If r_1 and r_2 are the radii to the contact point between the gear teeth, the velocity of the contact point can be expressed as either $r_1\omega_1$ or $r_2\omega_2$, which leads to

$$\omega_1 = \frac{r_2}{r_1}\omega_2 \tag{2-42}$$

Similarly, the simplest useful expression for the contact force is either τ_1/r_1 or τ_2/r_2, where τ_1 and τ_2 are the torques applied to the shafts, as shown in Fig. 2-23a. This idea leads to

$$\frac{r_2}{r_1}\tau_1 = \tau_2 \tag{2-43}$$

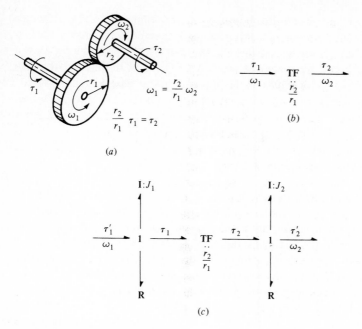

Figure 2-23 Models of a gear pair.

These relations are clearly transformer constitutive laws, i.e., two efforts proportional to each other and two flows proportional to each other in such a way that the power flowing in on one port always equals the power flowing out of the other. Thus, the gear pair can be represented as in Fig. 2-23b as a transformer with modulus r_2/r_1 and with a "through" power-sign convention.

Actually, a gear box with several sets of gears having an overall ratio of shaft speeds of $m:1$ can also be modeled as a TF, which means that the torque relationship will also have a modulus m that takes the place of r_2/r_1 in Eqs. (2-42) and (2-43). Because of power conservation, gearboxes that step torque up must step velocity down, and vice versa.

In our considerations of transformers so far we have implicitly neglected some additional physical effects that are always present to some degree. If we want to model these effects, we can supplement the basic transformer model. Figure 2-23c shows an example of this type of extended model. Two 1-junctions now have been added, and I and R elements have been bonded to the junctions. The I elements represent the rotary inertias of the two gears, and the R elements represent bearing friction. Because the 1-junctions have common angular velocities on all their bonds, really only two angular velocities are involved, ω_1 and ω_2, and they are still related, by Eq. (2-42). In other words, we assume that the gears are rigid bodies.

The torques τ_1' and τ_2' that appear on the external bonds are no longer the same as τ_1 and τ_2. They differ from τ_1 and τ_2 because some amount of torque is required to overcome friction and to accelerate the rotary inertia elements. Thus, although the TF neither stores nor loses energy, the expanded model of the gears shown in Fig. 2-23c dissipates power because of the R's and can store kinetic energy in the I's. The step from the highly idealized gear model of Fig. 2-23b to the more realistic model of Fig. 2-23c is typical of a reasonable approach to modeling a real system.

You can perhaps see that it often is a matter of engineering judgment which storage and dissipation to include in modeling a particular device or system. As a rule, a simple but adequate model is to be preferred to a more complex and detailed one, since it is more easily understood and costs less to study. Experience is a great help in building your modeling judgment.

Example 2-7 Figure 2-24 illustrates the use of a transformer to model coupling in a rack-and-pinion device which couples rotary to translatory motion. The input is rotational. The bearings holding the inertias J_1 and J_2 about the axis are not shown. The bond graph can be constructed by establishing 1-junctions for significant velocities and angular velocities such as V, ω_1, and ω_2. Then inertia elements are appended where appropriate, as well as a torque source, and resistance and capacitance elements are added for the rack and spring. Note that the rack, spring, and damper elements are governed by absolute velocity (V in this case). Finally, the torsional spring twists at a rate $\omega_2 - \omega_1$, so that a C element attached to a 0-junction is inserted between the two angular-velocity 1-junctions.

The transformer coupling comes from observing that

$$r\omega_2 = V \tag{2-44}$$

which represents one of the transformer constitutive laws; i.e., flow is proportional to flow. The other must be a relationship between a torque and a force, and it should have the same proportionality constant as Eq. (2-44). Can you draw a free-body

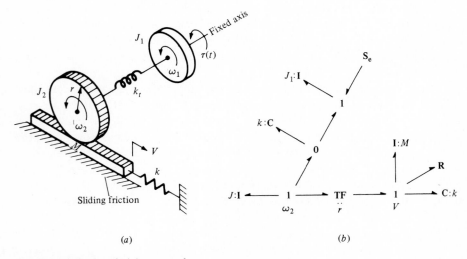

(a) (b)

Figure 2-24 Rack-and-pinion example.

diagram to convince yourself that the bond graph TF in Fig. 2-24b does indeed properly relate the tooth-contact force to the torque it causes on the pinion? If so, you see that a transformer effectively represents the conversion of rotary to translational power using a rack and pinion.

2-4 SOME CASES OF PLANE MOTION

An important example of the interaction between translation and rotation is the plane motion of rigid bodies. A major result proved in every first course in dynamics is that if the center-of-mass position is used to describe the translation of the body, the equations for translation are decoupled from those of rotation. That is, as far as translation is concerned, the body behaves as if all the mass were concentrated at the center of mass, and the torques around the center of mass are related to the rate of change of angular momentum about the center of mass as if the center of mass were a fixed point. We now examine these ideas in a bond-graph context.

In Fig. 2-25a, V_x and V_y are the x and y components of the velocity of the center of mass of the body B with respect to inertial x and y directions, and ω is the angular velocity around the z axis. For plane motion there is no velocity in the z direction, nor are there any other angular-velocity components. Then the equations of motion can be written

$$\dot{P}_x = \sum_i F_{xi} \qquad (2\text{-}45)$$

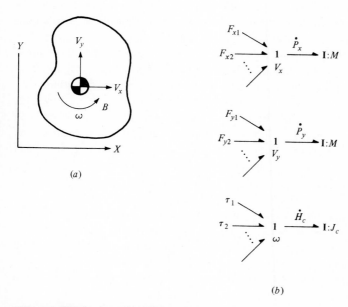

(a)

(b)

Figure 2-25 The dynamics of plane motion.

$$\dot{P}_y = \sum_i F_{yj} \tag{2-46}$$

$$\dot{H}_c = \sum_k \tau_k \tag{2-47}$$

where P_x, P_y = linear momenta
H_c = angular momentum
F_{xi}, F_{yj} = x and y forces
τ_k = torques about center of mass

In addition, we can write

$$P_x = MV_x \tag{2-48}$$

$$P_y = MV_y \tag{2-49}$$

$$H_c = J_c \omega \tag{2-50}$$

where M is the mass of the body and J_c is the moment of inertia about the center of mass. Figure 2-25*b* sums up these relations in an elegant way, representing Eqs. (2-48) to (2-50) with three I elements of exactly the same character as before. Equations (2-45) to (2-47) are represented by the 1-junctions, which properly sum the efforts (forces and torques).

Example 2-8 Although the use of the center of mass to describe motion decouples the dynamic equations of rigid bodies in a system, other force-generating elements may provide coupling between rotation and translation. Figure 2-26 provides an example. The pulley is supported in a gravity field by a flexible cable, which is not supposed to slip on the pulley. The pulley can bounce up and down and also simultaneously rotate back and forth, but we assume that no other motions occur. The center-of-mass velocity is V, and the angular velocity is ω. The free-body diagram of Fig. 2-26*b* shows forces F_1, F_2, and mg, which act on the body. Two torques about the center are implied, RF_2 in the $+\omega$ direction and RF_1 in the $-\omega$ direction.

Figure 2-26*c* shows a stage in the body-graph construction. Two 1-junctions have been established to represent the flows V and ω, and I elements have been appended, following the concept of Fig. 2-25. Then a piece of *junction structure,* involving a 1-junction for V_2, a 0-junction for F_2, and a transformer, has been attached that links the $F_2 V_2$ port to the pulley dynamics. To see where this came from note from pulley geometry that V_2 depends upon V and ω; in fact, it is the sum $V + R\omega$. The TF creates the velocity $R\omega$ from ω, and the 0-junction does the summation. Note that the sign on the $F_2 V_2$ port stems from the free-body diagram, in which $F_2 V_2$ represents power *supplied* to the pulley. Read the 0-junction signs to convince yourself that $+V_2 - R\omega - V = 0$, which is the proper relation. The bond graph of Fig. 2-26*c* is smart enough to imply that F_2 appears on all bonds attached to the 0-junction. Thus, F_2 is applied to the translation I element, and RF_2 is applied to the rotation I element after F_2 is passed through the transformer.

The $F_1 V_1$ port is connected through another junction structure, as was done in Fig. 2-26*c*, with the form shown in Fig. 2-26*d*. The power direction for the TF is

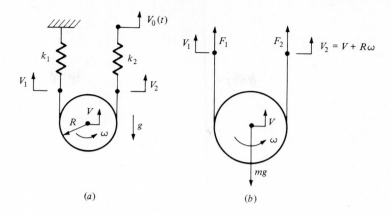

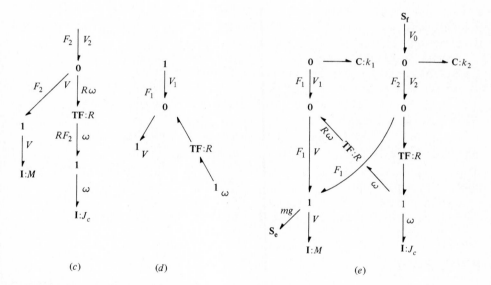

Figure 2-26 Plane-motion example.

reversed from that for V_2 because $V_1 = V - R\omega$, and the bond arrows around the F_1 0-junction enforce this relation. Since the force of gravity mg acts through the center of mass, a force source is appended to the translational I element. This is shown in Fig. 2-26e. Note the power direction on the S_e bond.

We usually insert springs between pairs of 1-junctions representing key velocities by means of 0-junctions. This is done in Fig. 2-26e, but some of the 1-junctions have been eliminated. The right-hand spring sees the relative velocity $V_0(t) - V_2$, but the left-hand spring is attached at one end to the (immovable) wall, so we argue that the spring's relative velocity is $V_{\text{wall}} - V_1 = 0 - V_1 = -V_1$. The extra 0-junction for the left-hand spring allows us to have the $F_1 V_1$ sign-convention arrow pointing down

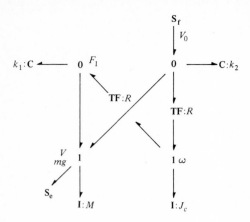

Figure 2-27 Simplified bond graph for the system of Fig. 2-26.

and also to have the C-sign arrow pointing into the C, as we always do. The role of that particular 2-port 0-junction is to enforce the idea that F_1 is the spring force and the cable force but that the spring extension is $-V_1$.

In two places on the graph we can simplify by combining two 0-junctions that are directly bonded. Notice that in each such simplification an interior bond disappears; thus what once was a distinct flow variable is no longer explicit in the model. If we simplify Fig. 2-26e, the velocities V_1 and V_2 are no longer available. The final graph is shown in Fig. 2-27.

If you followed this last example, you have come a long way. We have used the nine basic elements, C, I, R, S_e, S_f, 0, 1, TF, and GY, to represent inertia effects, springs, dampers, and various other effects, including rotational-translational coupling. Next steps include modeling more complex mechanical systems, as well as the generation and analysis of their associated differential equations. First, however, we consider the modeling of electric circuits, which is relatively straightforward.

PROBLEMS

2-1 Consider a mass moving on a horizontal surface with velocity V. If the mass experiences a friction force of magnitude $|F_f| = \mu N = \mu Mg$, where μ is a friction coefficient and $N = mg$ is the normal force, show in two free-body diagrams the direction of the friction force for positive and negative velocities. Sketch the shape of the friction law as in Fig. 2-2, including the effect of the sign of the velocity. Give an algebraic expression for F_f as a function of V.

2-2 Give the appropriate SI units for the following parameters: (a) the dashpot coefficient in Eq. (2-1), (b) the spring constant in Eq. (2-5), (c) the compliance in Eq. (2-6), and (d) the mass in Eq. (2-13).

2-3 Suppose that a real velocity source is intended to supply a constant velocity V_0 to a system but when it encounters a resisting force it slows down (Fig. P2-3a). Show that the bond graph in Fig. P2-3b, composed of an ideal source of velocity V_0, a 0-junction, and a resistor with damping coefficient b, can be used to model the real source. Relate the α of the real constitutive law to the damping coefficient b.

2-4 Make bond graphs for the systems illustrated.

2-5 Bond-graph this system, assuming a frictionless pivot but including the gravity force.

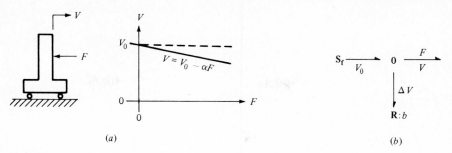

(a)

(b)

Figure P2-3

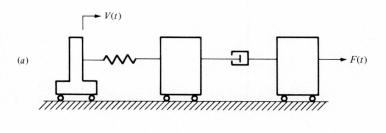

(a)

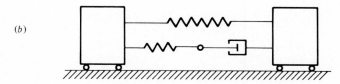

(b)

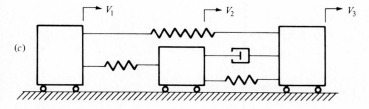

(c)

Figure P2-4

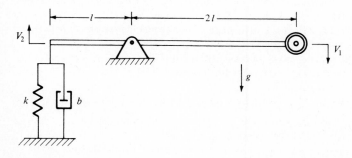

Figure P2-5

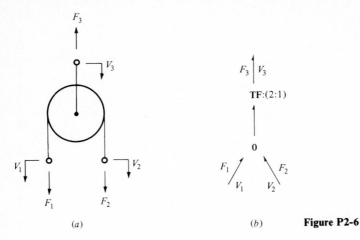

(a) (b) **Figure P2-6**

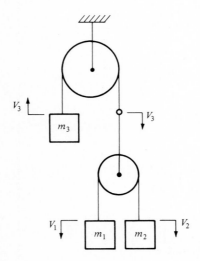

Figure P2-7

2-6 Show that the massless, frictionless pulley in Fig. P2-6a can be represented by the bond-graph junction structure in Fig. P2-6b.

2-7 Using the results of Prob. 2-6, construct a bond-graph for the system illustrated. All pulleys are massless and frictionless.

2-8 Suppose each floor of the building just moves sideways and the columns act as springs exerting shear forces that tend to line the floors up. Make a bond-graph model of this structure.

2-9 Convert the schematic into bond-graph form.

2-10 Make a tetrahedron of state like Fig. 2-7, using rotational variables instead of translational ones.

2-11 Give the SI units for the following parameters for rotary elements: (a) torsional spring constant, (b) torsional compliance, (c) torsional damper, (d) moment of inertia.

2-12 Make bond-graph representations for these fixed-axis rotary systems. (a) the shafts between fly-wheels have compliance, bearings have friction, and torque $\tau_0(t)$ is applied directly to flywheel $\alpha 1$.

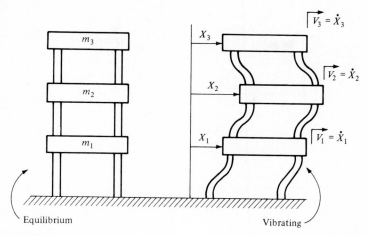

Figure P2-8

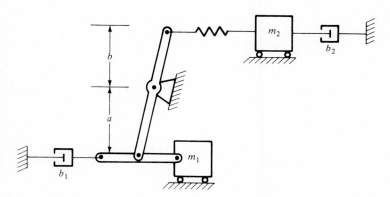

Figure P2-9

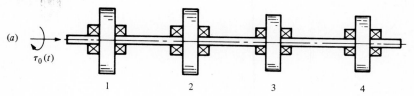

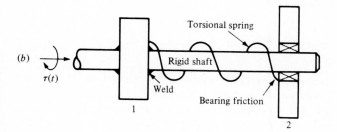

Figure P2-12

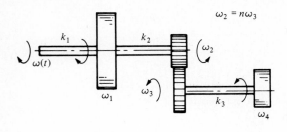

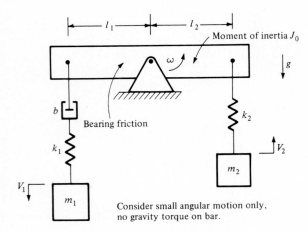

Figure P2-14

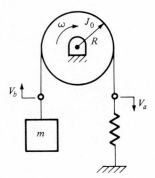

Figure P2-15

2-13 Model this geared system using a bond graph. Neglect frictional torques. Include inertia of both gears and both flywheels. Include compliance effects for all three shafts.

2-14 Make a bond graph for this system. You will need both rotary and translatory inertia elements. Consider small angular motions only and no gravity torque on the bar.

2-15 Make this system model into a bond graph. The cable does not slip, and there is bearing friction.

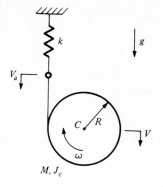

Figure P2-16

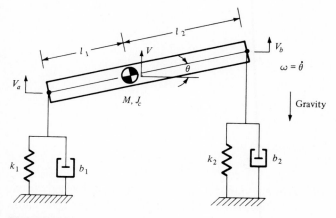

Figure P2-17

2-16 Bond-graph this system, in which a drum will fall with vertical velocity only, influenced by the cable wrapped around it (a Yo-Yo, more or less).

2-17 Consider small oscillations of the rigid body in the vertical direction, so that $V_a \approx V - l_1\omega$ and $V_b = V + l\omega_2$. Show a bond-graph model including gravity.

2-18 Modify your bond graph for Prob. 2-17 to allow motion inputs (velocity sources) at the bottom of both spring-damper suspension units.

THREE

ELECTRIC CIRCUITS AND NETWORKS

3-1 INTRODUCTION

Electrical systems are relatively simple to model because the well-established circuit symbols usually represent physical devices such as resistors, capacitors, and inductors rather well and the lines on a circuit diagram represent the connection pattern of conductors. Actually, a circuit graph is a mathematical model of a physical system, and, like all such models, it is accurate only for certain amplitude and frequency ranges. In fact, accurate modeling of even simple physical devices may require the use of several circuit symbols. For example, coils have not only inductance, but also resistance; resistors may have important inductance and capacitance effects, and so forth. In addition, nonlinear devices such as transistors may have small-signal equivalent-circuit representations that change as the operating point of the transistor is changed. This means that considerable skill and judgment may be required in making a circuit-graph representation of a real electrical system, although the simplest useful low-frequency model of an electrical system may be rather obvious.

In this chapter we concentrate on developing methods for converting circuit and network diagrams for electrical systems into bond graphs. This approach turns out to be more productive at the system level than using direct circuit-analysis techniques, even though direct circuit methods are highly developed and widely used by electrical engineers. In addition, bond graphs represent certain types of nonlinearities that arise in mechanical and hydraulic problems more efficiently than any other graphical language. Thus we choose to convert circuit graphs into bond graphs to achieve a unified approach to system dynamics.

3-2 ELECTRICAL VARIABLES AND BONDS

Figure 3-1 shows the basic electric bond, consisting of a pair of wires (ideal conductors) through which a single current i flows and across which a voltage e appears. In Fig. 3-1a, the arrows indicate the positive directions for current flow and voltage. The e arrow in this notation points to the higher potential when e is positive. Often a small plus sign is used near the point of the e arrow to signify the same thing. In Fig. 3-1b, the A and B parts of the circuit are shown as 1-ports bonded together by the wire pair, and in Fig. 3-1c the equivalent bond graph is shown. Note carefully how the sign conventions work: if ie is positive, power is flowing from A to B, so the bond-graph arrow points from A to B. If every current and every voltage on a circuit diagram has an indicated sign convention, every bond can be given the equivalent sign convention.

The e, f, p, and q variables for circuits are identified in Table 3-1. Since e is commonly used for both voltage and effort and q is commonly used for both charge and displacement, they should be easy to remember. That current i should be a flow-variable type probably seems natural, but you may be a little mystified by the magnetic-flux-linkage variable λ, which is the (electrical) momentum. However, as we use λ, it will become more familiar and natural.

Since voltage times current at a port is power, these are a natural pair of power variables, but the choice of which to take as effort and which as flow is not so obvious. The choice of voltage as effort seems natural in most respects. After all, voltage is electromotive *force*, so it is usually thought of as analogous to mechanical force, which we have called effort. Also, we tend to speak of current as *flowing*, and the

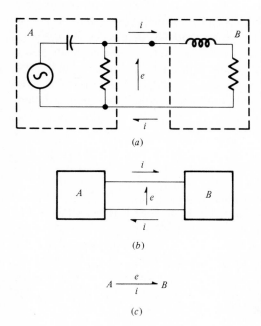

(a)

(b)

$$A \xrightarrow[\ i\ ]{\ e\ } B$$

(c)

Figure 3-1 The electric bond in circuit-diagram and bond-graph form.

Table 3-1 Electrical variables

Variable	General	Electrical	SI unit
	Notation		
Effort	$e(t)$	e, voltage	V
Flow	$f(t)$	i, current	A
Momentum or impulse	$p = \displaystyle\int^t e\, dt$	λ, flux linkage	Wb
Displacement	$q = \displaystyle\int^t f\, dt$	q, charge	C
Power	$\mathcal{P}(t) = e(t)f(t)$	$e(t)i(t)$	W
Energy	$E(p) = \displaystyle\int^p f\, dp$	$\displaystyle\int^\lambda i\, d\lambda$, magnetic	J
	$E(q) = \displaystyle\int^q e\, dq$	$\displaystyle\int^q e\, dq$, electric	J

analogy to velocity is immediate. Hence we shall define voltage as effort and current as flow.

One consequence of this choice of effort and flow identification is that series connections in electrical diagrams (in which two elements have the same current) are analogous to parallel connections in simple mechanical-translation diagrams (in which two elements have the same relative velocity). On occasion an engineer might want *connection* patterns for electrical and mechanical diagrams to be analogous. Then our definitions of effort and flow (and of course p and q) are reversed. Since the two analogies are equally useful in bond graphs, we adopt the traditional effort-flow assignments defined in Table 3-1. Hence a capacitor and a mechanical spring are analogous, and both are represented by C. Also, a mass is analogous to an inductor, and both are modeled by I elements. As far as the junctions go, you only need to remember that 0-junctions have a common effort (voltage) and 1-junctions have a common flow (current). With this in mind, series and parallel connections will take care of themselves.

3-3 ELECTRICAL 1-PORTS

It is easy to see how to represent the three basic electric-circuit elements (resistor, capacitor, and inductor) as 1-ports. All that needs to be done is to bend their leads around so that the voltage e across them and the single current i through them, appear on a conductor pair, as shown in Fig. 3-2. The bond-graph elements R, C, and I represent these circuit symbols by implying the appropriate constitutive laws between the proper variables. Resistors relate efforts (voltages) to flows (currents); capacitors relate efforts (voltages) to displacements (charges); and inertias relate flows (currents) to momenta (flux linkages). Both linear and nonlinear constitutive laws are sketched in Fig. 3-2. Note that an R, C, or I does not imply a linear law. We write these constitutive laws here in equation form for later reference

$$\text{Resistor:} \quad e = \begin{cases} Ri & \text{linear} & (3\text{-}1) \\ e(i) & \text{nonlinear} & (3\text{-}2) \end{cases}$$

$$\text{Capacitor:} \quad e = \begin{cases} \dfrac{q}{C} & \text{linear} & (3\text{-}3) \\ e(q) & \text{nonlinear} & (3\text{-}4) \end{cases}$$

$$\text{Inductor:} \quad i = \begin{cases} \dfrac{\lambda}{L} & \text{linear} & (3\text{-}5) \\ i(\lambda) & \text{nonlinear} & (3\text{-}6) \end{cases}$$

Equations (3-1), (3-3), and (3-5) appear as straight lines in Fig. 3-2. For a linear element we also speak of the *parameters* R, C, and L. Common practice is to mea-

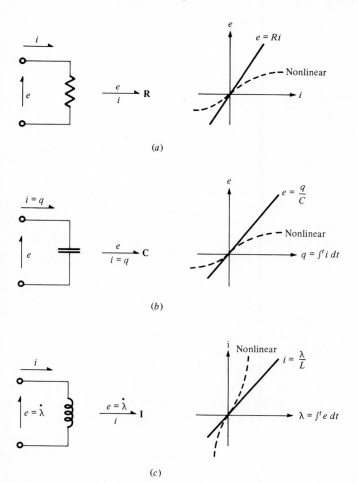

Figure 3-2 Electrical 1-port elements R, C, and I.

sure R in ohms (Ω), C in farads (F), and L in henrys (H). Because of their great practical importance, these units rightly deserve their special names in the SI system. (You can see from the constitutive laws above that an ohm is a volt per ampere, a farad is a coulomb per volt, and a henry is a weber per meter. The beauty of the SI is that if you stick to the basic units, you do not have to remember all the equivalences.) The notation in Eqs. (3-2), (3-4), and (3-6) is meant to imply a general relationship between the variables, which may or may not be linear.

The only electrical variable that may be unfamiliar is the flux linkage λ, which is analogous to mechanical momentum. Recall that in mechanics the common law $F = ma$ was integrated once in time to yield $P = mV$. In the same way, if you recall the equation for a linear inductor as

$$e = L\frac{di}{dt} \tag{3-7}$$

After consulting Table 3-1, you will see that one time integration with constant L produces

$$\lambda = \int^t e\, dt = Li \tag{3-8}$$

which is just Eq. (3-5) rearranged.

The time integral of any force is called the *impulse* of that force; when the force is also the net force on a mass initially at rest, the same time integral is equivalent to the mass momentum. We thus have used the general name "momentum" to stand for the integral of a force. In a similar way we call the time integral of a voltage a *flux-linkage variable*. If e is actually the net voltage on an inductor, then λ is the flux linkage for the inductor. Faraday's law states that $e = \dot{\lambda}$; that is, the voltage is equal to the time rate of change of the amount of flux linked by a current. An example of the use of this law is Eq. (3-7). It is not necessary here to describe in any detail what magnetic flux is or how it is linked by a current path in an inductor. It suffices for now to say that an inductor relates λ to i and by differentiating this relation we can find the basic relation between e and di/dt.

Since the properties of the basic 1-ports were presented at length in Chap. 2, here we remind you of them and point out how they describe electrical elements. Electric resistors dissipate power, and with the normal sign convention this means that the constitutive law for a resistor will lie in the first and third quadrant of the e-i plane, as shown in Fig. 3-2a. Ohm's law is just the linear version of a resistor law, and with our convention the resistance is positive. (Later you may encounter the idea of *negative resistance*, but this refers only to a locally negative slope of a nonlinear resistor relation which will still be in the first and third quadrants.)

Capacitors and inductors are conservative. The energy stored is

$$E(t) = \int^t ei\, dt \tag{3-9}$$

which can be written in two possible forms. One, associated with q, is

$$E(q) = \int^q e(q)\, dq \tag{3-10}$$

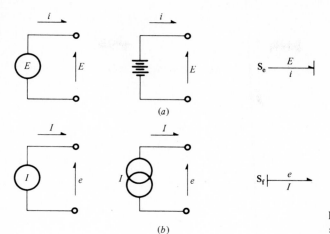

Figure 3-3 Voltage and current sources.

where $dq = i\ dt$. Here $E(q)$ is the electric energy stored in a capacitor; it can be thought of as the area under the constitutive law curve of Fig. 3-2b. Another form, associated with λ, is

$$E(\lambda) = \int^{\lambda} i(\lambda)\ d\lambda \qquad (3\text{-}11)$$

where $d\lambda = e\ dt$. Here $E(\lambda)$ is the magnetic energy and is represented as the area under the constitutive-law curve for the inductor shown in Fig. 3-2c.

For the common linear models of C and I the respective energies are

$$E(q) = \frac{q^2}{2C} \qquad (3\text{-}12)$$

and

$$E(\lambda) = \frac{\lambda^2}{2L} \qquad (3\text{-}13)$$

Thus q and λ are called the electric-energy variables.

Figure 3-3 shows the effort and flow sources that appear in circuit diagrams as voltage and current sources. The capital letters E and I on the bonds are reminders that these are the given sources variables, while e and i are the complementary bond variables determined by the system to which the sources are connected. Generally, E and I are functions of time, written $E(t)$ and $I(t)$; one important special case is when they are constant.

3-4 ELECTRICAL 3-PORTS

The common-effort and common-flow junctions in electric circuits are visualized easily as the connection patterns of ideal conductors in Fig. 3-4, where Fig. 3-4a shows how three pairs of terminals can be connected in such a way that three voltages are equal. The voltage equations are

$$e_1 = e_2 = e_3 \qquad (3\text{-}14)$$

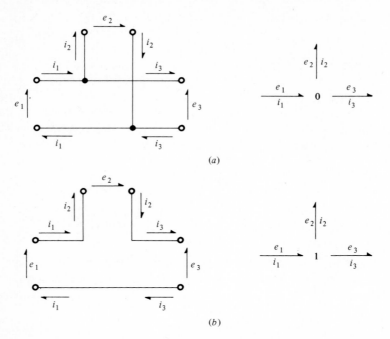

Figure 3-4 Circuit connections for 0- and 1-junctions.

By noting that the sum of the three currents entering either node in the circuit must add to zero and by taking into account the assumed positive directions shown for the currents we derive the relation

$$i_1 - i_2 - i_3 = 0 \qquad (3\text{-}15)$$

Equations (3-14) and (3-15) are incorporated in the 0-junction shown. Notice that there are three equations.

The circuit diagram and the bond graph of Fig. 3-4a contain the assumption that the same current flows out of one terminal and into the other at each terminal pair. This will be the case, for example, if 1-ports are connected across the terminals. The 0-junction is a representation of a *parallel electric connection*. Under similar assumptions, the connection pattern of Fig. 3-4b is represented by the 1-junction. Clearly

$$i_1 = i_2 = i_3 \qquad (3\text{-}16)$$

and by requiring the voltage drops encountered in a trip around the circuit to add to zero we have

$$-e_1 + e_2 + e_3 = 0 \qquad (3\text{-}17)$$

As you can see, 1-ports connected to a 1-junction are connected in *series* electrically. In summary, for electric circuits, the 0- and 1-junctions represent parallel and series connections, respectively. They embody local versions of the well-known *Kirchhoff current laws* and *Kirchhoff voltage laws* as part of their constitutive relations; see Eqs. (3-15) and (3-17).

Example 3-1 For simple circuits, a translation to a bond graph often can be made by inspection, requiring only that you learn to see series and parallel connections as shown in Fig. 3-4. An example of this approach appears in Fig. 3-5. The original circuit is distorted in Fig. 3-5b to show the terminal pairs of the 1-ports and 3-ports which are bonded together in Fig. 3-5c.

The sign-convention half arrows on the bond graph come from decisions about positive directions for voltages and currents. Since we *always* point the arrows into the passive 1-ports, only a few bonds need to be considered. The source voltage E and current i_a shown in Fig. 3-5b indicate that Ei_a represents power flowing *out* of the source, so the bond half arrow points *out* of S_e. Once we have decided on a direction for i_a, i_b must be as shown since the 1-junction implies that $i_a = i_b$. When we decide on the sense of e_b, the $e_b i_b$ bond must be oriented as in Fig. 3-5c. We similarly translate the $e_c i \cdot_c$ direction in Fig. 3-5b to the bond-graph sign in Fig. 3-5c.

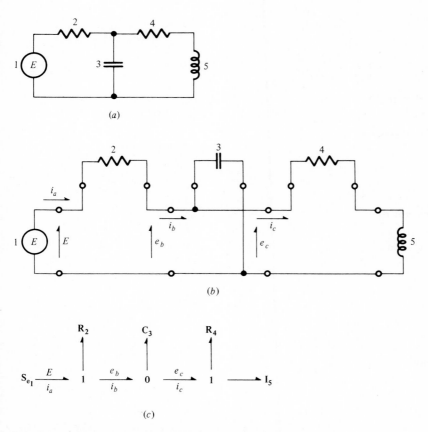

(a)

(b)

(c)

Figure 3-5 Circuit-to-bond-graph conversion for Example 3-1.

With some practice you can avoid writing circuit distortions to see 0- and 1-junctions. When two 1-ports have the same voltage across them, they are connected to a 0-junction. When two or more currents are the same, a 1-junction is involved. If you are clever enough, many circuits that seem complex can be broken down into series and parallel connections directly in this way. In Sec. 3-7 we present an organized procedure for doing this to any circuit.

Example 3-2 Figure 3-6 shows a slightly more complex circuit than the previous example. First, we assume that the parts of the circuit to the left of port *a* and to the

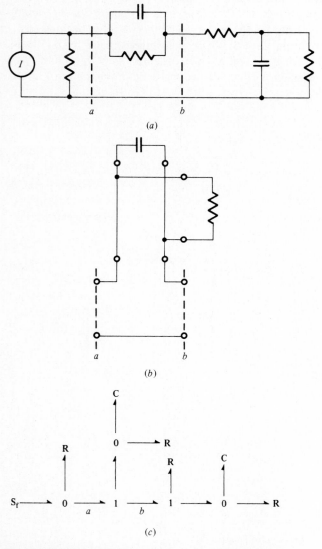

Figure 3-6 Circuit-to-bond graph conversion for Example 3-2.

right of port b can be translated directly into bond-graph form without circuit distortion. The center section, however, takes some thought. The R and C elements are clearly connected in parallel (since they feel the same voltage) and can be attached to a 0-junction. Also, the current at port a is equal to the current at port b, so the a and b ports should be attached to a 1-junction. We have just about argued ourselves from the circuit in Fig. 3-6a directly to the bond graph of Fig. 3-6c. The distorted piece of circuit in Fig. 3-6b should serve as a check on the 0-1, parallel-series connection pattern represented in the bond graph.

The two circuit examples we have just studied are quite easy to convert into bond graphs, mainly because each has a fairly simple series-parallel construction. For more complex circuits it may not be possible to make a basically sequential set of conversions from a circuit diagram to the bond graph. Before discussing a reliable procedure for doing any conversion we introduce the 2-ports that generalize *circuits* into *networks*.

3-5 ELECTRICAL 2-PORTS

The circuit symbols shown in Fig. 3-7 have the same meaning as the bond-graph symbols TF and GY. The circuit symbols and the bond-graph elements are useful in modeling real devices such as electric transformers and Hall-effect gyrators, but these elements are highly idealized and normally must be combined with other elements to model physical devices accurately. Since the constitutive laws of these elements have already been given for mechanical systems, we merely need to convert to the appropriate efforts and flows for electrical systems. The transformer laws are

$$e_1 = ne_2 \tag{3-18}$$

$$ni_1 = i_2 \tag{3-19}$$

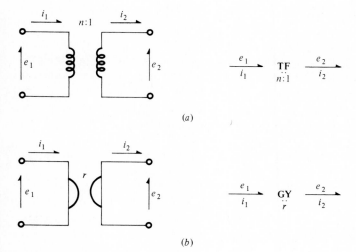

(a)

(b)

Figure 3-7 Ideal electrical 2-ports.

which together imply power conservation

$$\mathcal{P}_1 = e_1 i_1 = e_2 i_2 = \mathcal{P}_2 \tag{3-20}$$

Note that, as usual, we have the problem of deciding where the transformer ratio n belongs in Eqs. (3-18) and (3-19). If we turned the transformer end for end and still called the left end 1 and the right end 2, $1/n$ would appear in the equations. Since n is dimensionless, it is often safest both in a circuit and in a bond graph simply to write either a voltage or a current relation with n in its proper place next to the transformer. (A computer program may require a specific convention for specifying n, but the convention may be annoying when a person is doing the equation formulation. People normally are more imaginative than computers and more comfortable with ambiguous rules.)

The gyrator laws are easily remembered because the gyrator parameter r has dimensions of ohms, just like a resistor

$$e_1 = r i_2 \tag{3-21}$$

$$r i_1 = e_2 \tag{3-22}$$

Of course, power is also conserved

$$\mathcal{P}_1 = e_1 i_1 = e_2 i_2 = \mathcal{P}_2 \tag{3-23}$$

The 2-ports allow more complex electrical systems to be modeled than the basic 1-ports and the 0- and 1-junctions. It is sometimes useful to make the distinction between *circuits*, which consist only of series and parallel connections of 1-ports, and the more general *networks*, which may contain 2-ports as well as other elements. Before showing how the 2-ports enter bond-graph models of networks, we consider some other common network elements.

3-6 CONTROLLED SOURCES, ACTIVE BONDS, AND SIGNALS

Many electric-network models really represent linearized small-signal models of nonlinear devices. In addition to the elements already studied these networks often contain special elements called *controlled sources*, i.e., sources of voltage or current influenced by other variables at remote locations in the system. The influences are of a *signal*, or one-way, nature; e.g., a voltage source may provide an output voltage proportional to a voltage elsewhere in the system without having any back effect on the controlling voltage.

An *active bond* is like a *signal* in a block diagram or signal-flow graph, as discussed briefly in Chap. 1. An active bond is shown with a *full arrow:*

$$A \longrightarrow B$$

a notation borrowed from block diagrams. A normal bond always has two signals, effort and flow, but an active bond communicates only one of the two possible signals in a single direction.

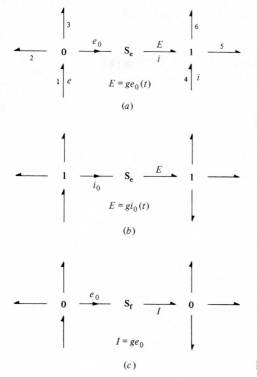

Figure 3-8 Controlled sources and active bonds.

The examples shown in Fig. 3-8 may make this clearer. The voltage source in Fig. 3-8*a* is controlled by a voltage somewhere in the system. The active bond impinging on S_e indicates that E is controlled by the voltage e_0. For example, E might be given by

$$E(t) = ge_0(t) \qquad (3\text{-}24)$$

where g might be a constant gain. More complex relationships are also possible.

Since the active bond comes from a 0-junction, the voltage e_0 is the same as e_1, e_2, and e_3. However, the active bond has *no power* because the current corresponding to e_0 is assumed to be negligible. The current sum at the 0-junction involves only i_1, i_2, and i_3. On the other hand, the source has finite power Ei. Therefore the controlled source has an implied source of power internally.

Figure 3-5*b* and *c* shows a current-controlled voltage source and a voltage-controlled current source, respectively. The current signal comes from a 1-junction but has no back voltage, just as the voltage signal comes from a 0-junction and has no reaction current.

Active bonds and controlled sources are indispensible for modeling certain systems, but one must use them with care since they imply power supplies, just as normal S_e and S_f elements do.

3-7 A PROCEDURE FOR BOND GRAPHING ELECTRICAL SYSTEMS

The following procedure considers each circuit node (or place at which two or more branches join) to be at a voltage defined with respect to an initially unspecified ground. Elements which connect between any two nodes then experience a relative voltage. If each node is represented by a 0-junction, 1-junctions between two 0-junctions exhibit relative voltages and the appropriate elements can be appended. A particular node is then chosen to be the ground node; it is eliminated, and the graph is simplified. The procedure is simple because the initial bond graph follows the circuit topology. We call this procedure the *circuit-construction method*.

Circuit-Construction Method

1. For each node in the circuit with a distinct voltage establish a 0-junction.
2. Insert each 1-port circuit element between the appropriate pair of 0-junctions by attaching the 1-port to a 1-junction and then bonding the 1-junction to the two 0-junctions.
3. Assign sign conventions using a "through" sign convention for the 0—1—0 sections; i.e.,

$$0 \longrightarrow 1 \longrightarrow 0$$

This assures that the 1-ports will see voltage differences defined consistently with respect to implied ground.
4. Choose a particular node to be ground. This will be the zero-voltage node so that all bonds connected to it can be eliminated.
5. Simplify the graph by replacing any 2-port 0- or 1-junctions that happen to have "through" power half arrows with simple bonds. For example

$$\longrightarrow 1 \longrightarrow \; = \; \longrightarrow \quad \text{and} \quad \longrightarrow 0 \longrightarrow \; = \; \longrightarrow$$

but leave alone

$$\longrightarrow 0 \longleftarrow \quad \text{and} \quad \longleftarrow 1 \longrightarrow$$

Example 3-3 Suppose we try the procedure on an example that could also be done by inspection. The circuit of Fig. 3-9a clearly has a single current, so we would expect the three 1-ports to be joined to a 1-junction. We put down 0-junctions for nodes a, b, and g in Fig. 3-9b. Each 1-port is attached to a 1-junction and is strung between the appropriate 0-junctions in Fig. 3-9c. You can see now that the bond graph in Fig. 3-9c follows the circuit-diagram pattern closely. The signs on the bonds between a and b carry the implication

$$e_{ab} = e_a - e_b \tag{3-25}$$

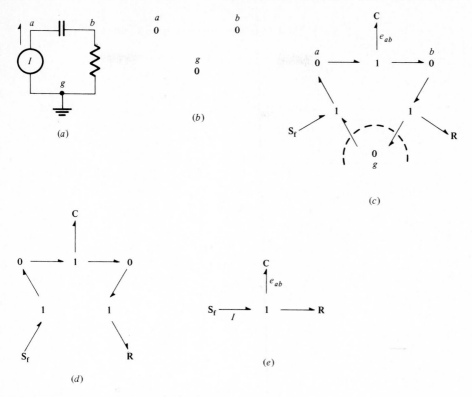

Figure 3-9 Circuit construction for Example 3-3.

If we wished, we could point the half arrows from b to a through the 1-junction, implying

$$e_{ab} = e_b - e_a \qquad (3\text{-}26)$$

If voltage and current signs are given on the circuit, they can be transferred to the bond graph.

Now we pick g to be the zero-voltage point and eliminate g's bonds in Fig. 3-9d. This could be a final bond graph, but we note that our sign conventions result in several 2-port junctions that can be simplified, yielding the result we anticipated (Fig. 3-9e). The next example is chosen to be more practical as well as more complex.

Example 3-4 Fig. 3-10a shows a Wheatstone-bridge circuit. We have shown four resistors in the bridge, which might represent strain gages that change resistance and unbalance the bridge when some mechanical device is loaded. The voltage e is zero when the bridge is balanced, and we mean to record e by means of an instrument that draws negligible current.

The four nodes a, b, c, and d are represented by four 0-junctions in Fig. 3-10b,

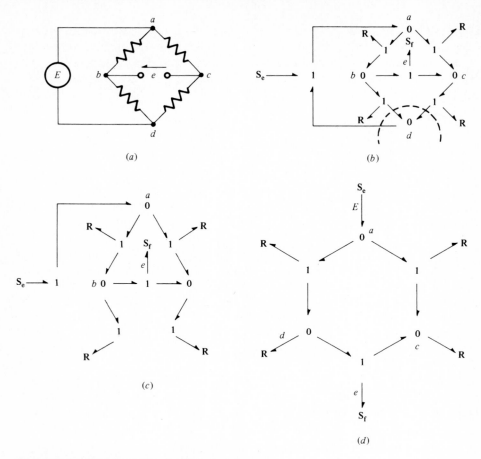

Figure 3-10 Wheatstone bridge for Example 3-4.

and we put the 1-ports in the graph on 1-junctions. Note that the relative voltage e appears on the bond of a current source. The current source (with current equal to zero) models the high-impedance instrument. The sign conventions have been applied with an eye to the future, since the circuit itself had few signs for voltages and currents. Since we plan to use d for the ground node, the signs are planned so that after the d 0-junction and its bonds have been removed in Fig. 3-10c, the maximum amount of simplification will occur. The result in Fig. 3-10d is a pattern reminiscent of the chemical formula for benzene.

This is an example in which the choice of ground node is not obvious. For every choice of ground node, a different bond graph results (although in all cases the benzene-ring type of pattern results) because each bond graph concerns itself with voltages defined with respect to a different ground node. They all are equivalent dynamically, of course.

An important point is that a ground node *must* be chosen. The bond graph of Fig. 3-10b may look all right to you, but it cannot be used to compute voltages at a, *b*, *c*, and *d* since these voltages are with respect to a floating, undefined ground. Equations from such a graph turn out to be incomplete or contradictory. By choosing *d* to be the ground node we defined voltages at *a*, *b*, and *c* with respect to *d*, and these voltages are computable. This point is related to the fact in circuit-analysis methods that for *n* nodes there are *n* − 1 independent Kirchhoff current-sum relations.

Network-Construction Methods

The circuit-construction procedure just given works for networks with only a few modifications. For 2-ports and controlled sources we need to ask which voltages and currents appear at the ports; using 0- and 1-junctions when necessary to relate these quantities to the node voltages, we can insert these multiports into the system.

Example 3-5 The first example is shown in Fig. 3-11 and, while it is fairly complicated, most of the steps in the construction should be familiar from the circuit-construction procedure. The new part of the system is the *isolating* transformer, which removes the hard-wire connection between the two parts of the system.

When a system is thus isolated, we need to choose *two* ground nodes, one on each side of the transformer. One way to see why is by observing that the voltage relation for a transformer, Eq. (3-18), is really a relation between relative voltages. Thus when we consider the four voltages at the four transformer terminals (points *d*, *e*, *f*, and *j* in Fig. 3-11) the transformer provides only a relation between difference voltages ($e_{de} = e_d - e_e$ and $e_{fj} = e_f - e_j$). This leaves us free to set two of the four voltages to zero. In practice, we need to pick a ground for every isolated circuit part. Let us choose two grounds as indicated in Fig. 3-11.

Notice how the 2-port transformer element is connected using 1-junctions as if each port were a 1-port. The 1-junctions ensure that the current entering from node *d* will flow through the transformer and into node *e*. It also enforces the constraint that the transformer voltage be the relative voltage $e_{de} = e_d - e_e$. Similar statements can be made for the other side of the transformer. In general, any port of any device can be modeled by a pair of 0-junctions with a 1-junction.

In the final bond graph of Fig. 3-11*d* we have noted that several voltages are defined with respect to one or the other of the ground nodes. Also, we were purposely not clever at assigning sign conventions on the right-hand side and are stuck with some nonremovable 2-port 1-junctions. You might like to go back and modify sign conventions to make the junctions removable.

As you may imagine, we do not have to choose two grounds except when two parts of the circuit are connected *only* by a transformer, i.e., when the transformer is an isolating transformer. When other circuit elements also connect the parts of the system on both sides of the transformer, voltages throughout the system are related, so that only a single ground need be chosen.

Example 3-6 Figure 3-12 shows a network containing a voltage-controlled current source. Most of the construction procedure is the same as it would be for a circuit

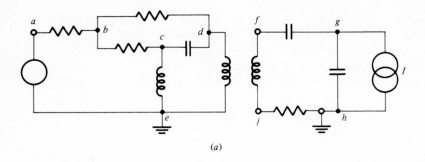

(a)

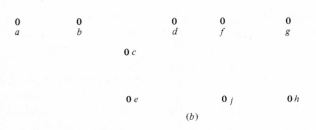

(b)

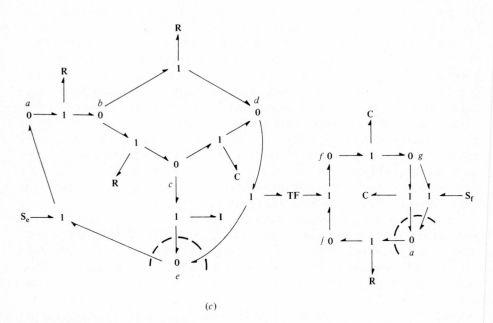

(c)

Figure 3-11 Network for Example 3-5.

(d)

(because most of the network is a circuit), but a word of explanation is in order for the structure surrounding the source. The controlled current source connects on a 1-junction between nodes d and f, just as a normal, uncontrolled source would. It receives a voltage signal e_0 on an active bond, which implies that current on the active bond vanishes. We know that the active bond carries a voltage signal since it emanates from a 0-junction and *all* bonds, active or not, have the same effort when connected to a 0-junction. The voltage e_0 is measured across nodes c and f, so the e_0 junction is connected to these nodes using a 1-junction.

The role of the 2-port 0-junction is important. First, the voltage at the junction is transmitted actively to the S_f. Second, the active bond has zero current and, by the laws of 0-junction, the algebraic sum of currents on all bonds vanishes. This means that the current on the normal bond must also vanish and further that there is no current on the 1-junction between nodes c and f. Thus the voltage e_0 which controls the current source is an *open-circuit* voltage.

The entire purpose of the extra 0-junction from which the active bond emanates is to create the relative voltage e_0 and to make sure that no current flows. When node f is chosen to be ground, the bond graph simplifies drastically, as shown in Fig. 3-12c. A number of 2-port 0- and 1-junctions can be eliminated, the extra 0-junction used to produce e_0 for the active bond being directly bonded to the 0-junction for node c. Thus these two 0-junctions can be combined into one.

As it happens, you probably could have produced Fig. 3-12c directly by inspection because node f is such an obvious place for the ground node. The extra structure necessary to produce e_0 in Fig. 3-12b is worth understanding since it illustrates a generally useful procedure. Just as we can simplify bond graphs containing 2-port 0- and 1-junctions with "through" sign conventions by changing them into simple bonds, so too can we insert extra 0- or 1-junctions in any bond when we want to take out an effort or flow signal on an active bond. The bond relationships are unaffected by the inserted junction and its active bond since the active bond draws no power

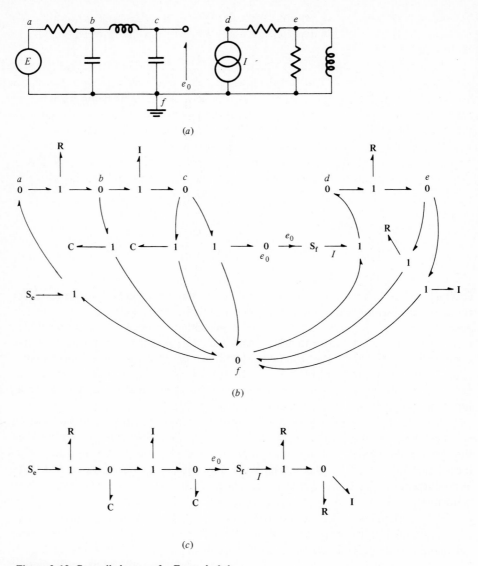

Figure 3-12 Controlled source for Example 3-6.

due to its lack of back effect. This type of modeling is particularly useful when instruments are coupled to power systems. Examples in Chap. 19 demonstrate the idea.

3-8 CIRCUIT AND BOND-GRAPH TOPOLOGY

The *topology* of a circuit refers to the pattern of connection of the elements, which determines how the circuit will behave. Circuit models do not change when the conduction paths are distorted and moved about as long as the connections are still the same at the nodes. This is true of physical circuits also, except at quite high frequencies, when field effects may depend upon the geometry of the circuit as well as its topology.

At the beginning of the circuit-construction procedure there is a close relationship between circuit and bond-graph topology, with every electric node having its counterpart in a bond-graph 0-junction and branch connections being mirrored by bond-graph connections through 1-junctions. After the ground-node choice and simplification, however, the bond graph often ceases to resemble the circuit.

Nonetheless, it is easy to associate bond graphs with circuits. To demonstrate we examine some typical circuit patterns and their bond-graph counterparts. Figure 3-13 shows some common connection patterns called *pi*, *tee*, and *ladder networks*.

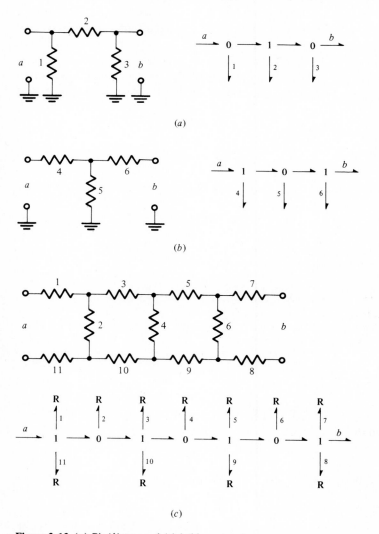

Figure 3-13 (a) Pi, (b) tee, and (c) ladder network.

Although the circuit graphs show resistors in each 1-port location, we really are interested in the connection patterns. The pi and tee networks in Fig. 3-13a and b look different in circuit-graph form, but the bond-graph junction structures are very similar, involving an interchange of 0's and 1's. This means that pi and tee networks are topological *duals;* i.e., the relationships for voltages in pi are the same as the relationships for currents in tee and vice versa. The bond graphs indicate this immediately if you recall that 0- and 1- are duals. This interesting relationship may be less evident in the circuit graphs. Sometimes it proves useful to obtain the topological dual of a circuit, and interchanging 0s and 1s on a bond-graph model may do the trick. If you convert a circuit into a bond graph and then dualize the bond-graph junctions, you can usually find a circuit equivalent to the dualized bond graph. You should be aware, however, that some circuits do not have duals; i.e., while every circuit can be converted into a bond graph (we have just shown you how), not every bond graph can be converted into a circuit. Do not assume that any random bond graph you happen to write represents a circuit.

Finally, note in Fig. 3-15c the ladder circuit and its associated bond graph. It appears to be a joining of two halves of a pi or tee. Such repeated structures are often used to model distributed electrical elements, such as power lines, by circuit elements.

PROBLEMS

3-1 One sometimes sees a nonlinear-inductor law written in terms of an inductance $L(i)$ which is a function of the current, so that

$$L(i)\frac{di}{dt} = e$$

We have restricted our use of the term "inductance" to the case in which L is constant. In the nonlinear case we say that λ is a nonlinear function of the current

$$\lambda = \phi(i)$$

How does this description of a nonlinear inductor relate to the $L(i)$ concept?

3-2 Show that a mechanical system which contains a spring and damper in series (Fig. P3-2a) has a bond graph different from, but related to, that of an electrical system containing a capacitor and resistor in series (Fig. P3-2b). Assume A and B are 1-ports.

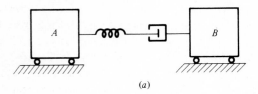

(a)

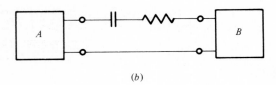

(b)

Figure P3-2 (a) Mechanical spring-damper; (b) electric capacitor-resistor.

3-3 Make a bond graph translation for the illustrated circuits using the inspection method.

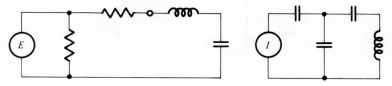

Figure P3-3

3-4 Make circuit shown into a bond graph and transfer the sign conventions.

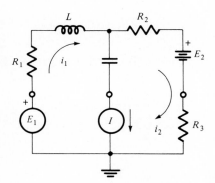

Figure P3-4

3-5 Use the circuit-construction procedure for creating an equivalent bond graph.

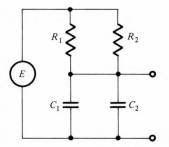

Figure P3-5

3-6 Make a bond-graph model for this network.

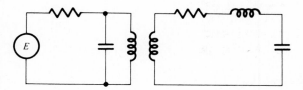

Figure P3-6

3-7 Convert the circuit shown into a bond graph.

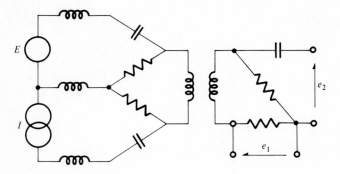

Figure P3-7

3-8 Convert this gyrator network into a bond graph.

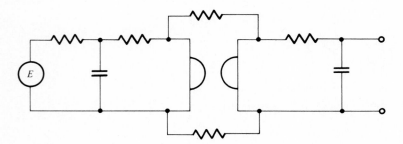

Figure P3-8

3-9 Make a bond graph by the inspection method for this network containing a voltage-controlled voltage source.

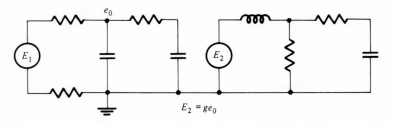

$$E_2 = g e_0$$

Figure P3-9

BASIC TECHNIQUES OF EQUATION FORMULATION

4-1 STANDARD FORMS FOR SYSTEM EQUATIONS

A bond graph, together with element constitutive equations, contains all the necessary mathematical information about a system model. One way to convince yourself that the model is complete is to write down a set of describing equations, usually differential, for the model. A remarkable feature of bond graphs is that the equation-formulation process can be studied without actually writing equations. If you do not like mathematical surprises, you will learn to appreciate the ability of bond-graph methods to let you know how the equations will turn out before you start.

All this is accomplished through *causality,* the topic we consider next. Before embarking on this journey, let us consider where we want to go. Bond graphs lend themselves to *state-space* methods, in which an nth-order dynamic system is represented by n first-order differential equations in n variables. This is a standard form popular with mathematicians, numerical analysts, and, more recently, automatic-control specialists. On the other hand, it is also possible to describe a system with a single nth-order equation involving a single unknown variable or with various combinations of equations and unknowns of appropriate orders.

As an example we consider the double-oscillator model shown in Fig. 4-1. There are a variety of ways to write equations for this mechanical system. One could use the coordinates X_1 and X_2 in Fig. 4-1a to write Newton's law for both masses. If X_1 is eliminated in favor of X_2 and its derivatives, a single fourth-order equation can be found:

$$\dddot{X}_2 + \left[\frac{k_2}{m_2} + k_1 \left(\frac{1}{m_1} + \frac{1}{m_2} \right) \right] \ddot{X}_2 + \frac{k_1 k_2}{m_1 m_2} X_2 = k_1 \left(\frac{1}{m_1} + \frac{1}{m_2} \right) g \quad (4\text{-}1)$$

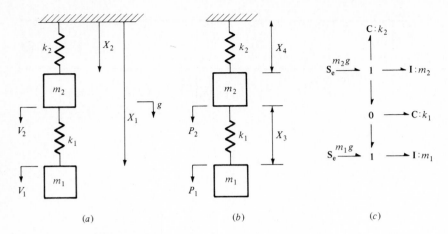

Figure 4-1 A double oscillator.

As you can imagine, obtaining this form required quite a bit of algebra. The physical parameters are combined in three coefficients, one for $\ddot{X}_2$, one for X_2, and one for g; but note that if m_1 is changed, for example, all three coefficients will change. This may be a disadvantage in design studies.

Another possibility might be to use X_3 and X_4 as generalized coordinates for Lagrange's equations (Fig. 4-1b). The result is two second-order equations

$$m_1\ddot{X}_3 + m_1\ddot{X}_4 + k_1X_3 = m_1g$$

$$m_1\ddot{X}_3 + (m_1 + m_2)\ddot{X}_4 + k_2X_4 = (m_1 + m_2)g \qquad (4\text{-}2)$$

Finally, using the bond graph of Fig. 4-1c, we find four first-order equations in momenta and displacements

$$\dot{P}_1 = +k_1X_3 + m_1g$$

$$\dot{P}_2 = -k_1X_3 - k_2X_4 + m_2g$$

$$\dot{X}_3 = -\frac{1}{m_1}P_1 + \frac{1}{m_2}P_2$$

$$\dot{X}_4 = \frac{1}{m_2}P_2 \qquad (4\text{-}3)$$

Now, the three equations (4-1) to (4-3) all represent the same physical model, and in this sense they are equivalent. However, we shall concentrate on the form represented by Eq. (4-3). As you can see, the n first-order equations are generally simpler in form than the smaller number of higher-order equations, and there will be an advantage in that the bond-graph equations are closely connected to the energy stored in the system. In fact, variables in bond-graph equation sets will always be the energy variables p and q, and the state equations themselves are effort ($\dot{p}$) and flow ($\dot{q}$) equations.

For general nonlinear models we shall write equations in the form

$$\dot{X}_1 = \phi_1(X_1, X_2, \ldots, X_n; u_1, u_2, \ldots, u_r)$$

$$\dot{X}_2 = \phi_2(X_1, X_2, \ldots, X_n; u_1, u_2, \ldots, u_r) \qquad (4\text{-}4)$$

$$\ldots\ldots\ldots\ldots\ldots\ldots\ldots\ldots\ldots\ldots\ldots\ldots\ldots$$

$$\dot{X}_n = \phi_n(X_1, X_2, \ldots, X_n; u_1, u_2, \ldots, u_r)$$

where X_i = *state variables* (energy variables)

$\dot{X}_i$ = derivatives with respect to time of state variables (efforts or flows)

u_j = efforts or flows from sources

ϕ_i = algebraic functions

For linear systems the ϕ's become (you guessed it) linear. Equations (4-4) become very orderly, although they seem harder to write out in longhand

$$\dot{X}_1 = a_{11}X_1 + a_{12}X_2 + \cdots + a_{1n}X_n + b_{11}u_1 + b_{12}u_2 + \cdots b_{1r}u_r$$

$$\dot{X}_2 = a_{21}X_1 + a_{22}X_2 + \cdots + a_{2n}X_u + b_{21}u_1 + b_{22}u_2 + \cdots b_{2r}u_r \quad (4\text{-}5)$$

$$\ldots\ldots\ldots\ldots\ldots\ldots\ldots\ldots\ldots\ldots\ldots\ldots\ldots\ldots$$

$$\dot{X}_n = a_{n1}X_1 + a_{n2}X_2 + \cdots + a_{nn}X_n + b_{n1}u_1 + b_{n2}u_2 + \cdots b_{nr}u_r$$

The a and b parameters are constant if the system is *time-invariant* but depend upon time if the system is time-varying. The formidable appearance of Eq. (4-5) is easily eliminated if one is willing to use vector and matrix notation

$$\dot{X} = AX + Bu \qquad (4\text{-}6)$$

where X and u are column vectors and A and B are matrices, the components of which appear in Eq. (4-5). Despite the formal simplicity of Eq. (4-6), we do not require you to use vectors and matrices unless you want to. We shall simply write our equations in such form that if they happen to be linear, they can be arranged in matrix form.

Always remember that linear equations are a special case and nonlinear equations are more general. However, linear, constant-coefficient systems are so important in practice that we shall study them in some detail later.

4-2 CAUSALITY FOR BASIC MULTIPORTS

To organize component constitutive laws into sets of differential equations we need to make a series of cause-and-effect decisions. This is fundamentally what is done whenever algebraic relations are combined, but we shall provide an organized approach using bond graphs. It is an interesting fact that bond graphs and the phys-

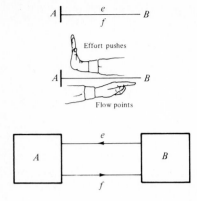

Figure 4-2 The meaning of the causal stroke.

ical systems they represent are full of interacting variable pairs. If two components are bonded together, we can think of the effort as causing one component to respond with the flow while the flow causes the first component to respond with the effort. Thus the cause-effect relations for efforts and flows are represented in opposite directions. A single mark on a bond, which we call the *causal stroke*, indicates how e and f simultaneously are determined causally on a bond.

Figure 4-2 shows A and B bonded together with the short perpendicular line (the causal stroke) on the end of the bond near A. What this means is that B determines $e(t)$ and impresses it on A, which causes A to respond with $f(t)$, which is returned to B. Of course, f causes B to respond with e—but let's stop going around this interaction loop. You may think of the stroke as a hand *pushing the effort* on A and the other end of the bond as a finger *pointing the flow* at B. Finally, you can split the bond into two signals, e and f, as in a block diagram. Then the causal stroke simultaneously shows the directions in which both the effort signal and the flow signal are determined. Everything is reversed if the causal stroke is moved to the b side.

What may be hard to see is that the sign-convention half arrow and the causal stroke are completely independent; i.e., all four cases

$$A \vdash B \qquad A \vdash B \qquad A \dashv B \qquad A \dashv B$$

are possible. Recall that the sign convention has to do with the choice of positive sense for e and f on a bond; the causal stroke indicates an input-output sense for e and f at each port of the two components together.

The basic bond-graph elements can be put into a pattern based upon how much flexibility they have with respect to causality. This is shown in Table 4-1, where the causally most restricted elements appear at the top and the least restricted at the bottom.

The source elements make sense only with one causality. Since effort sources are assumed to apply a predetermined effort on other components, the only possible causality is as shown. The flow source must impress its flow on other components, and so it also has a single causality.

The junction structure elements TF, GY, 0, and 1 have some flexibility regarding causality, but only a restricted set of patterns is possible. Consider the transformer, for example,

$$\overset{1}{\underset{}{\rightharpoonup}} \overset{m}{\underset{}{\text{TF}}} \overset{2}{\underset{}{\rightharpoonup}}$$

$$e_1 = me_2 \tag{4-7}$$

$$mf_1 = f_2 \tag{4-8}$$

If e_2 is impressed on the transformer, Eq. (4-7) could be used to find e_1, which then could be impressed on some other component. Using causal strokes, this causality would be written

$$\overset{1}{\underset{}{\vdash}} \overset{m}{\underset{}{\text{TF}}} \overset{2}{\underset{}{\vdash}}$$

Note that this causal assignment also implies that f_1 is impressed on TF. Then the TF computes f_2 and delivers f_2 to some other component. What the TF does in this causality can be expressed in computer-program form (input variables on right, outputs on left) by rearranging Eqs. (4-7) and (4-8)

$$e_1 = me_2 \tag{4-7}$$

and

$$f_2 = mf_1 \tag{4-8}$$

In other words, given e_2 and f_1 as inputs, the TF computes e_1 and f_2 as outputs. The only other causal possibility is that we could give the TF e_1 and f_2, and it could find

Table 4-1 Causal considerations for the basic multiports

Necessary causality	S_e —⊣ S_f ⊢—
Restricted causality	—⊣ TF —⊣ or ⊢— TF ⊢— —⊣ GY ⊢— or ⊢— GY —⊣ —⊣ 0 —⊣ or ⊢— 0 —⊣ or ⊢— 0 ⊢— ⊢— 1 ⊢— or —⊣ 1 ⊢— or —⊣ 1 —⊣
Integral causality	—⊣ I ⊢— C
Derivative causality	⊢— I —⊣ C
Arbitrary causality	—⊣ R or ⊢— R

e_2 and f_1

$$e_2 = \frac{1}{m} e_1 \tag{4-7a}$$

and

$$f_1 = \frac{1}{m} f_2 \tag{4-8a}$$

This corresponds to the other possible causal pattern

$$\overset{m}{\underset{1}{\dashv}} \overset{..}{\text{TF}} \overset{2}{\dashv}$$

Because of Eq. (4-7) two causal marks cannot be put on the ends of both bonds next to the TF since this would imply that two different efforts could be impressed on the TF, which would contradict Eq. (4-7) in general.

The idea behind the gyrator causal patterns is similar:

$$\overset{r}{\underset{1}{-}} \overset{..}{\text{GY}} \overset{2}{-}$$

$$e_1 = rf_2 \tag{4-9}$$

$$rf_1 = e_2 \tag{4-10}$$

If we are given f_2, we know e_1. Then also f_1 must be given, and e_2 can be found from Eq. (4-10). This causal pattern and the equations in input-output form are

$$\overset{r}{\underset{1}{\vdash}} \overset{..}{\text{GY}} \overset{2}{\dashv}$$

$$e_1 = rf_2 \tag{4-9}$$

$$e_2 = rf_1 \tag{4-10}$$

The only other causal possibility is

$$\overset{1}{\dashv} \text{GY} \overset{2}{\vdash}$$

$$f_1 = \frac{1}{r} e_2 \tag{4-10a}$$

$$f_2 = \frac{1}{r} e_1 \tag{4-9a}$$

Thus we see that the causal strokes tell us exactly how the implied equations should be written.

The 0- and 1-junctions have three possible causal patterns if they are 3-ports, as shown in Table 4-1. An n-port version of 0 or 1 would have n possible causal patterns. To see why consider the specific example

$$\overset{1}{\rule{0.5em}{0.4pt}} \overset{\displaystyle \mid 2}{0} \overset{}{\underset{3}{\longrightarrow}}$$

$$e_1 = e_2 = e_3 \qquad\qquad (4\text{-}11)$$

$$f_1 - f_2 - f_3 = 0 \qquad\qquad (4\text{-}12)$$

Note that Eq. (4-11) is really two independent equations and that the plus and minus signs for a 0-junction all occur in the flow relations, Eq. (4-12). (The net sum of flows is zero, where we have taken arrows in as positive.) Suppose e_1 is impressed on the 0. Then e_2 and e_3 will be known from the effort identity. We put e_1 on the right

$$e_2 = e_1 \qquad e_3 = e_1 \qquad\qquad (4\text{-}11a)$$

Also, f_1 can be found from f_2 and f_3 (since e_1 was input) by

$$f_1 = f_2 + f_3 \qquad\qquad (4\text{-}12a)$$

Again, we put f_2 and f_3 (inputs) on the right. The causal pattern is

$$\overset{1}{\dashv} \overset{\displaystyle \overset{2}{\mid}}{0} \overset{3}{\vdash}$$

The key property of causal assignment which allows us to organize Eqs. (4-11) and (4-12) into (4-11a) and (4-12a) is the bilateral nature of effort and flow causalities: if e_1 is an input to an element, f_1 is an output and vice versa. There are exactly three possible causal patterns for a 3-port 0-junction because there is only a single effort, and once it has been impressed on one bond by putting the stroke next to the 0, the other bonds must regard the effort as known and must communicate this effort outward. Alternatively, given two flow inputs, the third flow must be an output because of Eq. (4-12). The other causal patterns of Table 4-1 come by changing the role of bond 1 in our example to other bonds. For example, let e_2 be input effort

$$\overset{1}{\vdash} \overset{\displaystyle \mid 2}{0} \overset{3}{\vdash}$$

This means that

$$e_1 = e_2 \qquad e_3 = e_2 \qquad\qquad (4\text{-}11b)$$

and

$$f_2 = f_1 - f_3 \qquad\qquad (4\text{-}12b)$$

The 1-junction causal patterns are found by the same kind of reasoning, with efforts and flows interchanged. As an example, the 1-junction

$$\overset{1}{\rule{0.5em}{0.4pt}} \overset{\displaystyle \mid 2}{1} \overset{3}{\rule{0.5em}{0.4pt}}$$

$$f_1 = f_2 = f_3 \qquad\qquad (4\text{-}13)$$

$$-e_1 + e_2 + e_3 = 0 \qquad\qquad (4\text{-}14)$$

might have f_1 defined as input. Then Eq. (4-13) requires that f_2 and f_3 be imposed on other components

$$f_2 = f_1 \qquad f_3 = f_1 \tag{4-13a}$$

$$e_1 = e_2 + e_3 \tag{4-14a}$$

This also shows that if two of the three efforts are imposed on a 1-junction, the third must be an output because of Eq. (4-14). Here is one more example, in which the role of bond 1 is exchanged with bond 3

$$f_1 = f_3 \qquad f_2 = f_3 \tag{4-13b}$$

$$e_3 = e_1 - e_2 \tag{4-14b}$$

In this case, the 1-junction receives f_3, e_1, and e_2 as input quantities and responds with the complementary set e_3, f_1, and f_2 as output quantities.

Returning to Table 4-1, we see that the next level concerns the energy-storing elements I and C. Although these elements can accept either possible causality, there is a major difference between the two choices. In one case, the effort-flow relation involves integration; in the other it involves differentiation.

Consider the C element. The constitutive law for a C always relates e to q, and q is the integral of f. Thus, if f is the input to a C, it is first integrated to find q and then e can be an output related to q. This situation in graph and equation form is

$$\begin{matrix} e \\ \vdash C \\ \dot{q} = f \end{matrix}$$

$$e = \begin{cases} e(q) = e\left(\int^t f\, dt \right) & \text{nonlinear} \tag{4-15} \\[3mm] \dfrac{q}{C} = \dfrac{\int^t f\, dt}{C} & \text{linear} \tag{4-16} \end{cases}$$

where C is the (constant) capacitance. Reversing the causality causes the same constitutive law to be expressed in quite a different form:

$$\begin{matrix} e \\ \dashv C \\ \dot{q} = f \end{matrix}$$

$$f = \begin{cases} \dfrac{d}{dt}\, q = \dfrac{d}{dt}\, q(e) & \text{nonlinear} \tag{4-17} \\[3mm] \dfrac{d}{dt}\, q = \dfrac{d}{dt}\, Ce = C\dfrac{de}{dt} & \text{linear} \tag{4-18} \end{cases}$$

In Eq. (4-15) the function $e(q)$ means that the effort is expressed as a function of displacement, while in Eq. (4-17) $q(e)$ means that displacement is expressed as a function of effort, but $e(q)$ and $q(e)$ are really just two ways of expressing *the same constitutive law*. Maybe this is easier to see in the linear case, where $e(q)$ is just $e = q/C$ and $q(e)$ is just $q = Ce$. Each causality indicates how to use the constitutive relation and whether integration or differentiation is implied.

The same concepts apply to the I element. The I element relates f to p, and p is the time integral of e. Integral causality then is

$$e = \dot{p}$$
$$\dashv I$$
$$f$$

$$f = \begin{cases} f(p) = f\left(\displaystyle\int^{t} e\, dt \right) & \text{nonlinear} & (4\text{-}19) \\[3ex] \dfrac{p}{I} = \dfrac{\int^{t} e\, dt}{I} & \text{linear} & (4\text{-}20) \end{cases}$$

where I is the (constant) inertance. Derivative causality is

$$e = \dot{p}$$
$$\vdash I$$
$$f$$

$$e = \begin{cases} \dfrac{d}{dt} p = \dfrac{d}{dt} p = \dfrac{d}{dt} p(f) & \text{nonlinear} & (4\text{-}21) \\[3ex] \dfrac{d}{dt} p = \dfrac{d}{dt} If = I \dfrac{df}{dt} & \text{linear} & (4\text{-}22) \end{cases}$$

Again, $p(f)$ and $f(p)$ represent the same constitutive law expressed in two ways.

There are some old friends among these causal variations. For example, if you remember $F = kx$ as the law of a linear mechanical spring, this is just Eq. (4-16) with $k = 1/C$ and $X = \int^{t} V\, dt$. Also, $F = ma$ is just Eq. (4-22) with $m = I$ and $a = dV/dt$. Our way or organizing system equations will involve trying to have as many of the I and C elements as possible in integral causality.

The final level in Table 4-1 contains only the R element, which takes whatever causality it is given. It does not matter much whether we impose *resistance causality*

$$e$$
$$\vdash R$$
$$f$$

$$e = \begin{cases} e(f) & \text{nonlinear} & (4\text{-}23) \\[2ex] Rf & \text{linear} & (4\text{-}24) \end{cases}$$

or *conductance causality*

$$
\begin{array}{c}
e \\
\dashv\ R \\
f
\end{array}
$$

$$
f = \begin{cases}
f(e) & \text{nonlinear} \qquad\qquad (4\text{-}25) \\[2mm]
\dfrac{e}{R} = Ge & \text{linear} \qquad\qquad (4\text{-}26)
\end{cases}
$$

Note that the inverse of the resistance, $G = 1/R$, is called the conductance. If an R element is nonlinear, it is possible that one of Eqs. (4-23) or (4-25) might not be as useful as the other. Even in the linear case, for a resistance approaching zero (e.g., short circuit), Eq. (4-24) is preferred to (4-26).

Throughout this discussion of causality, we have assumed that the constitutive laws for the 1-ports are bi-unique functions; i.e., output can be determined uniquely from input in two ways, as Eqs. (4-23) and (4-25). There are, however, many nonlinear functions that are unique in one form but not in the inverse form. For example,

$$ y = x^2 $$

gives a unique value of y for a given x, but

$$ x = \pm\ \sqrt{y} $$

yields two possible values of x, given a value of y. (An additional problem, if we mean to have x and y be real variables, is that we should not try to give y as a negative number.) We see that certain elements, even the normally causally indifferent R's, may be easy to handle in one causality but not in another. Such problems do occur in practice (coulomb friction in mechanical systems causes such problems), but for now we shall consider linear elements and bi-unique nonlinear ones only. We are now ready to use causality to organize component constitutive laws in an orderly fashion so that system state equations can be derived.

4-3 ASSIGNMENT OF CAUSALITY

Bond-graph models are usually developed by building up the structure and adding some bond orientations for power. Once all bonds are directed by power signs, the graph implies a complete mathematical model. For all but the simplest problems it will repay us to add more information to the graph, as described next.

Before equations can be written in an orderly way, the bond graph needs to be completely *augmented*, by which we mean that each bond should be numbered, should possess a sign-convention half arrow, and should be given a causal stroke. The bond numbering allows one to refer to every variable and element by bond number; force F_1 would be the effort on bond 1; I_3 would be the I element attached to bond 3; spring constant k_2 would be the parameter of C_2. All the elements can be conve-

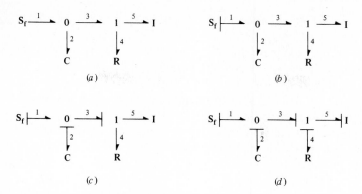

Figure 4-3 Causality assignment for Example 4-1.

niently referred to by the type and the numbers of their bonds. $TF_{4,5}$ is a transformer with bonds numbered 4 and 5, for example. And junction $0_{7,9,12}$ could designate a specific junction in the graph by its attached bond set.

The bond numbering is basically arbitrary provided each bond label is unique. One reasonable convention is to number the bonds sequentially, starting with 1 and not skipping numbers. The sign convention has been discussed previously. It is the assignment of causality which we now discuss.

The causality assignment procedure works from the top of Fig. 4-3 down. The sources *must* have appropriate causality. The 0, 1, TF, and GY elements have *restricted* possibilities and often can be used to propagate causal strokes in the system. We seek where possible to impose *integral* causality on I and C elements and finally (when necessary) to assign R elements arbitrarily if they have not already been assigned. In order to predict how the final equations will turn out and to avoid inconsistency, the procedure below must be followed in strict order.

Causality-Assignment Procedure

1. Choose any S_e or S_f and assign its required causality. Immediately extend the causal implications, using all 0, 1, TF, and GY restrictions that apply.
2. Repeat step 1 until all sources have been causally assigned.
3. Choose any C or I and assign integral causality. Again extend the causal implications of this action, using 0, 1, TF, and GY restrictions.
4. Repeat step 3 until all C and I elements have been causally assigned.
5. Choose any R that is unassigned and give it an arbitrary causality. Extend the causal implications, using 0, 1, TF, and GY.
6. Repeat step 5 until all R elements have been causally assigned.

In most cases all bonds in the graph will have been assigned a causal stroke during step 3. The examples we study now will terminate by step 4, since this is the most common case. If one has to continue to step 5, it means that some algebraic relations must be solved if the equations are to be put in standard form. We defer discussion of this case until Part Four.

When the graph is causally complete, you will often find that you have been able to put all C and I elements in integral causality. Then the state equations can be written down by a very direct process of substitution. This case is discussed next. If you were forced by 0, 1, TF, or GY elements to extend causality in such a way that some C or I elements end up in derivative causality, some algebraic manipulation will be required to achieve the standard state-equation form. This situation will be explained in Part Four. We start with the simplest case, which fortunately is quite common in engineering systems.

Example 4-1 Figure 4-3 shows a bond graph that could have come from either an electric circuit or a mechanical schematic diagram. In Fig. 4-3a the graph has been augmented by both bond numbers and sign half arrows. In the subsequent parts of the figure causality is assigned following the causality-assignment procedure. Step 1 is shown in Fig. 4-3b, in which the flow source has been assigned its required causality. Although one bond of the attached 0-junction now has a flow-input causality, causal implication does not propagate since a 0-junction has two flow-input causalities; therefore we cannot decide on the causality of bonds 2 and 3 yet. Since there is only one source, we move on to step 3.

Figure 4-4c shows the initial results of step 3. The element C_2 now has integral causality, involving that an effort has been assigned to the 0-junction; thus there is no choice but to put the causal mark on bond 3 as shown. Note that the 0-junction has a permissible causal pattern of the type shown in Table 4-1.

Next we check the effect of the causality of bond 3 on the 1-junction. The 1-junction has an effort input (stroke near) on a single bond, but since a 1-junction will end up with two effort inputs, we cannot say at this stage what the causality on bonds 4 and 5 will be. We therefore continue putting the causal mark on bond 5 so that I_5 is in integral causality, as shown in Fig. 4-3d. Then the only choice for bond 4 that gives the 1-junction a permissible causal pattern is the one shown.

For this example we see that C and I have integral causality and all bonds have been assigned a causality. Such a graph will yield its state equations to us with little resistance, as we shall soon see, but first let us causally orient some other graphs for experience.

Example 4-2 In Fig. 4-4 the sign half arrows are left off until the end for clarity in the causality-assignment procedure. This should also serve to emphasize the independence of the causal strokes and the half arrows. Figure 4-4a is an *acausal graph*. It represents a model of the system, but how the constitutive laws of the elements interact has not been studied. In Fig. 4-4b all three effort sources have been assigned their necessary causality. No propagation of causality was possible in any case since this merely resulted in a single bond of a 4-port 1-junction having an effort-input causality. Since each 4-port 1-junction will have three effort-in causalities and a single flow-in causality, the causal orientation of the three remaining bonds on the 1-junction was not determined.

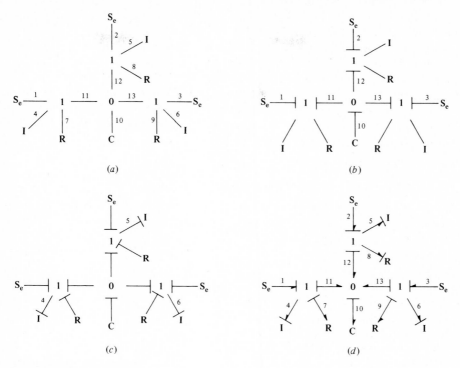

Figure 4-4 Example 4-2.

Also in Fig. 4-4b the C element on bond 10 was given integral causality, immediately implying that bonds 11, 12, and 13 were required to have the causality shown. Why? Any 0-junction has only one effort-input causality bond, and that role was assigned to bond 10. The remaining bonds therefore must have effort-outpuut (flow-input) causality to the 0-junction. Even with this additional causality, the 1-junctions are not able to propagate causality farther in Fig. 4-4b. They are 4-ports and have two of the three required effort-input causality bonds, but we do not yet know which bonds will be the single flow-input bonds for each 1-junction.

The causality is completed in Fig. 4-4c by orienting bonds 4, 5, and 6 so that the I elements are in integral causality. At each 1-junction this leaves no causal freedom for the R-element bonds 7, 8, and 9, since the 1-junctions can have only a single flow-input bond. Finally, Fig. 4-4c shows the completely augmented bond graph, which again terminates after step 4 with all energy-storage elements in integral causality and all bonds causally oriented.

Example 4-3 The last example we consider contains a transformer. The acausal graph is shown in Fig. 4-5a, and in Fig. 4-5b the two effort sources have causal marks. Bonds 1 and 2 impinge on 1-junctions and, as you should now begin to see, no propagation of causal marks is possible. In Fig. 4-5c, however, we orient bond 3, giving C_3 integral causality, as in step 3. Now a lot happens. Because bond 3 now

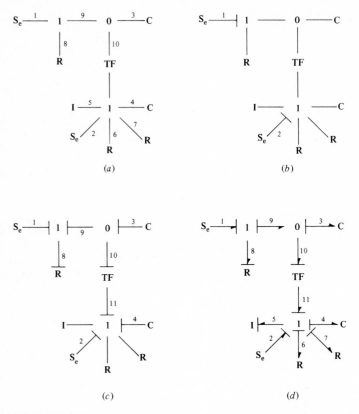

Figure 4-5 Example 4-3.

determines the effort on the 0-junction, bonds 9 and 10 must have the causality shown. But if bonds 1 and 9 give two effort inputs to the upper 1-junction, bond 8 must give the single flow input as shown. Also, as soon as we put a causal mark on bond 10, we must orient bond 11 so that the TF will have one of its possible causal patterns. Hence choosing causality on bond 3 caused us to propagate causality to bonds 9, 8, 10, and 11. Figure 4-5c also shows C_4 in integral causality, but this choice impinges on a 6-port 1-junction and does not lead to any further causal choices.

The final result comes when we put I_5 into integral causality, since then bond 5 provides the single flow input to the 1-junction, meaning that bonds 6 and 7 must have the causality shown. Also in this example, all I and C elements are in integral causality and all bonds had causal marks after step 4.

You may already have observed that, for the examples, the final causality of a graph does not depend upon the order of element bonds chosen for causal assignment. This is an accurate observation that needs only a little modification in cases where we are not finished after step 4 or are forced into derivative causality at some stage. For the cases we study at this stage, which terminate after step 4, you may assume that there is only one correct causal pattern for the entire system.

4-4 BASIC FORMULATION AND REDUCTION

A fully augmented bond graph can be processed in an orderly way to yield state equations. In contrast to less organized methods of deriving equations, bond-graph methods are so predictable that computers can be instructed to write equations from bond graphs. This is essentially what the ENPORT programs do. Do not feel degraded by this when you formulate equations by hand. It only means that we have an organized way of producing equations rather than a haphazard way.

Before we start to write equations, let us see what we know by having a fully augmented bond graph available.

1. The state equations will be in the form of Eq. (4-4) if nonlinear elements are involved or of Eq. (4-5) if the system model is linear.
2. The state variables $x_1, x_2, \ldots, x_n$ will be energy variables, i.e., a momentum p for each I and a displacement q for each C.
3. For a graph in which every I and C has integral causality, the order of the system n (or, equivalently, the number of state equations) is the number of I and C elements.
4. The system input variables $u_1, u_2, \ldots, u_r$ will be efforts from effort sources and flows from flow sources.
5. Therefore the number of input variables r is the number of sources.

With some practice, you will often be able to write down state equations almost by inspection, but initially it is best to proceed in a more detailed way. Here is a basic procedure.

1. Identify three key sets of variables: as inputs use $E(t)$ and $F(t)$ associated with sources; as state variables use p and q associated with I and C elements having integral causality; and as coenergy variables use f and e on I and C elements, respectively.
2. Formulate some initial system equations using these variables.
3. Reduce these equations to final state-space form.

It is often useful to write the key variables next to their bonds on the bond graph. We try to write efforts above or to the left of bonds and flows below or to the right. The input E's and F's are written near their source bonds. The list of input variables will be called U.

The state variables p and q themselves are not written on the bond graph. Instead write $\dot{p}$ and $\dot{q}$, which are efforts and flows. Coenergy variables f for I and e for C may be also indicated. It is these variables which will be eliminated in the final equations. Thus typical elements will look like

$$
\begin{array}{cc}
\dot{p}_3 & e_4 \\
\dashv I & \vdash C \\
f_3 & \dot{q}_4
\end{array}
$$

if $\rightharpoonup I$ were on bond 3 and $\rightharpoonup C$ were on bond 4, for example. The list of state (energy) variables will be called X.

Example 4-4 In Fig. 4-6 the key variables for Example 4-1 have been identified and labeled on the bond graph. The state variables are q_2 and p_5, and the input variable is $F_1(t)$. The variable lists are then

$$X = \begin{bmatrix} q_2 \\ p_5 \end{bmatrix} \qquad U = [F_1]$$

Since q_2 is a displacement and $\dot{q}_2$ is the flow on bond 2, q_2 is placed where a flow should be. Similarly, p_5 is a momentum and $\dot{p}_5$ is the effort on bond 5, and so $\dot{p}_5$ is placed next to bond 5 in the effort position. To help formulation, the coenergy variables e_2 and f_5 are also indicated. These variables will appear during the formulation process, as will other efforts and flows such as e_4, and f_4, e_3, and f_3, but all will be eliminated in the final result.

The general form of the equations is completely known at this point: if R_4, C_2, and I_5 are linear, we shall have

$$\dot{q}_2 = a_{11}q_2 + a_{12}p_5 + b_1 F_1$$
$$\dot{p}_5 = a_{21}q_2 + a_{22}p_5 + b_2 F_1 \qquad (4\text{-}27)$$

while if one or more elements are nonlinear, the general form will be

$$\dot{q}_2 = \phi_1(q_2, p_5; F_1)$$
$$\dot{p}_5 = \phi_2(q_2, p_5; F_1) \qquad (4\text{-}28)$$

For simplicity, we shall concentrate first on linear systems.

The plan is first to write down the constitutive relations between the coenergy and the energy variables. Then we write down equations for the derivatives of the energy variables, using the causal marks to guide us until these derivatives can be expressed in terms of input and coenergy variables. Finally, we eliminate the coenergy variables using the C and I constitutive laws. The results will be state equations. In the equations that follow, we put the element's output on the left and its input(s) on the right.

The constitutive laws for C_2 and I_5 are

$$e_2 = \frac{q_2}{C_2} \qquad (4\text{-}29)$$

and

$$f_5 = \frac{p_5}{I_5} \qquad (4\text{-}30)$$

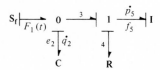

Figure 4-6 Key variables for Fig. 4.3d.

Using these relations, we can express e_2 and f_5 in terms of q_2 and p_5. (Remember that C_2 would be an electric capacitance or mechanical compliance and I_5 would be an inductance or a mass.)

Now comes the essential part. We seek an expression for $\dot{q}_2$ which is a flow. The causal mark shows that $\dot{q}_2$ is determined by the 0-junction, and we write a relation giving $\dot{q}_2$ in terms of F_1 and f_3, since these are inputs to the 0-junction. The *causality* says that we can write

$$\dot{q}_2 = \pm F_1(t) \pm f_3$$

but the *half arrows* determine the signs. The correct result is

$$\dot{q}_2 = F_1(t) - f_3 \tag{4-31}$$

[Equation (4-31) is arranged as it is because of the causal marks. You see how causal strokes and sign half arrows play distinct roles.]

The next step is to work on Eq. (4-31) until $\dot{q}_2$ is expressed in terms of inputs and coenergy variables. F_1 is an input, but f_3 is nothing special. The causal mark indicates that f_3 is determined by the 1-junction. The flow input to the 1-junction is f_5, so we write f_5 in place of f_3,

$$\dot{q}_2 = F_1(t) - f_5 \tag{4-32}$$

(True, f_3, f_4, and f_5 are all equal since they all are on the same 1-junction, but the causality says that we find f_3 from f_5, not f_4.)

Now we can use Eq. (4-30) to find f_5 and the result is a state equation

$$\dot{q}_2 = F_1(t) - \frac{p_5}{I_5} \tag{4-33}$$

The derivative of a state variable $\dot{q}_2$ is now only a function of the input F_1 and the state variable p_5.

The other state equation comes by considering the effort $\dot{p}_5$, which is determined by a 1-junction. It is composed of the two effort inputs to the 1-junction, e_3 and e_4. Using the sign convention, we find the actual relation to be

$$\dot{p}_5 = e_3 - e_4 \tag{4-34}$$

Since neither e_3 nor e_4 is a coenergy variable, we follow the causal strokes to see what determines e_3 and e_4; e_3 is an effort coming from a 0-junction. What determines the effort? The causal strokes say e_2, and

$$\dot{p}_5 = e_2 - e_4 \tag{4-35}$$

and e_2 is one of our coenergy variables. On the other hand, since e_4 comes from R_4 with input f_4, we write

$$\dot{p}_5 = e_2 - R_4 f_4 \tag{4-36}$$

The variable f_4 is determined by the 1-junction which has the flow input f_5. We therefore substitute

$$\dot{p}_5 = e_2 - R_4 f_5 \qquad (4\text{-}37)$$

which gets us to two coenergy variables. Using Eqs. (4-29) and (4-30), we have the final state equation

$$\dot{p}_5 = \frac{q_2}{C_2} - \frac{R_4 p_5}{I_5} \qquad (4\text{-}38)$$

Let us arrange Eqs. (4-33) and (4-38) in the form of Eq. (4-27)

$$\dot{q}_2 = 0 q_2 - \frac{1}{I_5} p_5 + 1 F_1(t)$$

$$\dot{p}_5 = \frac{1}{C_2} q_2 - \frac{R_4}{I_5} p_5 + 0 F_1(t) \qquad (4\text{-}39)$$

in which the predicted form appears but the a and b parameters are expressed in terms of the physical parameters R_4, C_2, and I_5.

Example 4-5 Consider the circuit of Fig. 4-7a and its fully augmented bond graph in Fig. 4-7b. You could have made this graph, don't you think? Since we plan to use generalized variables and parameters, a table relating the electrical quantities to those in the bond graph is included. The key variables have been shown on the bond graph of Fig. 4-7c. The two state variables will be p_2 and q_5, and the single input is

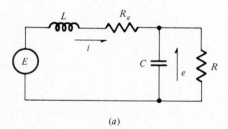

(a)

(b)   (c)

Figure 4-7 A circuit example.

$E_1(t)$. First, the constitutive laws for I_2 and C_5 are written to relate the coenergy variables f_2 and e_5 to the state variables p_2 and q_5

$$f_2 = \frac{p_2}{I_2} \tag{4-40}$$

$$e_5 = \frac{q_5}{C_5} \tag{4-41}$$

Next, we express the voltage $\dot{p}_2$ in terms of E_1, e_3, and e_4, as the causal marks instruct, giving

$$\dot{p}_2 = E_1(t) - e_3 - e_4 \tag{4-42}$$

where the signs are found from the half arrows. E_1 is an input, but we follow the causal mark to find how e_3 and e_4 are determined

$$\dot{p}_2 = E_1(t) - R_3 f_3 - e_5 = E_1(t) - R_3 f_2 - e_5 \tag{4-43}$$

Since f_2 and e_5 are coenergy variables, we find the final form by using Eqs. (4-40) and (4-41)

$$\dot{p}_2 = E_1(t) - \frac{R_3 p_2}{I_2} - \frac{q_5}{C_5} \tag{4-44}$$

In the same way, we express the flow $\dot{q}_5$ as

$$\dot{q}_5 = f_4 - f_6 = f_2 - \frac{e_6}{R_6} = f_2 - \frac{e_5}{R_6} \tag{4-45}$$

Again we have followed the causal strokes back to coenergy variables. (Can you see why f_4 was replaced by f_2? The causal stroke told us that the flows on the 1-junction were all determined by f_2.) For the final form, the laws of Eqs. (4-40) and (4-41) are used

$$\dot{q}_5 = \frac{p_2}{I_2} - \frac{q_5}{R_6 C_5} \tag{4-46}$$

Lining up Eqs. (4-44) and (4-46) in the standard form, we have

$$\dot{p}_2 = -\frac{R_3}{I_2} p_2 - \frac{1}{C_5} q_5 + 1 E_1(t) \tag{4-47}$$

and

$$\dot{q}_5 = +\frac{1}{I_2} p_2 - \frac{1}{R_6 C_5} q_5 + 0 E_1(t)$$

Here
$$X = \begin{bmatrix} p_2 \\ q_5 \end{bmatrix} \quad \text{and} \quad U = [E_1(t)]$$

which we expected from the start.

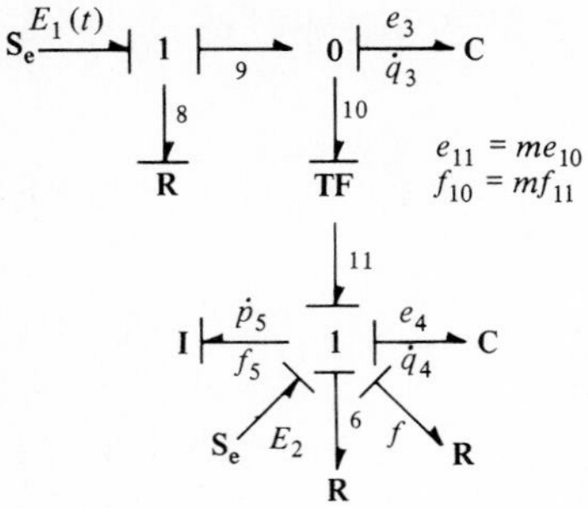

Figure 4-8 Key variables for Fig. 4.5.

Example 4-6 We write equations for Example 4-3. The bond graph, with key variables, is reproduced in Fig. 4-8. Note that

$$X = \begin{bmatrix} q_3 \\ q_4 \\ p_5 \end{bmatrix} \quad \text{and} \quad U = \begin{bmatrix} E_1(t) \\ E_2(t) \end{bmatrix}$$

In writing the constitutive laws for C_3, C_4, and I_5 we assume that the first two elements are nonlinear and hence represented by general functions

$$e_3 = \phi_3(q_3) \tag{4-48}$$

$$e_4 = \phi_4(q_4) \tag{4-49}$$

but since I_5 is linear,

$$f_5 = \frac{1}{I_5} p_5 \tag{4-50}$$

Because it is sometimes a bother to remember just where the transformer modulus m goes, we have written equations in Fig. 4-8 in the forms causality tells us will be needed; i.e., we shall need e_{11} given e_{10} and f_{10} given f_{11}. Now we are ready to express the derivatives of the state variables in terms of efforts and flows by following the causal strokes until we get back to the inputs and coenergy variables. We simply write down the steps

$$\dot{q}_3 = f_9 - f_{10} = f_8 - mf_{11} = \frac{1}{R_8} e_8 - mf_5$$

$$= \frac{1}{R_8} [E_1(t) - e_9] - mf_5 = \frac{1}{R_8} [E_1(t) - e_3] - mf_5 \tag{4-51}$$

$$\dot{q}_4 = f_5 \tag{4-52}$$

$$\dot{p}_5 = E_2 + e_{11} - e_6 - e_7 - e_4 = E_2(t) + me_{10} - R_6 f_6 - R_7 f_7 - e_4$$

$$= E_2(t) + me_3 - R_6 f_5 - R_7 f_5 - e_4 \tag{4-53}$$

The final form follows by using Eqs. (4-48) to (4-50) to eliminate the coenergy variables

$$\dot{q}_3 = -\frac{\phi_3(q_3)}{R_8} + 0q_4 - \frac{mp_5}{I_5} + \frac{1}{R_8} E_1(t) \qquad (4\text{-}54)$$

$$\dot{q}_4 = 0q_3 + 0q_4 + \frac{p_5}{I_5} \qquad (4\text{-}55)$$

$$\dot{p}_5 = m\phi_3(q_3) - \phi_4(q_4) - (R_6 + R_7)\frac{p_5}{I_5} + E_2(t) \qquad (4\text{-}56)$$

Even though these are nonlinear equations, they still have considerable internal structure. They can be arranged to show the separate influence of each state variable, although the influence may be in the form of a function rather than a coefficient. (Generally, however, matrices cannot be used for nonlinear equations.)

In a physical sense, each state equation has a meaning that may help to check the correctness of the equations. The circuit equations (4-47) for Example 4-5 can readily be interpreted as a voltage-balance equation when traversing a loop and a current-sum equation at a node. Even a check of units can be helpful. Every $\dot{p}$ equation must contain only terms which are efforts, and terms $\dot{q}$ equations must have the dimensions of flow. Generalizing a bit, we observe that using energy variables as state variables leads to state equations with the following characteristics: (1) in electrical systems every state equation is either a loop-voltage summation $\dot{\lambda}$ or a node-current summation $\dot{q}$; (2) in mechanical-translation systems, every state equation is either a force-balance summation $\dot{p}$ or a description of geometric compatibility $\dot{q}$. Similar observations can be made for every physical domain.

4-5 OUTPUT EQUATIONS

State equations involve the energy state variables and the input variables. If the equations are solved somehow, we get information about the state variables. In addition, we can get information about every other system variable. When we *simulate* the system, we usually start from given initial values of the state variables, choose specific time functions for the input variables, and compute time histories of the state variables over a time span of interest. But there are many other variables in the system that may be of interest besides the state variables. They are usually called *output variables.*

Typical output variables include efforts and flows in the system which were eliminated during state-equation formulation. This elimination provides a clue to how they can be found; they can be expressed, *following the causal marks,* as functions of the state variables and input variables. Thus during a simulation, if we know the instantaneous values of the state variables and input variables, the output variables can be found by algebraic relationships. If we call the output variables $y_1, y_2, \ldots, y_m$, the output equations will have the following form in the general case:

$$y_1 = f_1(x_1, x_2, \ldots, x_n; u_1, u_2, \cdots u_r)$$

$$y_2 = f_2(x_1, x_2, \ldots, x_n; u_1, u_2, \cdots u_r) \qquad (4\text{-}57)$$

$$\ldots\ldots\ldots\ldots\ldots\ldots\ldots\ldots\ldots\ldots\ldots\ldots\ldots\ldots\ldots$$

$$y_m = f_m(x_1, x_2, \ldots, x_n; u_1, u_2, \cdots u_r)$$

In the linear case

$$y_1 = c_{11}x_1 + c_{12}x_2 + \cdots + c_{1n}x_n + d_{11}u_1 + d_{12}u_2 + \cdots + d_{1r}u_r$$

$$y_2 = c_{21}x_1 + c_{22}x_2 + \cdots + c_{2n}x_n + d_{21}u_1 + d_{22}u_2 + \cdots + d_{2r}u_r \quad (4\text{-}58)$$

$$\ldots\ldots\ldots\ldots\ldots\ldots\ldots\ldots\ldots\ldots\ldots\ldots\ldots\ldots\ldots$$

$$y_m = c_{m1}x_1 + c_{m2}x_2 + \cdots + c_{mn}x_n + d_{m1}u_1 + d_{m2}u_2 + \cdots + d_{mn}u_n$$

As in Eq. (4-5 a), the use of matrices makes it easier to represent the linear version. We write

$$y = Cx + Du \qquad (4\text{-}58a)$$

which simply collects all the coefficients in Eq. (4-58) in the matrix symbols C and D.

If you compare Eqs. (4-57) and (4-58) with the state-equation forms of Eqs. (4-4) and (4-5), you will see that state equations are differential equations while output equations are algebraic equations.

When you know how to write state equations, writing output equations is easy. Suppose we want to find e_3 and e_4 in Fig. 4-7. We just look at the causality to see how these efforts are determined. The effort e_3 is determined by the 0-junction, which has its effort determined by bond 2. Thus

$$e_3 = e_2 = \frac{q_2}{C_2} \qquad (4\text{-}59)$$

That was easy because e_3 is determined only by the state variable q_2. It is not much harder to find e_4. It is determined by R_4, which reacts to f_4, which in turn comes from the 1-junction, whose flow is determined by f_5. The sequence of substitutions, following causality, is

$$e_4 = R_4 f_4 = R_4 f_5 = \frac{R_4 p_5}{I_5} \qquad (4\text{-}60)$$

Putting these outputs in a list Y, we have

$$Y = \begin{bmatrix} e_3 \\ e_4 \end{bmatrix}$$

where Y can be as large or as small as we wish since it depends upon how many variables beyond the state variables happen to be of interest. Finally we define three output variables for the example of Fig. 4-9.

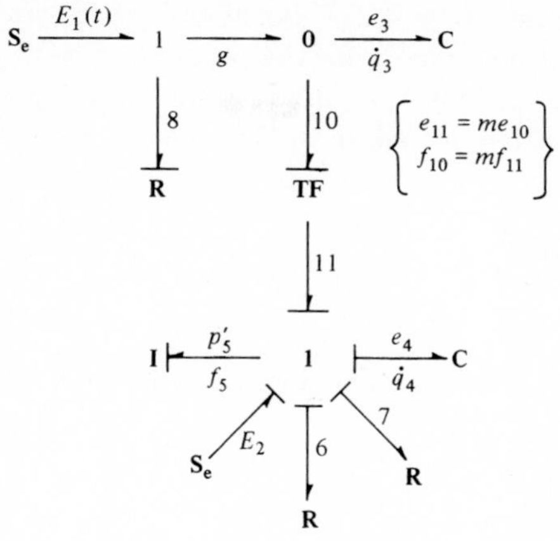

Figure 4-9 Key variables for the example of Fig. 4-6.

$$Y = \begin{bmatrix} f_8 \\ e_{11} \\ e_6 \end{bmatrix}$$

The corresponding output equations are

$$f_8 = \frac{e_8}{R_8} = \frac{E_1(t) - e_9}{R_8} = \frac{E_1(t) - e_3}{R_8} = \frac{E_1(t) - q_3}{C_3 R_8} \qquad (4\text{-}61)$$

$$e_{11} = me_{10} = me_3 = \frac{mq_3}{C_3} \qquad (4\text{-}62)$$

$$e_6 = R_6 f_6 = R_6 f_5 = \frac{R_6 p_5}{I_5} \qquad (4\text{-}63)$$

In each case, we followed the causal marks until we came to either a state variable or an input variable. Notice that a little rearranging of Eq. (4-61) would be required to obtain the C and D matrix entries directly.

Let us pause a moment to see how far we have come. We have completed our first pass at identifying basic elements, modeling systems, and organizing the component relationships into state equations and output equations suitable for analytical or computational study. We have restricted ourselves to relatively simple electrical and mechanical systems and have avoided major algebraic complexities to give you a better chance to see where we are going.

In Part Three we shall use the standard bond-graph elements to model more complicated systems in several energy domains. In many practical cases considerable judgment needs to be brought to bear on the modeling process, since many physical effects, while clearly present to some degree, may not be that important. It is no virtue to make the most complex model you can think of. Of course, we know that you do not expect real devices to have little circuit diagrams or mass-spring-dashpot diagrams stamped on them to tell you how to make a bond graph from them. What you will learn to do in practice is to make several models of varying degrees of sophistication and to compare them—both with each other and, if possible, with experimental evidence—to see which model is most appropriate for the situation at hand.

Thus we remind you once again that the successful practice of system dynamics requires a beautiful blending of science and art, of procedure and intuition, of theory and experience.

PROBLEMS

4-1 Convert $m\ddot{x} + b\dot{x} + kx = F(t)$ into state-variable form using the two state variables x and p. Let $p = mv = m\dot{x}$.

4-2 Convert

$$\dot{p} = -\frac{bp}{m} - kx + bV(t) + F(t)$$

$$\dot{x} = \frac{p}{m} - V(t)$$

into a single second-order equation for x.

4-3 For the example bond graph of Fig. 4-3, show a mechanical schematic diagram and electric circuit. Relate the physical parameters mass m, damper constant b, spring constant k, inductance L, resistance R, and capacitance C to the generalized parameters C_2, R_4, and I_5. What are the meanings of e_2 and f_5 in the two physical systems?

4-4 Apply causal strokes to the system illustrated. Identify both state (X) and input (U) lists.

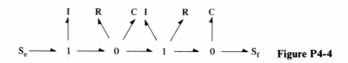

Figure P4-4

4-5 Make a bond graph and apply causal strokes. Identify X and U.

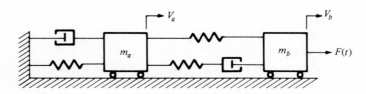

Figure P4-5

4-6 Make a bond graph and apply causality. Identify X and U.

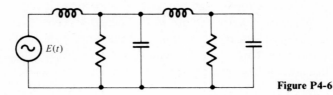

Figure P4-6

4-7 For this rack-and-pinion mechanism the relationship $V = r\omega$ is represented by a transformer. Show that if the rack has mass m and the pinion has moment of inertia J_a, the bond graph for this system must have derivative causality on either the mass I element or the pinion I element.

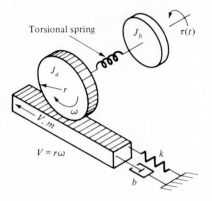
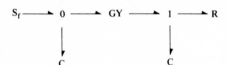

Figure P4-7

4-8 Apply causal strokes. Identify X and U.

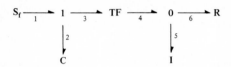

Figure P4-8

4-9 Apply causal strokes. Identify X and U.

$$S_f \longrightarrow 0 \longrightarrow GY \longrightarrow 1 \longrightarrow R$$

$$C \qquad\qquad C$$

Figure P4-9

4-10 The vehicle model shown has a body mass M, main spring K, damper constant B, tire mass m, and tire spring k. The input is the velocity $V_0(t)$ at the bottom of the tire, and gravity acts on both masses.

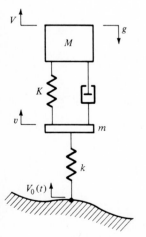

Figure P4-10

(*a*) Make a bond graph showing the physical variables v, V, $V_0(t)$, Mg, and mg. Also show the physical parameters M, K, B, m, and k.

(*b*) Convert to a bond graph using generalized variables e, f, p, q. Fully augment this graph. Include a table relating physical parameters to generalized parameters; e.g., if C_3 represents the main suspension spring, $C_3 = 1/K$.

4-11 Three flywheels are coupled by torsional springs, and each is subject to a viscous friction torque $\tau = b\omega$. One of the torsion springs has an angular-velocity input $\tau(t)$. One flywheel has an input torque $\tau(t)$. Make a bond graph and apply causality. Identify X and U.

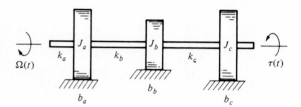

Figure P4-11

4-12 The lever has negligible mass and friction. Make a fully augmented bond graph. Identify X and U.

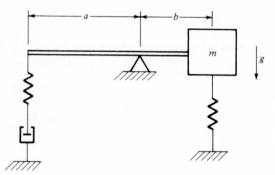

Figure P4-12

4-13 Apply causality. Identify X and U.

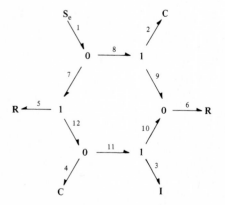

Figure P4-13

4-14 Make two bond graphs for this system, one using physical variables such as $V_0(t)$, $F_0(t)$, V and the physical parameters m, b, k, and the other using generalized variables and parameters and numbered bonds. Write state equations for both versions.

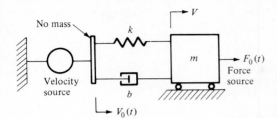

Figure P4-14

4-15 Make a bond graph for this system and write equations using generalized variables. Assume linear components.

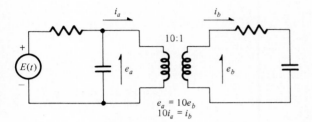

Figure P4-15

4-16 Write equations for the system of Prob. 4-11 in generalized variables. The six state equations fall into two groups of three. What are the physical meanings of the two groups of equations?

4-17 Make two bond graphs, one with physical variables and parameters and one with generalized variables and numbered bonds. Write state equations for either one, but relate m, b, k, V_a, V_b, and mg to generalized quantities.

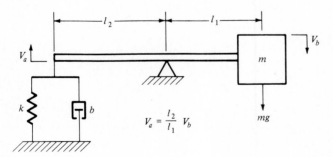

Figure P4-17

4-18 Write state equations in generalized variables for the system of Prob. 4-10.

4-19 Continue Prob. 4-18 to write the acceleration of the upper mass as an output related to state variables and inputs. Note that the acceleration $a = \dot{V}$ is proportional to the net force on the mass or the rate of change of the momentum.

4-20 Express the damper force in Prob. 4-14 as an output variable and write an output equation for this force.

4-21 Write an output equation for the generalized variable corresponding to i_a in Prob. 4-15.

4-22 The system shown has two nonlinear springs and coulomb friction represented by a signum or sign function. Write the nonlinear state equations in the physical variables force F, velocity V, momentum P, and deflection α.

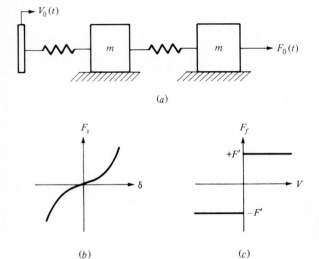

(a)

(b)   (c)

Figure P4-22 (*b*) Spring force $F = A\delta^3$ and deflection δ; (*c*) friction force $F = F'$ sgn V.

4-23 The suspension point of the spring-mass oscillator is moved with velocity $V_0(t)$, and the mass stays submerged in fluid. Make a bond-graph model and write two sets of state equations, assuming (1) viscous damping force on the mass, $F = BV$, and (2) quadratic damping force, $F = \beta V|V|$. Note that the resistor law (2) always represents power dissipation, no matter what sign V has.

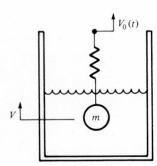

Figure P4-23 Gravity forces mg and buoyant force $\rho \overline{V}$, where $\overline{V} = $ volume of mass.

TIME RESPONSE OF LOW-ORDER LINEAR SYSTEMS

5-1 INTRODUCTION

In this chapter we consider the dynamic response of low-order linear constant-coefficient systems. The models are described by a set of state-space equations together with a set of initial conditions. Our dual tasks are to extract insight and to obtain formal results concerning the dynamic behavior of the models in question.

Our principal method will be the use of Laplace transforms, which enable us to convert a problem in linear differential equations expressed in the *time domain* into a set of linear algebraic equations expressed in the *complex frequency domain*. The transformation is well worth knowing and using because it yields both insight and analytical results in an efficient manner.

Example 5-1 To motivate our discussion of the technical details that follow in this chapter, we first consider a familiar example. A mass-spring configuration is shown in Fig. 5-1*a*. The system bond graph is given in Fig. 5-1*b*, where the *C, R,* and *I* elements represent the spring, damper, and mass, respectively. The S_e element represents the gravity force on the mass. Here are some questions we wish to address:

1. Under what conditions will the system oscillate?
2. What are the steady-state values of the system variables?
3. Exactly how does the system move from its initial state to its final state?

We shall see that the amount of work required to answer each of these questions differs, and our transform method allows us direct control over the effort invested.

111

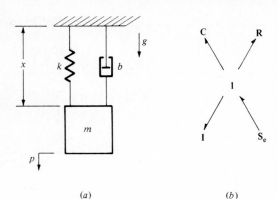

Figure 5-1 System for Example 5-1: (*a*) mechanical schematic and (*b*) bond-graph model.

(*a*)   (*b*)

Do not try to understand each step in detail now. Just try to follow the pattern. Later you can return to this example when you understand the details of the operations.

We begin by writing the state equations for the system from the bond graph of Fig. 5-1*b*

$$\dot{p}(t) = -\frac{b}{m}p(t) - kx(t) + mg \qquad (5\text{-}1a)$$

$$\dot{x}(t) = \frac{1}{m}p(t) \qquad (5\text{-}1b)$$

where b = damping constant
m = mass
k = spring constant
g = acceleration of gravity

and we have indicated which variables are functions of time. Next, we obtain transformed equations

$$sP(s) - p(0) = -\frac{b}{m}P(s) - kX(s) + \frac{mg}{s} \qquad (5\text{-}2a)$$

$$sX(s) - x(0) = \frac{1}{m}P(s) \qquad (5\text{-}2b)$$

Each equation above is the transform of one of the state equations. The unknowns $P(s)$ and $X(s)$ are the Laplace transforms of $p(t)$ and $x(t)$, respectively. For now, it is sufficient to notice that Eqs. (5-2) are linear algebraic equations and the independent variable is not t (time) but s (generalized frequency). The initial conditions are incorporated into Eqs. (5-2) in the terms $p(0)$ and $x(0)$, which are values of p and x at $t = 0$.

Next we solve for the particular unknown of interest. In this case, let us choose $X(s)$, corresponding to both spring deflection and mass position. From Eqs. (5-2) we obtain

$$X(s) = \frac{(s + b/m)\, x(0) + 1/m\, p(0) + g/s}{s^2 + (b/m)\, s + k/m} \tag{5-3}$$

This is a transformed solution for $x(t)$ obtained by solving the simultaneous algebraic equations (5-2a) and (5-2b).

Our first equation can be answered from a study of the denominator of Eq. (5-3). We calculate the roots of the denominator polynomial; i.e., we solve

$$s^2 + \frac{b}{m}\, s + \frac{k}{m} = 0 \tag{5-4}$$

to get

$$s_{1,2} = -\frac{b}{2m} \pm \frac{1}{2} \sqrt{\left(\frac{b}{m}\right)^2 - 4\frac{k}{m}} \tag{5-5}$$

If

$$\left(\frac{b}{m}\right)^2 - 4\frac{k}{m} > 0 \tag{5-6}$$

the roots are real and the system will not oscillate. If

$$\left(\frac{b}{m}\right)^2 - 4\frac{k}{m} < 0 \tag{5-7}$$

the roots are complex and the system will execute damped oscillations.

In physical terms, we see that as the damping b increases relative to the stiffness k, there is less tendency for oscillation. This type of insight is very useful in the preliminary stages of design and analysis.

To answer the second question, we can apply the final value theorem, (FVT), which is derived in appendix B, to the Laplace solution for $X(s)$, provided the roots s_1 and s_2 both have negative real parts (which is the case if b, k, and m are all positive). Applying the FVT yields

$$x(t \to \infty) = \lim_{s \to 0} sX(s) = \left. \frac{s[(s + b/m)\, x(0) + 1/m\, p(0) + g/s]}{s^2 + (b/m)s + k/m} \right|_{s \to 0}$$

or

$$x(t \to \infty) = \frac{mg}{k} \tag{5-8}$$

This result tells us that the spring stretches enough to generate a force $kx(t \to \infty)$ to counterbalance the weight mg. That is reasonable.

We have answered two of our original questions from the transformed solution $X(s)$ rather than from the complete time solution $x(t)$, and the computational effort was quite small. To obtain the complete time response, we must transform back from the frequency s domain to the time t domain. This is done most effectively by the use of partial fractions and a table of Laplace transforms.

Suppose $m = 1$ kg, $b = 5$ N$\cdot$s/m, and $k = 4$N/m. Let $p(0) = 0$ and $x(0) = 0$. Find $x(t)$. Observe that

$$X(s) = \frac{g}{(s^2 + 5s + 4)s} = \frac{g}{(s + 4)(s + 1)s} \tag{5-9}$$

This can be expressed as

$$X(s) = \frac{g/12}{s + 4} + \frac{-g/3}{s + 1} + \frac{g/4}{s} \tag{5-10}$$

which yields the inverse

$$x(t) = \frac{g}{12} e^{-4t} - \frac{g}{3} e^{-1t} + \frac{g}{4} \tag{5-11}$$

Check that the initial condition is correct: $x(0) = 0$. Observe that the steady-state $x(t \rightarrow \infty) = g/4$, which corresponds to Eq. (5-8). Finally, note that the solution does not oscillate, i.e., contains no sine or cosine terms.

The Laplace Procedure

The next sections develop the details for each of the steps in the Laplace-transform approach to state-equation study and solution. Now we summarize the major steps for future reference:

1. Laplace-transform the state equations.
2. Solve for the unknown(s) of interest.
3. Find the roots of the transform denominator.
4. Separate the transform solution by partial fractions.
5. Invert to get the complete time solution.

5-2 THE LAPLACE TRANSFORM

Transform methods were introduced by Oliver Heaviside in the late 1800s in the form of an *operational calculus* and have been the subject of much mathematical study since. We shall restrict our attention to those features of the Laplace transform which make it useful for our problems. More complete discussions are available.[1-3]†

Definition The Laplace transform of a given time function $x(t)$ is denoted by $X(s)$ and written

$$X(s) = \mathcal{L}\{x(t)\}$$

The defining relation is

$$\mathcal{L}\{x(t)\} = \int_0^\infty x(t)e^{-st}\, dt \tag{5-12}$$

† Numbered references appear at the end of the chapter.

This transformation converts a function of time t into a function of generalized frequency s, where the units of s are the reciprocal of the units of time. The variable s is a complex quantity.

Although it is of limited practical use to us, the inverse transform is defined by the complex inversion integral

$$x(t) = \begin{cases} \lim_{T \to \infty} \dfrac{1}{2\pi j} \displaystyle\int_{\sigma - jT}^{\sigma + jT} e^{st} X(s) \, ds & t > 0 \\ 0 & t < 0 \end{cases} \tag{5-13}$$

Our principal approach is to learn about a number of common transform pairs, $x(t)$ matched with $X(s)$, and to use the pairs to solve our problems. Although our working table of pairs (Table 5-1) is small, it covers a large percentage of typical linear engineering problems. Furthermore, extensive tables of transform pairs are available[2,3] if

Table 5-1 Short table of transform relations

Entry	$X(s)$	$x(t), \, t > 0$	Type
1	$\dfrac{1}{s}$	1	Unit step
2	$\dfrac{1}{s^2}$	t	Unit ramp
3	$\dfrac{1}{s + \sigma}$	$e^{-\sigma t}$	Exponential
4	$\dfrac{1}{(s + \sigma)^2}$	$te^{-\sigma t}$	Repeated root
5	$\dfrac{\omega}{s^2 + \omega^2}$	$\sin \omega t$	Sine
6	$\dfrac{s}{s^2 + \omega^2}$	$\cos \omega t$	Cosine
7	$\dfrac{\omega}{(s + \sigma)^2 + \omega^2}$	$e^{-\sigma t} \sin \omega t$	Damped sine
8	$\dfrac{s + \sigma}{(s + \sigma)^2 + \omega^2}$	$e^{-\sigma t} \cos \omega t$	Damped cosine
9	$sX(s) - x(0)$	$\dfrac{dx(t)}{dt}$	First derivative
10	$\dfrac{1}{s} X(s)$	$\displaystyle\int_0^t f(t)\, dt$	Integral
11	$X(s + \sigma)$	$e^{-\sigma t} x(t)$	Frequency shift
12	$e^{-ts} X(s)$	$x(t - t)$	Time shift
13	$\dfrac{1}{a} X\left(\dfrac{s}{a}\right)$	$x(at)$	Scale change

the need arises. Thus, our objective is to use the defining relation (5-12) only enough to become familiar with some common pairs.

Some Common Transform Pairs

Consider the time function $x_1(t) = 1$, a constant. Then, from Eq. (5-12),

$$X_1(s) = \int_0^\infty (1)e^{-st}\, dt = \frac{-1}{s} e^{-st} \Big|_0^\infty = -\frac{1}{s}\left(e^{-st}\Big|_{t\to\infty} - e^{s0} \right) = \frac{1}{s}$$

provided that $e^{-st}\Big|_{t\to\infty}$. We shall assume that s is restricted to ensure this situation. Then we have computed the first entry in Table 5-1.

Now try $x_2(t) = e^{-at}$. Again from Eq. (5-12) we get

$$X_2(s) = \int_0^\infty e^{-at}e^{-st}\, dt = \int_0^\infty e^{-(s+a)t}\, dt = \frac{-1}{s+a} e^{-(s+a)t} \Big|_0^\infty = \frac{1}{s+a}$$

provided that s is restricted to make the first limit $(t \to \infty)$ converge to zero. This is entry 2 in the transform table.

One more pair should suffice at this point. Suppose $x_3(t) = \sin \omega t$. Then

$$
\begin{aligned}
X_3(s) &= \int_0^\infty (\sin \omega t)e^{-st}\, dt \\
&= \int_0^\infty \frac{1}{2j}(e^{j\omega t} - e^{-j\omega t})e^{-st}\, dt \\
&= \frac{1}{2j}\left[\int_0^\infty e^{-(s-j\omega)t}\, dt - \int_0^\infty e^{-(s+j\omega)t}\, dt \right] \\
&= \frac{1}{2j}\left(\frac{1}{s-j\omega} - \frac{1}{s+j\omega} \right) \\
&= \frac{1}{2j}\frac{2j\omega}{s^2+\omega^2} = \frac{\omega}{s^2+\omega^2}
\end{aligned}
$$

This pair is listed as entry 3 in the table. In most cases of interest, $X(s)$ is a polynomial ratio; $X(s)/D(s)$.

Clearly, a large table of transform pairs could be built up by patiently computing the transforms of a series of time functions. Before considering the problem of inverting transforms, we investigate some useful properties of the Laplace transform.

Some Transform Properties

In the study of constant-coefficient state equations, several operations occur repeatedly. The transform properties we introduce in this section are related to those operations.

Superposition

Let $X(s)$ be the Laplace transform of $x(t)$. Let us investigate the Laplace transform of $Cx(t)$, where C is a constant

$$\mathcal{L}\{Cx(t)\} = \int_0^\infty Cx(t)e^{-st}\,dt = C\int_0^\infty x(t)e^{-st}\,dt$$

or
$$\mathcal{L}\{Cx(t)\} = CX(s) \tag{5-14}$$

That is, multiplying a time function by a constant multiplies its transform by the same constant. Now consider that we have two transform pairs $x_1(t)$, $X_1(s)$ and $x_2(t)$, $X_2(s)$. Suppose we wish to evaluate the transform of $x_3(t)$, where

$$x_3(\mathrm{t}) = C_1x_1(t) + C_2x_2(t)$$

is an arbitrary linear combination. We have

$$
\begin{aligned}
\mathcal{L}\{x_3(t)\} &= \mathcal{L}\{C_1x_1(t) + C_2x_2(t)\} \\
&= \int_0^\infty [C_1x_1(t) + C_2x_2(t)]e^{-st}\,dt \\
&= \int_0^\infty [C_1x_1(t)]e^{-st}\,dt + \int_0^\infty [C_2x_2(t)]e^{-st}\,dt \\
&= C_1\int_0^\infty x_1(t)e^{-st}\,dt + C_2\int_0^\infty x_2(t)e^{-st}\,dt
\end{aligned}
$$

or
$$\mathcal{L}\{C_1x_1(t) + C_2x_2(t)\} = C_1X_2(s) + C_2X_2(s) \tag{5-15}$$

Thus, we say that the Laplace transform obeys the superposition property.

Differentiation Suppose that we know the pair $x(t)$ and $X(s)$, its transform. We wish to evaluate the transform of $\dot{x}(t)$, the first derivative. Using integration by parts, we get

$$\mathcal{L}\left\{\frac{dx(t)}{dt}\right\} = \int_0^\infty \frac{dx(t)}{dt}e^{-st}\,dt = x(t)e^{-st}\Big|_0^\infty - \int_0^\infty x(t)(-s)e^{-st}\,dt$$

or
$$\mathcal{L}\left\{\frac{dx}{dt}\right\} = sX(s) - x(0) \tag{5-16}$$

provided s is restricted to make the first limit ($t \to \infty$) vanish. Thus, the first derivative can be obtained directly from the original transform by multiplying by s and subtracting the initial condition.

For example, let $x(t) = 2 \sin \omega t$. Then $X(s) = 2\omega/(s^2 + \omega^2)$. To find the transform of $\dot{x}(t)$ let

$$\mathcal{L}\{\dot{x}(t)\} = sX(s) - x(0) = \frac{2\omega s}{s^2 + \omega^2} - 2\sin(0) = \frac{2\omega s}{s^2 + \omega^2}$$

On the other hand, $\dot{x}(t)$ is given directly by $\dot{x}(t) = 2\omega \cos \omega t$. From Table 5-1 the transform is found to be

$$\mathcal{L}\{\dot{x}\} = \frac{2\omega s}{s^2 + \omega^2}$$

when we include the constant 2ω. So the formula yields the correct result.

Initial-value theorem (IVT) It is often useful to check a solution by investigating the initial value associated with the transform solution. We state here the initial-value theorem (IVT) for that purpose.

$$x(t \to 0) = \lim_{s \to \infty} sX(s) \qquad (5\text{-}17)$$

For example, let $x(t) = 6 \cos 2t + 3e^{-4t}$. Then

$$X(s) = \frac{6s}{s^2 + 4} + \frac{3}{s + 4}$$

and

$$x(t \to 0) = \lim_{s \to \infty} sX(s)$$
$$= \lim_{s \to \infty} \left(\frac{6s^2}{s^2 + 4} + \frac{3s}{s + 4} \right)$$
$$= 6 + 3 = 9$$

Clearly, this is what $x(t = 0)$ yields directly.

Final-value theorem (FVT) A related theorem with considerable usefulness is the final-value theorem (FVT), derived in Appendix B,

$$x(t \to \infty) = \lim_{s \to \infty} sX(s) \qquad (5\text{-}18)$$

This theorem must be applied with caution since it will yield the correct value only if there is a unique finite limit for $x(t \to \infty)$. More precisely, all roots of the denominator of $X(s)$ must have negative real parts except for a single root at the origin.

For example, let $X(s) = 2/(s + 4)(s + 1)$. The roots are -4 and -1; the FVT applies

$$x(t \to \infty) = \lim_{s \to 0} sX(s) = \lim_{s \to 0} \frac{2s}{(s+4)(s+1)} = 0$$

Since the time function is

$$x(t) = -\tfrac{2}{3}e^{-4t} + \tfrac{2}{3} e^{-1t}$$

and $x(t \to \infty) = 0$. Now suppose $X(s) = 6/(s + 7)s$. The roots are -7 and 0. The FVT applies. Thus

$$x(t \to \infty) = \lim_{s \to 0} sX(s) = \lim_{s \to 0} \frac{6s}{(s + 7)s} = \frac{6}{7}$$

Since the time function is

$$x(t) = -\tfrac{6}{7}e^{-7t} + \tfrac{6}{7}$$

we see that the FVT predicted correctly.

To see why it is important to check first the applicability of the FVT, consider this case. Let $x(t) = \cos \omega t$. Then

$$X(s) = \frac{s}{s^2 + \omega^2}$$

Now the FVT predicts

$$x(t \to \infty) = \lim_{s \to 0} sX(s) = \lim_{s \to 0} \frac{s^2}{s^2 + \omega^2} = 0$$

But the roots are $\pm j\omega$, which do not have negative real parts, and the time function $x(t)$ does not have a unique "final value."

5-3 LAPLACE SOLUTION OF STATE EQUATIONS

In this section we show how to transform constant-coefficient state equations in a systematic manner. Then we discuss the insight into the system dynamics obtainable from a Laplace solution by means of the characteristic polynomial. Finally, we offer a formal solution method based on determinants.

Transformed State Equations

Begin with the state equations in general form for a second-order system

$$\dot{x}_1(t) = a_{11}x_1(t) + a_{12}x_2(t) + b_1u(t) \tag{5-19a}$$

$$\dot{x}_2(t) = a_{21}x_1(t) + a_{22}x_2(t) + b_2u(t) \tag{5-19b}$$

and
$$x_1(0) = x_{10}, \quad x_2(0) = x_{20} \tag{5-19c}$$

Now apply the Laplace transform to each side of Eq. (5-19 a)

$$\mathcal{L}\{\dot{x}_1(t)\} = \mathcal{L}\{a_{11}x_1(t) + a_{12}x_2(t) + b_1u(t)\}$$

Denoting the transforms of $x_1(t)$, $x_2(t)$, and $u(t)$ by $X_1(s)$, $X_2(s)$, and $U(s)$, respectively, we get

$$sX_1(s) - x_1(0) = a_{11}X_1(s) + a_{12}X_2(s) + b_1U(s)$$

If the unknowns are collected on the left and the known data are collected on the right, the equation becomes

$$(s - a_{11})X_1(s) - a_{12}X_2(s) = x_1(0) + b_1U(s) \tag{5-20a}$$

Apply the same procedure to Eq. (5-19 b). The result is

$$-a_{21}X_1(s) + (s - a_{22})X_2(s) = x_2(0) + b_2U(s) \tag{5-20b}$$

We now have two linear algebraic equations (5-20 a) and (5-20 b) in two unknowns $X_1(s)$ and $X_2(s)$. The inputs $U(s)$ and initial conditions $x_1(0)$ and $x_2(0)$ are incorporated into the equations.

This procedure can be applied to a set of state equations of any order, giving

$$(s - a_{11})X_1(s) - a_{12}X_2(s) - \cdots - a_{1n}X_n(s)$$
$$= X_1(0) + b_{11}U_1(s) + b_{12}U_2(s) + \cdots + b_{1m}U_m(s) \quad (5\text{-}21a)$$

. . . .

$$-a_{n1}X_1(s) - a_{n2}X_2(s) \cdots + (s - a_{nn})X_n(s)$$
$$= x_n(0) + b_{n1}U_1(s) + \cdots + b_{nm}U_m(s) \quad (5\text{-}21\,b)$$

There are n equations. The coefficients on the left are the *negatives* of the A array. Added to each diagonal entry, i.e., the negative of the coefficient of $X_i(s)$ in the ith equation is s. On the right appear the initial conditions, one for each equation, plus the transformed inputs weighted by the B array. Chapter 15 examines these equations based on array notation, which is a very efficient way to study such systems.

Let us apply the Laplace transformation to a specific set of equations. Suppose we have derived

$$\dot{x}_1 = -2x_1 + 4x_2 + 7u(t) \qquad \dot{x}_2 = -3x_1 + 2u(t) \qquad (5\text{-}22)$$

and

$$x_1(0) = -2 \qquad x_2(0) = 0$$

Applying the Laplace-transform procedure one step at a time, we get

$$sX_1(s) - x_1(0) = -2X_1(s) + 4X_2(s) + 7U(s)$$
$$sX_2(s) - x_2(0) = -3X_1(s) + 0X_2(s) + 2U(s)$$

A zero entry has been written in for element a_{22}. Now collect unknowns on the left

$$(s + 2)X_1(s) - 4X_2(s) = x_1(0) + 7U(s) = -2 + 7U(s) \quad (5\text{-}23a)$$

$$3X_1(s) + (s - 0)X_2(s) = x_2(0) + 2U(s) = 0 + 2U(s) \quad (5\text{-}23b)$$

Close examination of the equations shows that $-A$ components appear on the left, plus s on the diagonal. Initial conditions and inputs appear on the right with the inputs weighted by the components of B.

Solution for $X_i(s)$

The next step is to solve for a particular unknown of interest. Say we wish to find $X_2(s)$ from Eqs. (5-23). We can solve Eq. (5-23 b) for $X_1(s)$ in terms of $X_2(s)$

$$X_1(s) = \tfrac{1}{3}[-sX_2(s) + 2U(s)] \qquad (5\text{-}24)$$

Substitute this result into Eq. (5-23 a) to get

$$(s + 2)(\tfrac{1}{3})[-sX_2(s) + 2U(s)] - 4X_2(s) = -2 + 7U(s)$$

or, after the necessary algebra has been carried out

$$X_2(s) = \frac{6 + (2s - 17)U(s)}{s^2 + 2s + 12} \tag{5-25}$$

The result is in the form of a polynomial ratio, with $U(s)$ yet to be specified. The numerator includes a contribution from the initial conditions (the 6 term) and the input.

Now let us solve for $X_1(s)$, the other unknown. Referring to Eq. (5-24) and substituting, we get

$$X_1(s) = \frac{1}{3} - s\,\frac{6 + (2s - 17)\,U(s)}{s^2 + 2s + 12} + 2U(s)$$

or

$$X_1(s) = \frac{-2s + (7s + 8)U(s)}{s^2 + 2s + 12} \tag{5-26}$$

Behold, the same denominator appears! This denominator, called the *characteristic polynomial* of the system, contains the necessary information to gain much insight into the system dynamics. Notice that the characteristic polynomial does not depend upon either the initial conditions or the inputs. As we shall see next, it can be derived directly from the A array by a suitable determinant operation.

Cramer's rule Consider again the general second-order system in the transformed s domain

$$(s - a_{11})X_1(s) - a_{12}X_2(s) = x_1(0) + b_1U(s) \tag{5-20 a}$$

$$-a_{21}X_1(s) + (s - a_{22})X_2(s) = x_2(0) + b_2U(s) \tag{5-20 b}$$

Using Cramer's rule, we can find any unknown by forming a ratio of determinants. The denominator is the determinant of the array of coefficients of all the unknowns. Here we have a 2×2 array. The numerator array is obtained from the denominator array by replacing the column corresponding to the unknown of interest by the (entire) right-hand side, which is a column itself. For example, if we want $X_j(s)$, we put the right-hand-side column into the jth column in the coefficient array.

Thus, to solve for $X_1(s)$ from Eqs. (5-20)

$$X_1(s) = \frac{\begin{vmatrix} x_1(0) + b_1U(s) & -a_{12} \\ x_2(0) + b_2U(s) & s - a_{22} \end{vmatrix}}{\begin{vmatrix} s - a_{11} & -a_{12} \\ -a_{21} & s - a_{22} \end{vmatrix}} \tag{5-27}$$

The vertical lines indicate the determinant of the array contained within. The denominator determinant will yield a polynomial in s. In fact, it will yield a second-order polynomial in this case but an nth-order polynomial for an nth-order system. This polynomial is the *characteristic polynomial* of the system. Notice from the Cramer's-rule solution that it will be the same for all system unknowns. Only the numerator changes, as different columns are replaced.

In particular, the second-order characteristic polynomial (CP) from Eq. (5-27) is

$$CP(s) = (s - a_{11})(s - a_{22}) - a_{12}a_{21}$$

or $$CP(s) = s^2 + (-a_{11} - a_{22})s + (a_{11}a_{22} - a_{12}a_{21}) \qquad (5\text{-}28)$$

This is an extremely useful result, as we shall see.

If Eq. (5-28) is applied directly to the state equations in (5-22), we get

$$CP(s) = s^2 + (2 + 0)s + (-2)(0) - 4 \cdot (-3)$$

or $$CP(s) = s^2 + 2s + 12$$

This agrees with Eq. (5-25).

The general nth-order $CP(s)$ is given by

$$CP(s) = \begin{vmatrix} s - a_{11} & -a_{12} & \cdots & -a_{1n} \\ -a_{21} & s - a_{22} & \cdots & -a_{2n} \\ & & & \\ -a_{n1} & -a_{n2} & \cdots & s - a_{nn} \end{vmatrix} \qquad (5\text{-}29)$$

which yields an nth-order polynomial

$$CP(s) = s^n + (-a_{11} - a_{22} \cdots - a_{nn})s^{n-1} + \cdots |-A| \qquad (5\text{-}30)$$

The first coefficient is unity; the second is called the *trace* of $-A$; the constant term is the determinant of $-A$.

While the Cramer's-rule approach is useful for low-order systems, it is also valuable for higher-order problems because typical engineering problems of higher order have many zeros in the A array. Hence the determinant of the modified A array in Eq. (5-29), as well as the derived numerator array, can be expanded about a sparse row or column and reduced by that method. Appendix B explains the procedure.

We have shown two ways to obtain the Laplace solution for any unknown $X_i(s)$. The specific denominator obtained from either Cramer's rule or ad hoc reduction before the $U(s)$ functions are substituted in is the characteristic polynomial, which plays a crucial role in subsequent analysis. It is time to consider the time solution and other information that can be derived from such a solution.

5-4 TIME RESPONSE

The key steps in obtaining a time response from a Laplace-transform solution are:

1. Find the roots of the denominator polynomial.
2. Separate the Laplace solution by partial fractions.
3. Invert each fraction by table look-up.

Each of these steps is discussed and illustrated in a series of examples. The final part of this section shows how valuable information can be developed from the final-value theorem and how the initial-value theorem can be used to check the Laplace solution.

The Roots of the Denominator $D(s)$ Consider $X(s)$ in the form

$$X(s) = \frac{N(s)}{D(s)} = \frac{n_1 s^m + n_2 s^{m-1} + \cdots + n_m s^1 + n_{m+1}}{s^n + d_1 s^{n-1} + d_2 s^{n-2} + \cdots + d_n} \qquad (5\text{-}31)$$

where m is the degree of the numerator, n is the degree of the denominator, and $n > m$. We can rewrite Eq. (5-31) in factored form as

$$X(s) = \frac{N(s)}{(s + p_1)(s + p_2) \cdots (s + p_n)} \qquad (5\text{-}32)$$

where $-p_1, -p_2, \ldots, -p_n$ are called the *poles* of $X(s)$. The poles are found by solving the equation $D(s) = 0$. They are values of s which make the polynomial vanish and expressions such as Eq. (5-32) blow up. Note that there are exactly n roots of $D(s)$. In general, factoring an nth-order polynomial, i.e., finding its roots, is difficult manually. Computers are very helpful for large problems, and the structure of a problem frequently simplifies the factorization.

The part of $D(s)$ that is obtained from the system itself before the specific inputs are entered is, of course, the characteristic polynomial $CP(s)$. The roots of this polynomial are called the *eigenvalues* of the system. They are values of s for which the *characteristic equation* $CP(s) = 0$ is satisfied. These roots offer great insight into the system's dynamic-response tendencies or the natural or unforced dynamics of the system. In Eq. (5-32) some poles are eigenvalues and some come from the specific forcing functions $U(s)$.

In defining factors we want to keep real coefficients. Hence, if a quadratic factor has complex-conjugate roots, we do not reduce it to two first-order factors but preserve it in σ, ω form. The motivation comes from the entries in Table 5-1, specifically the denominator forms. Notice that they include first-order forms $s + \sigma$ and second-order forms $(s + \sigma)^2 + \omega^2$ but no higher forms.

Example 5-2

$$D(s) = s^2 + 6s + 8$$

By inspection, $D(s) = (s + 4)(s + 2)$, and the roots are -4 and -2.

Example 5-3

$$D(s) = s^2 + 6s + 10$$

Solve for the roots

$$s_{1,2} = \frac{-6 \pm \sqrt{6^2 - 4(10)}}{2}$$

or $s_{1,2} = -3 \pm j1$, a complex-conjugate pair. In this case, we shall not factor the quadratic form further.

Example 5-4

$$D(s) = s^3 - 1s^2 - 2s$$

Since $D(s)$ is a cubic, it has three roots. Write $D(s)$ as $s(s^2 - 1s - 2)$, from which it can be written by inspection as $(s+0)(s+1)(s-2)$. The roots are 0, -1, and $+2$.

Example 5-5

$$D(s) = (s^3 + 4s^2 + 9s + 10)(s^2 + 4)$$

This polynomial is already partially factored. By guessing at a (real) root for the cubic, we find that $D(s) = 0$ for $s = -2$. (Check it.) Hence, we can remove the factor $s + 2$ from the cubic by division

$$
\begin{array}{r}
s^2 + 2s + 5 \\
s + 2\overline{)s^3 + 4s^2 + 9s + 10} \\
\underline{s^3 + 2s^2} \\
2s^2 + 9s \\
\underline{2s^2 + 4s} \\
5s + 10 \\
5s + 10
\end{array}
$$

Now we have $D(s) = (s + 2)(s^2 + 2s + 5)(s^2 + 4)$. Study the middle factor; its roots are $-1 \pm j2$. The roots of the last factor are $\pm j2$. Now we have the complete story on $D(s)$; all five of its roots are known. Do not reduce $D(s)$ further at this point.

Partial-fraction expansion

Partial-fraction expansion is a key step in preparing a Laplace-transform function $X(s)$ for inversion from s to t. Separate $X(s)$ into a set of fractions, i.e., polynomial ratios, each of whose denominators is either a first- or a second-order factor, depending upon the roots. Construct a numerator that is a polynomial one degree less than the denominator, with constants to be determined. Now comes the work. Determine the constants by cross-multiplying the separate factors to reconstruct the proper numerator. By equating the coefficients of like powers of s in the numerator a set of equations from which to find the constants can be derived.

Example 5-6

$$X(s) = \frac{4}{s^2 + 6s + 8} = \frac{4}{(s + 4)(s + 2)}$$

Then

$$X(s) = \frac{C_1}{s + 4} + \frac{C_2}{s + 2}$$

or
$$X(s) = \frac{C_1(s + 2) + C_2(s + 4)}{(s + 4)(s + 2)}$$

but $C_1(s + 2) + C_2(s + 4) = 4$, which leads to

$$(C_1 + C_2)s + (2C_1 + 4C_2) = 4$$

or, by equating powers of s,

$$C_1 + C_2 = 0$$
$$2C_1 + 4C_2 = 4$$

Since $C_1 = -2$ and $C_2 = 2$ from above,

$$X(s) = \frac{-2}{s + 4} + \frac{2}{s + 2}$$

Example 5-7

$$X(s) = \frac{4s^2 + 5}{s(s^2 + 4)}$$

Then
$$X(s) = \frac{C_1}{s} + \frac{C_2 s + C_3}{s^2 + 4}$$

or
$$X(s) = \frac{C_1(s^2 + 4) + s(C_2 s + C_3)}{s(s^2 + 4)}$$

which leads to the polynomial equation

$$(C_1 + C_2)s^2 + (C_3)s + (4C_1) = 4s^2 + 5$$

Then
$$C_1 + C_2 = 4$$
$$C_3 = 0$$
$$4C_1 = 5$$

Solving these equations and substituting yields

$$X(s) = \frac{1.25}{s} + \frac{2.75s}{s^2 + 4}$$

Example 5-8 An important special case arises when a root is repeated in $D(s)$. We illustrate the treatment for a repeated root here. Let

$$X(s) = \frac{s + 3}{(s + 4)^2}$$

Then
$$X(s) = \frac{C_1}{(s + 4)^2} + \frac{C_2}{s + 4}$$

Note the special form of the first numerator. Recombine to find

$$C_1 + C_2(s + 4) = s + 3$$

or

$$C_2 = 1$$
$$C_1 + 4C_2 = 3$$

Thus

$$X(s) = \frac{-1}{(s + 4)^2} + \frac{1}{s + 4}$$

More general information on treatment of repeated quadratic factors and factors repeated more than twice is available.[2,3] The case shown in the example, a real root of multiplicity 2, is the most common practical case by far.

We have illustrated only the most basic techniques, but there are some useful tricks. For example, we can find C_1 in the first example with little effort as follows:

1. Multiple the original $X(s)$ and the partial-fraction expansion by C_1's denominator $s + 4$.
2. Cancel $s + 4$ out of $X(s)$, leaving

$$\frac{4}{s + 2} = C_1 + \frac{C_2(s + 4)}{s + 2}$$

3. Now make $s + 4 = 0$ by setting $s = -4$. This makes C_2's factor vanish and

$$C_1 = \frac{4}{-4 + 2} = -2$$

A generalization of this idea can often be used to find constants one at a time. You might try the idea on the other examples.

Inversion from $X(s)$ to $x(t)$

Our approach to inversion from $X(s)$ to $x(t)$ is based on knowing the table entries for a very common and important set of transform pairs. The last step in obtaining $x(t)$ from $X(s)$ is to put the separate fractions of $X(s)$ into table format.

Example 5-9

$$X(s) = \frac{-2}{s + 4} + \frac{2}{s - 2}$$

By direct table look-up we find

$$x(t) = -2e^{-4t} + 2e^{-2t}$$

Example 5-10

$$X(s) = \frac{1.25}{s} + \frac{2.75 s}{s^2 + 4}$$

By direct table look-up we get

$$x(t) = 1.25 + 2.75 \cos 2t$$

Example 5-11

$$X(s) = \frac{s + 5}{s^2 + 6s + 10}$$

The roots of $D(s)$ are $-3 \pm j1$ (from a previous example), so we write $X(s)$ in the form

$$X(s) = \frac{s + 5}{(s + 3)^2 + 1^2}$$

Now we reformat the numerator as

$$X(s) = \frac{(s + 3) + 2(1)}{(s + 3)^2 + 1^2}$$

or

$$X(s) = \frac{s + 3}{(s + 3)^2 + 1^2} + \frac{2(1)}{(s + 3)^2 + 1^2}$$

The inverse is

$$x(t) = e^{-3t} \cos 1t + 2e^{-3t} \sin 1t$$

Example 5-12

$$X(s) = \frac{-3}{s + 2} + \frac{2(s + 6)}{s^2 + 2s + 5} + \frac{-(s - 8)}{s^2 + 4}$$

The second and third terms need to be reformatted as

$$X(s) = \frac{-3}{s + 2} + \frac{2[(s + 1) + 2.5(2)]}{(s + 1)^2 + 2^2} + \frac{-1s + 4(2)}{s^2 + 2^2}$$

Then, inverting on a term-by-term basis, we get

$$x(t) = -3e^{-2t} + 2e^{-1t} \cos 2t + 5e^{-1t} \sin 2t - 1 \cos 2t + 4 \sin 2t$$

Remark on a Useful Equivalence

The inversion procedure we have been using often generates sine and cosine terms of the same frequency. It is convenient to combine such terms into a single term as follows. Let

$$x(t) = C_1 \sin \omega t + C_2 \cos \omega t$$

where C_1, C_2, and ω are given. But

$$x(t) = M \sin (\omega t + \phi) = M \sin \omega t \cos \phi + M \cos \omega t \sin \phi$$

$$= (M \cos \phi) \sin \omega t + (M \sin \phi) \cos \omega t$$

Then, since C_1 and C_2 are known, we have

$$M = \sqrt{C_1^2 + C_2^2} \qquad \text{and} \qquad \tan \phi = \frac{C_2}{C_1}$$

The quantity M is the magnitude of the signal $x(t)$, and ϕ is the phase angle.

Example 5-13

$$x(t) = -1 \cos 2t + 4 \sin 2t$$

Then $C_1 = 4$ and $C_2 = -1$, so that

$$M = \sqrt{4^2 + (-1)^2} = \sqrt{17} = 4.123$$

and $\tan \phi = -\frac{1}{4}$ so $\phi = -0.245$ rad. Thus

$$x(t) = 4.123 \sin (2t - 0.245)$$

Example 5-14

$$x(t) = 5e^{-1t} \sin 2t + 2e^{-1t} \cos 2t$$

Then $C_1 = 5$, $C_2 = 2$, and e^{-1t} can be factored out. We find

$$M = \sqrt{5^2 + 2^2} = \sqrt{29} = 5.385$$

and $\tan \phi = \frac{2}{5}$, so $\phi = 0.381$ rad. Then

$$x(t) = 5.385e^{-1t} \sin (2t + 0.381)$$

Checking the Solution

One way to check the time solution for a problem is to apply the initial-value theorem to the Laplace transform. We see whether $x(t)$ for $t = 0$ and $X(s)$ subject to the IVT agree.

Example 5-15

$$X(s) = \frac{-2s^4 + 14s^3 + 28s^2 + 102s + 116}{(s + 2)(s^2 + 25 + 5)(s^2 + 4)}$$

By the IVT,

$$x(t = 0) = \lim_{s \to \infty} sX(s) = -2$$

The solution to this $X(s)$ is given in Example 5-12 as

$$x(t) = -3e^{-2t} + 2e^{-1t} \cos 2t + 5e^{-1t} \sin 2t - 1 \cos 2t + 4 \sin 2t$$

Clearly, $\qquad\qquad x(0) = -3 + 2 - 1 = -2$

In essence, we have checked the leading numerator coefficient of $X(s)$ and our time solution. In a problem derived from physical conditions, the initial condition will be known and both $X(s)$ and $x(t)$ can be checked independently.

A second important check that can sometimes be made is for a constant steady-state value. This check should be applied only when the roots of the denominator have negative real parts or there is at most one root at zero. It involves the use of the final-value theorem.

Example 5-16

$$X(s) = \frac{4s + 1}{s(s + 2)(s + 4)}$$

Predict $x(t \rightarrow \infty)$ by $\lim_{s \to 0} sX(s)$:

$$x(t \rightarrow \infty) = \frac{1}{2(4)} = 0.125$$

The time solution is

$$x(t) = 0.125 + 1.75e^{-2t} - 1.875e^{-4t}$$

from which we see that $x(t \rightarrow \infty) \rightarrow 0.125$. (Also notice that the initial condition checks at zero.)

Nature of the Response

The partial-fraction expansion shows what types of time functions are involved in a response. For example, Table 5-1 shows that poles (or eigenvalues) which are real and negative correspond to decaying exponentials, while only complex poles (or eigenvalues) lead to damped sinusoidal solution terms. A system which has purely sinusoidal response terms must have a factor $s^2 + \omega^2$ in its $D(s)$ polynomial, while one which oscillates in a decaying manner will have a factor $(s + \sigma)^2 + \omega^2$. These correspond to poles at $s = \pm j\omega$ and $s = -\sigma \pm j\omega$, respectively. Remember that $D(s)$ comes partly from the characteristic polynomial and partly from the forcing functions. Therefore, a system with complex eigenvalues will have damped oscillatory parts in its response even if the forcing does not contribute any oscillatory response through its own transform. Thus we say that the eigenvalues determine the natural response dynamics of the system independent of the particular forcing.

Eigenvalue Interpretation

The eigenvalues are the roots of the characteristic polynomial $CP(s)$, as previously described. The actual response then depends upon the initial conditions. In fact, the initial conditions determine only the numerator polynomial $N(s)$. Furthermore, all system variables have the same denominator, except in certain special cases when one or more common factors are canceled from both $N(s)$ and $CP(s)$. In this section we exploit the information and insight contained in the eigenvalues of first- and second-order systems. Since all higher-order system responses are composed of sums of first- and second-order response, our considerations are general enough to be very useful in practice.

First-order systems

Any first-order system without forcing can be put in the form

$$\dot{x}(t) = -ax(t) \qquad x(t_0) = x_0 \tag{5-33}$$

The characteristic polynomial is $s + a$, the eigenvalue is $-a$, and the time response has the factor e^{-at} in it. The initial condition provides the particular weighting of the time factor. The solution is

$$x(t) = x_0 e^{-at} \tag{5-34}$$

Provided the initial condition is not zero and a is finite, we can nondimensionalize this result as

$$\frac{x}{x_0} = e^{-t/\tau} \tag{5-35}$$

where x/x_0 and t/τ are dimensionless and τ is defined as $1/|a|$, the time constant. Only *one* curve is needed to represent the nondimensionalized first-order response of Eq. (5-35), provided that a is positive. That curve has initial value 1 for $t/\tau = 0$ and decays exponentially to 0 as $t/\tau \to \infty$.

The exponential response curve in dimensionless form is shown in Fig. 5-2. Several key points from the graph are labeled, and their values are displayed, in Table 5-2. Notice that the curve starts at 1.0 and drops off by a factor of $1/e$ for each unit of t/τ. Thus τ is the critical constant in assessing first-order transient response. Recall that τ is given as $1/a$ (a positive), minus where $-a$ is the eigenvalue. Hence, any first-order response can easily be estimated from its equation by inspecting the coefficient a.

Example 5-17 Let $\dot{x} = -4x$. When does x fall to within 1 percent of its final value? From the dimensionless response curve of Fig. 5-2 we see that x/x_0 falls to within 1

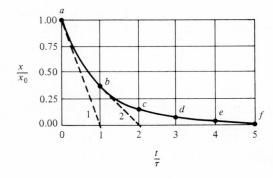

Figure 5-2 The exponential response curve in dimensionless form (see Table 5-2).

Table 5-2 Values for Fig. 5-2

Point	Time t/τ	x/x_0 Exact	Approximate	Percent
a	0	e^0	1.000 ...	100
b	1	e^{-1}	0.368	<37
c	2	e^{-2}	0.135	<14
d	3	e^{-3}	0.050	5
e	4	e^{-4}	0.018	<2
f	5	e^{-5}	0.007	<1

percent (of zero) by 5 time constants; i.e., when t/τ is 5, the ratio is down to 0.007. So the quick and easy answer is 5τ. The given equation tells us that since $a = 4$, $\tau = 0.25$. Therefore, by t equal to 1.25 the response is within 1 percent of final value.

Another property of the exponential responses is also depicted in Fig. 5-2. Observe the dashed lines. Curve 1 goes from unity to zero in a 1-time-constant interval. The response curve is tangent to this line at $t/\tau = 0$. The same property holds for curve 2 and in fact is true for every point on the curve. Can you prove that? At least try it for integer values of t/τ.

By knowing that an exponential curve approaches its final value by a factor of $1/e$ for every interval τ and by using the tangent property at key points (for example, $\tau, 2\tau, 3\tau, \ldots$) it should be easy for you to sketch an exponential response of the form $x_0 e^{-t/\tau}$ (or $x_0 e^{-at}$) given the data x_0 and a.

Second-order systems

A second-order system has a characteristic polynomial of the form

$$\text{CP}(s) = s^2 + Bs + C \tag{5-36}$$

which yields two roots, i.e., eigenvalues. Two distinct cases can arise. Consider the general solution expression

$$s_{1,2} = \frac{-B + \sqrt{B^2 - 4C}}{2} \tag{5-37}$$

where we assume that B and C are both positive. If $B^2 - 4C > 0$, there are two real roots. The solution can be written as

$$x(t) = C_1 e^{-\sigma_1 t} + C_2 e^{-\sigma_2 t} \tag{5-38}$$

where σ_1 and σ_2 are positive real constants. The solution is the sum of two exponentials. If $B^2 - 4C < 0$, there are complex-congugate roots. One way to write the roots is

$$s_{1,2} = -\sigma \pm j\omega \tag{5-39}$$

where σ and ω are positive real constants. This leads to a time response of the form

$$x(t) = C_1 e^{-\sigma t} \sin(\omega + \phi) \tag{5-40}$$

This is called an *exponentially damped sinusoidal response;* σ defines the decay envelope, while ω defines the oscillation frequency. C_1 determines the amplitude, while ϕ determines the phase; these are found from the initial conditions and do not concern us here. The fundamental dynamic nature of the response is governed by σ and ω, which are derived in turn from B and C in CP(s).

A second way to write the roots for complex conjugates is derived from CP(s) as follows:

$$CP(s) = s^2 + 2\zeta\omega_n s + \omega_n^2 \tag{5-41}$$

Then

$$s_{1,2} = -\zeta\omega_n \pm j\omega_n \sqrt{1 - \zeta^2} \tag{5-42}$$

where ζ is the *damping ratio* (a positive constant) and ω_n is the *natural frequency* (a positive constant).

In Fig. 5-3 we plot a series of response curves for a given second-order system. These curves are the solution to the equation $\ddot{x} + 2\zeta\omega_n \dot{x} + \omega_n^2 x = 0$ with the initial conditions $x(0) = x_0$ and $\dot{x}(0) = 0$. Start with the curve labeled by $\zeta = 0$. It corresponds to an undamped oscillation of period $2\pi/\omega N$. This oscillation persists indefinitely with its peak amplitude undiminished. As the damping ratio is increased, each curve has a more rapid decay of amplitude and a longer period. The amplitude-decay envelope is in fact governed by $e^{-\zeta\omega_n t}$, while the frequency is $\omega_n \sqrt{1 - \zeta^2}$. Thus the equivalence between σ and ω, on the one hand, and ζ and ω_n, on the other, is

$$\sigma = \zeta\omega_n \quad \text{and} \quad \omega = \omega_n \sqrt{1 - \zeta^2} \tag{5-43}$$

In engineering we normally speak of a system with $\zeta = 0$ as *undamped,* with $\zeta < 1$ as *underdamped,* with $\zeta = 1$ as critically damped (two real repeated roots), and with $\zeta > 1$ as *overdamped.*

In summary, we note that two parameters control the natural time response of

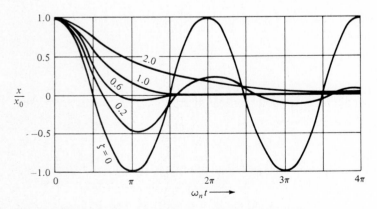

Figure 5-3 Characteristic response curves for the damped sinusoidal case.

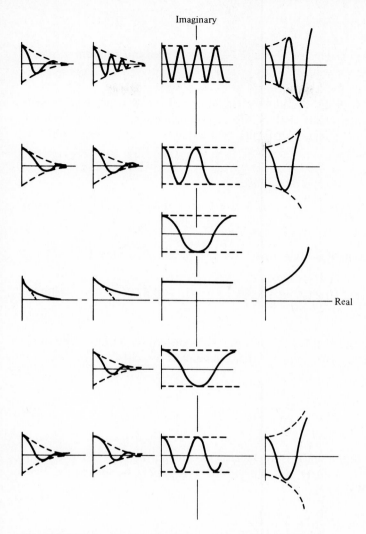

Figure 5-4 Time response correlated with eigenvalues in the *s* plane.

a second-order system; we can convert the CP(*s*) coefficients *B* and *C* into either σ, ω or ζ,ω$_n$ form if the system has complex-conjugate roots or into σ$_1$,σ$_2$ form if the system has two real roots. All this information is nicely represented by a correlation of typical time responses with the *s* plane, i.e., the complex plane showing eigenvalue locations (see Fig. 5-4). We remind you that all linear-system models with real constant coefficients lead to eigenvalues that are either real or paired as complex conjugates. Consequently, all natural time responses are sums of exponential curves and (damped) sinusoidal curves. The eigenvalues tell you what the time factors of the natural responses are, independent of initial conditions and input functions. We shall learn later how to exploit such information in analysis and design.

5-5 EXAMPLES

Now that we have developed tools sufficient to the task of predicting system response, we apply them to several examples.

Example 5-18 Refer to Fig. 5-1. Suppose the mass is 1 kg, the spring stiffness is 100 N/m, and the damper value is 20 N/m-s. The acceleration of gravity is 9.8 m/s^2. Find the position of the mass if the initial conditions on x and p are both zero. Does the system oscillate?

The state equations for the system are given by Eq. (5-1); with the parameter values given above for m, k, and b they become

$$\dot{p}(t) = -20p(t) - 100x(t) + 9.8 \tag{5-44a}$$

and
$$\dot{x}(t) = 1p(t) \tag{5-44b}$$

When Eq. (5-44) are Laplace-transformed and unknowns collected on the left-hand side, we get

$$(s + 20)P(s) + 100X(s) = p(0) + 9.8/s \tag{5-45a}$$

$$-1P(s) + sX(s) = x(0) \tag{5-45b}$$

Note that the quantity $9.8/s$ is the transform of the constant term (9.8). Letting the initial conditions be zero and solving for the transform of x yields

$$X(s) = \frac{\begin{vmatrix} s + 20 & 9.8/s \\ -1 & 0 \end{vmatrix}}{\begin{vmatrix} s + 20 & 100 \\ -1 & s \end{vmatrix}} = \frac{9.8/s}{s^2 + 20s + 100} \tag{5-46}$$

The denominator in Eq. (5-46) is the characteristic polynomial. To determine whether the system will oscillate, we calculate the eigenvalues by solving the characteristic equation $CP(s) = 0$.

$$s_{1,2} = \frac{-20 \pm \sqrt{(20)^2 - 4(100)}}{2} = -10 \pm 0 \tag{5-47}$$

This means that there are repeated roots. The system does not oscillate because the roots are real. Now write $X(s)$ in the form

$$X(s) = \frac{9.8}{s(s + 10)^2} \tag{5-48}$$

expand by partial fractions

$$X(s) = \frac{C_1}{s} + \frac{C_2}{(s + 10)^2} + \frac{C_3}{s + 10}$$

and solve for C_1, C_2, and C_3. The equations are

$$C_1 + C_3 = 0$$

$$20C_1 + C_2 + 10C_3 = 0$$
$$100C_1 = 9.8$$

Thus
$$X(s) = \frac{0.098}{s} + \frac{-0.98}{(s + 10)^2} + \frac{-0.098}{s + 10} \qquad (5\text{-}49)$$

Inverting term by term yields

$$x(t) = 0.098 - 0.98te^{-10t} - 0.098e^{-10t} \qquad (5\text{-}50)$$

Check the solution by the initial condition $x(t = 0) = 0.098 - 0.098 = 0$. Apply the FVT to Eq. (5-48). This yields

$$x(t \to \infty) = \lim_{s \to 0} sX(s) = \frac{9.8}{(10)^2} = 0.098$$

This result is the same as that derived from Eq. (5-50) with $t \to \infty$. (Observe that the limit of te^{-10t} as $t \to \infty$ is zero.)

We conclude that the system does not oscillate. Suppose we wished to induce oscillations at, say, a frequency of 5 rad/s. How should the spring stiffness k be adjusted to ensure this?

Return to Eq. (5-44 a) and (5-46). The value of 100 represents k. This could be shown more directly by retaining k in the state equations and working it through the transform procedure to the point of Eq. (5-47). If we did so, we should find the roots as

$$s_{1,2} = \frac{-20 \pm \sqrt{(20)^2 - 4k}}{2} \qquad (5\text{-}51)$$

The root form we seek is $\sigma \pm j\omega$ with $\omega = 5$. To achieve this, set

$$\sqrt{(20)^2 - 4k} = j10 \qquad \text{or} \qquad k = 125 \text{ N m}$$

The resulting roots are $-10 \pm j5$; the system will oscillate at a frequency of 5 rad/ s, as desired, but it will also be damped. You may wish to verify that the new $x(t)$ solution is

$$x(t) = 0.0784 - 0.0784e^{-10t} \cos 5t - 0.1568e^{-10t} \sin 5t$$

or
$$x(t = 0.0784 + 0.1753e^{-10t} \sin (5t - 2.678)$$

where the phase angle is in radians.

Example 5-18 Consider the electric-circuit diagram and bond graph in Fig. 5-5. We wish to find $v_{out}(t)$ as it depends upon $v_{in}(t)$. The state equation describing the circuit is

$$\dot{q}_4 = -0.15q_4 + 0.1v_{in} \qquad (5\text{-}52)$$

and the output equation is

$$v_{out} = 0.59q_4 \qquad (5\text{-}53)$$

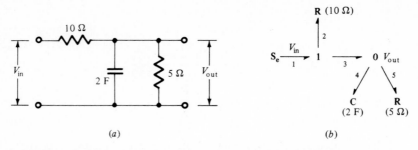

(a)   *(b)*

Figure 5-5 An electric-circuit example: (*a*) circuit diagram and (*b*) bond graph.

If we use Eq. (5-53) in Eq. (5-52), we arrive directly at the input-output equation

$$\dot{v}_{out} = -0.15v_{out} + 0.05v_{in} \tag{5-54}$$

We wish to find the output in response to an input of the form

$$v_{in}(t) = 110 \sin 0.1t \tag{5-55}$$

In particular, we are interested in the part of the response that persists after the influence of the initial condition effect is small. Let

$$V_{in}(s) = \mathcal{L}\{v_{in}(t)\} \quad \text{and} \quad V_{out}(s) = \mathcal{L}\{v_{out}(t)\}$$

Then Eq. (5-54) can be transformed to yield

$$(s + 0.15)V_{out}(s) = v_{out}(0) + 0.05V_{in}(s) \tag{5-56}$$

where $v_{out}(0)$ is the initial condition, derivable from Eq. (5-53) in terms of $q_4(0)$.

The general Laplace solution for $V_{out}(s)$ is

$$V_{out}(s) = \frac{v_{out}(0)}{s + 0.15} + \frac{0.05 \, V_{in}(s)}{s + 0.15} \tag{5-57}$$

The first term on the right can be inverted by inspection to yield $v_{out}(0)e^{-0.15t}$. We shall concentrate on the second term for now

$$V_2(s) = \frac{0.05}{s + 0.15} \frac{110(0.1)}{s^2 + (0.1)^2} \tag{5-58}$$

or

$$V_2(s) = \frac{C_1}{s + 0.15} + \frac{C_2s + C_3}{s^2 + (0.1)^2} \tag{5-59}$$

where $V_2(s)$ is the second part of the solution and we have used the transform of Eq. (5-55). Solving for the constants by cross-multiplying, we get

$$V_2(s) = \frac{16.923}{s + 0.15} + \frac{-16.923s + 2.538}{s^2 + (0.1)^2}$$

or

$$V_2(s) = \frac{16.923}{s + 0.15} + \frac{-16.923s}{s^2 + (0.1)^2} + \frac{25.38(0.1)}{s^2 + (0.1)^2}$$

This inverts to yield

$$v_2(t) = 16.923e^{-0.15t} - 16.923(\cos 0.1)t + 25.38(\sin 0.1)t \qquad (5\text{-}60)$$

As a partial check, we note that $v_2(0) = 0$, which corresponds to the result obtained by applying the IVT to Eq. (5-58).

Let us combine the sine and cosine terms of Eq. (5-60), obtaining

$$v_2(t) = 16.923e^{-0.15t} + 30.505 \sin (0.1t - 0.588)$$

The complete solution for $v_{out}(t)$ is

$$v_{out}(t) = (v_{out}(0) + 16.923e)^{-0.15t} + 30.505 \sin (0.1t - 0.588) \qquad (5\text{-}61)$$

After a certain period of time, the first term, involving the initial condition and the transient part of the solution, will become small compared with the persistent sinusoid. The magnitude of the sinusoidal response is 30.505, compared with an input magnitude of 110. Hence the circuit attenuates the sinusoidal input signal and delays it in phase relative to the input signal. If we studied input voltages at various forcing frequencies, we would conclude that the circuit is a low-pass filter which attenuates high frequencies and passes low ones with little attenuation and phase shift.

5-6 SUMMARY

In this chapter we have defined and applied the Laplace transform to the solution of state equations with constant coefficients. The method converts a set of differential equations in the time domain into a set of algebraic equations in the (complex) frequency domain.

Useful information was extracted directly from the Laplace transform solution. In particular the roots of the characteristic polynomial, called eigenvalues, determine what types of time functions will be found in the response no matter what input is involved.

The time solution was obtained by using partial-fraction expansion and comparison of terms to a standard table of Laplace transforms.

PROBLEMS

5-1 Show that the Laplace transform of a *unit impulse* at $t = 0$ is 1 by first finding the Laplace transform of $x(t)$ (Fig. P5-1). Then set $h\Delta = 1$, and let $h \to \infty$ as $\Delta \to 0$.

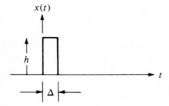

Figure P5-1

5-2 Derive entry 8 in Table 5-1 from entry 6 by using the frequency-shift relation.

5-3 (a) Derive entry 5 in Table 5-1 from entry 6 by using entry 10.

(b) Derive entry 5 in Table 5-1 from entry 6 by using entry 9.

5-4 Prove entry 13 in Table 5-1 directly from the defining Laplace-transform integral.

5-5 Find the Laplace transform of the following functions:

(a) $x(t) = 3e^{-7t}$ (b) $x(t) = 4e^{-t} + 7e^{-3t}$ (c) $x(t) = 3e^{-5t}\cos 2t - 1e^{-5t}\sin 2t$

(d) $x(t) = 8$ $t \geq 0$ (e) $\text{x(t)} = \begin{cases} 0 & t < 3 \\ A & t \geq 3 \end{cases}$ (f) $x(t) = 5\sin(2t + \frac{\pi}{4})$

(g) $x(t) = ct^2$ (h) $x(t) = -2te^{-5t}$

5-6 Find $x(t)$ for the following Laplace transforms:

(a) $X(s) = \dfrac{0.01}{2s + 1}$ (b) $X(s) = \dfrac{3}{s^2 + 3s + 2}$

(c) $X(s) = \dfrac{2s + 5}{s + 3}$ Hint: Divide to obtain a proper fraction. Then see the result of Prob. 5-1.

(d) $X(s) = \dfrac{7s + 1}{s^2 + 4s + 13}$

(e) $X(s) = \dfrac{2s^2 + 5}{s^3 + 10s^2 + 33s + 36}$ Hint: Investigate a root near -4.

(f) $X(s) = \dfrac{4e^{-s}}{s^2}$ (g) $X(s) = \dfrac{6s + 1}{s^2 + 9}$ (h) $X(s) = \dfrac{6s + 1}{s^2 - 9}$

5-7 Solve the following state equations. Check your answers.

(a) Find $x_1(t)$ and $x_2(t)$ for

$$\dot{x}_1 = -2x_1 + 4x_2 \quad \text{with } x_1(0) = -2$$
$$\dot{x}_2 = -1x_1 - 1x_2 \quad\quad\quad x_2(0) = 0$$

(b) Find k to give an oscillation frequency of $4v$

$$\dot{x}_1 = -kx_2$$
$$\dot{x}_2 = 2x_1$$

(c) Find b to yield repeated roots (note that $b > 0$):

$$\dot{x}_1 = \qquad\qquad - x_3$$
$$\dot{x}_2 = \qquad -2x_2 + 1x_3$$
$$\dot{x}_3 = 2x_1 \qquad\quad - bx_3$$

(d) Find $x(t)$ if

$$u(t) = 6\sin 2t$$
$$\dot{x} = -5x + 10u(t) \quad \text{with } x(0) = 1$$

(e) Repeat part (d) for

$$u(t) = \begin{cases} 0 & t < 2 \\ = 5 & t \geq 2 \end{cases} \quad x(0) = 0$$

What if $x(0) = 1$?

5-8 A classic mechanical system is shown in the figure. The parameters are $m_1 = m_4 = 1$ kg, $k_2 = k_5 = 1$ N/m, and $g = 9.8$ m/s^2.

(a) If $b_3 = b_6 = 0$ (no damping), what are the oscillation frequencies? Hint: The resulting quartic equation is a special form that you can solve.

(b) Now let $b_3 = b_6 = 1$ N·m/s. What is the steady-state configuration of the system? In particular, what is $x_5(t \rightarrow \infty)$, the spring deflection? Hint: You can apply the FVT to $X_5(s)$ since the roots all have negative real parts. It is not necessary to find the roots.

(c) If $b_3 = 0$, can b_6 be chosen so that the system does not oscillate? Explain.

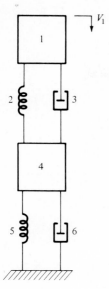

Figure P5-8

5-9 The circuit in Fig. 5-9 acts as a filter. We are interested in the output voltage $v_o(t)$ in response to the input voltage $v_i(t)$ under *no-load* conditions.

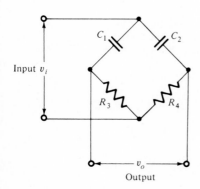

Input v_i

C_1 C_2

R_3 R_4

Output   **Figure P5-9**

v_o

(*a*) Make a bond-graph model of the circuit.

(*b*) Derive state equations and an output equation for $v_o(t)$.

(*c*) Laplace-transform the state and output equations and derive $V_o(s)$ in terms of $V_i(s)$. Assume that the initial conditions are zero.

(*d*) Let $R_3 C_1 = 0.5 \text{ s}^{-1}$ and $R_4 C_2 = 0.25 \text{ s}^{-1}$. Also let $v_i(t) = 10 \sin 3t$. Find $v_o(t)$.

5-10 A rotational mechanical load is driven by a motor that sets the driving speed (Fig. P5-10*a*). A bond-graph model is shown in Fig. P5-10*b*, and the parameters are

$$k = 1 \text{ N} \cdot \text{m} \qquad g = 20 \ (\omega_3 = 20\omega_4) \qquad J = 2 \text{ kg} \cdot \text{m}^2 \qquad b = 20 \text{ N} \cdot \text{m} \cdot \text{s}$$

with $T_L = 10 \text{ N} \cdot \text{m}$. The input $\Omega_1(t)$ is shown in Fig. P5-10*c*

$$\Omega_1 = \begin{cases} 100 \text{ rad/s} & 0 \le t < 1 \\ 0 & 1 \le t < \infty \end{cases}$$

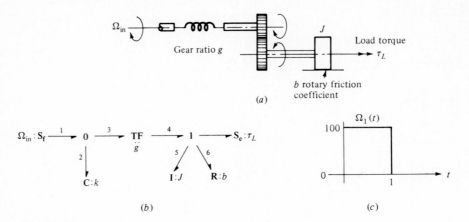

(a)

(b)

(c)

Figure P5-10

Find $\omega_5(t)$, the angular velocity of the load inertia. Initial conditions are zero. *Hint:* Divide the problem into two periods. Solve for ω_5 from $0 \leq t < 1$. Then use the *system state* at $t = 1$ as new initial conditions. Solve for ω_5 from $1 \leq t < \infty$.

REFERENCES

1. R. V. Churchill: *Operational Mathematics,* McGraw-Hill, New York, 1958.
2. W. T. Thomson: *Laplace Transformation,* Prentice-Hall, Englewood Cliffs, N.J., 1960.
3. M. R. Spiegel: *Laplace Transforms,* Schaum's Outline Series, McGraw-Hill, New York, 1965.

BLOCK DIAGRAMS

In previous chapters we have seen that many dynamic physical systems can be modeled by bond graphs. A set of state and output equations can be derived from a given bond graph using causality methods. For linear equations with constant coefficients the Laplace transform has proved to be extremely useful in analysis. This chapter extends the usefulness of earlier methods by introducing the *block diagram* for input-output analysis.

6-1 BLOCK DIAGRAMS AND TRANSFER FUNCTIONS

A block diagram is a graphical way to represent equation information. In this chapter we shall restrict our attention to linear models based upon Laplace-transformed variables, but block diagrams can also represent nonlinear systems. Each directed line in the diagram represents a single variable (voltage, pressure, angular momentum), while each node or *block* shows a particular type of input-output relation between variables.

A *transfer function* is the ratio of the Laplace transform of an output to the Laplace transform of an input with all initial conditions zero. It is important to remember the qualifier "all initial conditions zero."

Consider a system with the state equation

$$\dot{x}(t) = -2x(t) + 3u(t) \tag{6-1}$$

We define the transfer function $T(s)$ as

$$T(s) = \frac{X(s)}{U(s)} \tag{6-2}$$

where $X(s) = \mathcal{L}\{x(t)\}$ and $U(s) = \mathcal{L}\{u(t)\}$. Applying the Laplace transform to Eq. (6-2) and solving for $X(s)$ in terms of $U(s)$ with the initial condition set to zero, we obtain

$$sX(s) - x(0)^{0} = -2X(s) + 3U(s)$$

or
$$(s + 2)\,X(s) = 3U(s) \tag{6-3}$$

According to Eq. (6-2), the transfer function is

$$\frac{X(s)}{U(s)} = T(s) = \frac{3}{s + 2} \tag{6-4}$$

Observe that the transfer function itself does *not* depend upon the specific input. That is why it is so useful. It characterizes the system response (measured by x) to any input u.

A block diagram for Eq. (6-4) is just

Reading such a diagram, we write

$$X(s) = \frac{3}{s + 2}\,U(s) \tag{6-5}$$

The arrows on the signal lines indicate the direction of the flow of information. That is, $U(s)$ comes in and is multiplied by the transfer function to produce $X(s)$. Consider the system equations

$$\dot{x}_1 = -2x_1 + 3u(t) \tag{6-6a}$$

$$\dot{x}_2 = -4x_1 \tag{6-6b}$$

If all initial conditions are zero, we obtain the transformed equations

$$(s + 2)X_1(s) = 3U(s) \tag{6-7a}$$

$$4X_1(s) + sX_2(s) = 0 \tag{6-7b}$$

We wish to obtain the transform function between $X_2(s)$ and $U(s)$. One way is to solve Eq. (6-7) for $X_2(s)$ using Cramer's rule

$$X_2(s) = \frac{\begin{vmatrix} s + 2 & 3U(s) \\ 4 & 0 \end{vmatrix}}{\begin{vmatrix} s + 2 & 0 \\ 4 & +s \end{vmatrix}} = \frac{-12U(s)}{s^2 + 2s} \tag{6-8}$$

Then
$$\frac{X_2(s)}{U(s)} = T(s) = \frac{-12}{s^2 + 2s} \tag{6-9}$$

The block diagram for Eq. (6-9) is

$$U(s) \rightarrow \boxed{\dfrac{-12}{s^2 + 2s}} \rightarrow X_2(s)$$

We now try a slightly different approach with the same goal in mind, namely, to find $X_2(s)/U(s)$ using block diagrams. First, solve Eq. (6-7 a) for $X_1(s)$

$$X_1(s) = \frac{3}{s + 2} U(s) \tag{6-10}$$

The corresponding block diagram is

$$U(s) \rightarrow \boxed{\dfrac{3}{s + 2}} \rightarrow X_1(s)$$

Now, solve Eq. (6-7 b) for $X_2(s)$, temporarily treating $X_1(s)$ as an input

$$X_2(s) = -\frac{4}{s} X_1(s) \tag{6-11}$$

The block diagram for Eq. (6-11) is

$$X_1(s) \rightarrow \boxed{-\dfrac{4}{s}} \rightarrow X_2(s)$$

Next, we join the block diagram for Eq. (6-10) to that for Eq. (6-11) to get

$$U(s) \rightarrow \boxed{\dfrac{3}{s + 2}} \rightarrow X_1(s) \rightarrow \boxed{-\dfrac{4}{s}} \rightarrow X_2(s)$$

which can be reduced to

$$U(s) \rightarrow \boxed{\dfrac{-12}{s^2 + 2s}} \rightarrow X_2(s)$$

if $X_1(s)$ is eliminated as an explicit variable. The diagram with two blocks is called a *cascade connection*. Simplification of such a connection is obvious.

Only three basic building blocks are required in block diagrams for linear systems; they are shown in Table 6-1. We have already used the transfer-function block. Now we add a summation element and a distribution element. Sometimes a minus sign is associated with one or more inputs on a summation element; then the corresponding sign for that term is changed in the equation. To be completely specific, you can provide plus and minus signs for each input signal to a summation block. Observe that a summation has exactly one output. The distribution element is the counterpart; it has exactly one input, which is distributed to several output signal paths.

Table 6-1 Block-diagram elements†

Function	Element	Equation
Transfer		$X_2 = T(s)X_1$
Summation		$Y = X_1 + X_2 + X_3$
Distribution		$Y_1 = X$ $Y_2 = X$ $Y_3 = X$

† All variables are Laplace transforms: $X(s)$, $Y(s)$, $T(s)$.

6-2 BLOCK-DIAGRAM ALGEBRA

Although many types of connections show up in block diagrams, we have summarized the three most common in Table 6-2. The first entry is the cascade connection, which involves multiplication of transfer functions. The second entry involves parallel paths. Its equivalent transfer function for the reduced form is just the sum of the transfer functions in each path. The third entry in the table, an important one, is the classic feedback connection.

Let us prove the equivalence shown in the table for the feedback loop. In Fig. 6-1 a feedback connection is shown with each signal labeled. Three equations are implied

$$X_4(s) = X_1(s) - X_3(s) \tag{6-12a}$$

$$X_2(s) = T_1(s)X_4(s) \tag{6-12b}$$

$$X_3(s) = T_2(s)X_2(s) \tag{6-12c}$$

We wish to find $X_2(s)$ in terms of $X_1(s)$ by eliminating $X_3(s)$ and $X_4(s)$.

$$X_2(s) = T_1(s)[X_1(s) - X_3(s)]$$

or

$$X_2(s) = T_1(s)[X_1(s) - T_2(s) X_2(s)]$$

or

$$[1 + T_1(s) T_2(s)]X_2(s) = T_1(s)X_1(s)$$

Table 6-2 Some block-diagram equivalents

Connection	Expanded	Reduced
	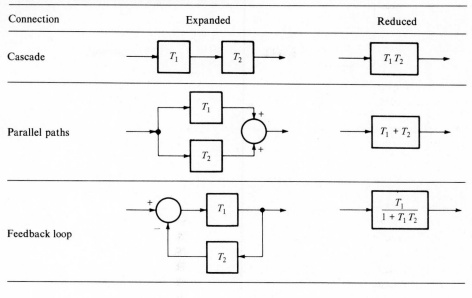	
Cascade		
Parallel paths		
Feedback loop		

$$\frac{X_2(s)}{X_1(s)} = \frac{T_1(s)}{1 + T_1(s)\ T_2(s)} \tag{6-13}$$

The result includes the effect of a minus sign in the feedback path, since that is the most common situation in physical systems and control systems. This classic result is stated in words as follows:

> The system transfer function is equal to the ratio of the forward transfer function to 1 minus the loop transfer function.

In Fig. 6-1 the forward function is $T_1(s)$. The loop transfer function is $-T_1(s)T_2(s)$. Hence the denominator is $1 + T_1(s)T_2(s)$.

Example 6-1 Consider the block diagram shown in Fig. 6-2a. The goal is to reduce it to a single block with u as input and y as output. We look for any reductions that can be made in terms of cascade, parallel-path, or feedback connections. Observe that transfer functions $T_1(s)$ and $T_2(s)$ are in a parallel-path configuration and can

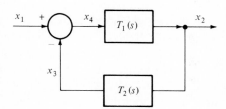

Figure 6-1 A feedback loop.

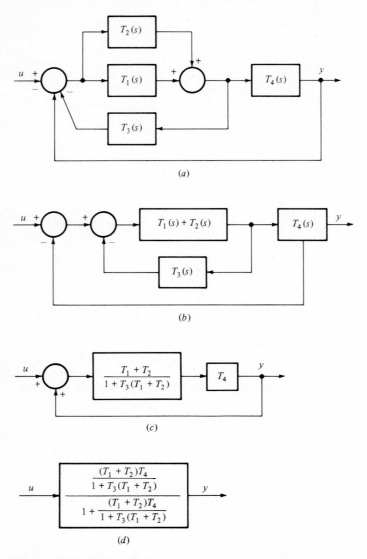

Figure 6-2 Block-diagram reduction: (a) block diagram; (b) partially reduced; (c) reduced further; (d) final block reduction.

be combined. This has been done in Fig. 6-2b. We have also separated the three-input summation into two two-input summers to facilitate further reductions.

Now a feedback loop appears. The result of reducing the loop involving $T_1 + T_2$ and T_3 is shown in Fig. 6-2c. At this point we sense victory, since only a cascade connection and another feedback loop remain. Note that a signal path with no block implies a transfer function of value unity. The result (Fig. 6-2d) can be checked by labeling each signal in the original figure, writing an equation for each block and summation, and using a direct algebraic approach. A simplified form of the result is

$$\frac{v(s)}{u(s)} = \frac{(T_1 + T_2)T_4}{1 + (T_3 + T_4)(T_1 + T_2)} \qquad (6\text{-}14)$$

This suggests another pattern of reduction from the diagram. Can you see what it is?

6-3 STATE EQUATIONS

State equations are organized in a particular fashion. When block diagrams are derived from them, it is possible to take advantage of the organization. Consider a second-order system governed by

$$\dot{x}_1 = a_{11}x_1 + a_{12}x_2 + b_1 u \qquad (6\text{-}15a)$$

$$\dot{x}_2 = a_{21}x_1 + a_{22}x_2 + b_2 u \qquad (6\text{-}15b)$$

and
$$y = c_1 x_1 + c_2 x_2 \qquad (6\text{-}16)$$

Assume that the initial conditions are zero. When the equations are transformed and X_1 and X_2 collected on the left, we get

$$(s - a_{11})X_1 - a_{12}X_2 = b_1 U \qquad (6\text{-}17a)$$

$$-a_{21}X_1 + (s - a_{22})X_2 = b_2 U \qquad (6\text{-}17b)$$

and
$$y = c_1 X_1 + c_2 X_2 \qquad (6\text{-}18)$$

Now solve Eq. (6-17 a) for X_1 and Eq. (6-17 b) for X_2; that is,

$$X_1 = \frac{a_{12}X_2 + b_1 U}{s - a_{11}} \qquad X_2 = \frac{a_{21}X_1 + b_2 U}{s - a_{22}} \qquad (6\text{-}19)$$

A block diagram can be constructed from Eq. (6-19) and (6-18), as shown in Fig. 6-3. There is one row from u to y that is the X_1 path and one that is the X_2 path. The paths are coupled by the terms a_{12} and a_{21}.

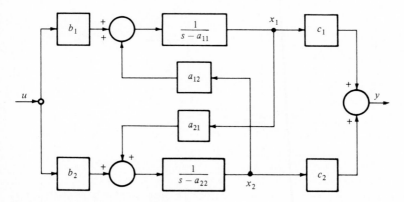

Figure 6-3 Second-order block diagram.

A block diagram for a state-space system of any order can be constructed in this manner. In large systems it is common for most of the coupling coefficients a_j to be zero, so that the diagram may not be so dense as might first appear. Also, many of the b_j and c_k terms are often zero, thus simplifying the block diagram further.

6-4 BOND GRAPHS

A bond graph that has power directions and causality assigned to all its bonds can be converted directly into an equivalent block diagram without writing equations. To see this, consider the augmented bond graph of Fig. 6-4a. Each bond corresponds to a pair of directed signals (e and f). In Fig. 6-4b we have separated each bond into its signal pairs. The causal marks indicate how the signal information flows and thus how the arrows on the signal lines in a block diagram are oriented. For example, bond 3 is diagrammed as having an effort input into the 1-junction and a flow input into the 0-junction.

The next step is taken in Fig. 6-4c, where each bond-graph element is replaced by its block-diagram equivalent set of operations. For example, the R element simply returns f_5 proportional to input e_5 by the constant R^{-1}. The 0-junction denotes a pair of operations. One is to sum the input flows f_3 and f_5 to get the output flow f_4. Note the signs on the summer block. The input effort e_4 is distributed to the outputs e_3 and e_5. The 1-junction is treated similarly, although the efforts sum and the flow is distributed. The effort source applies an effort signal to the system from some external agency, and the system responds by generating the flow f_1.

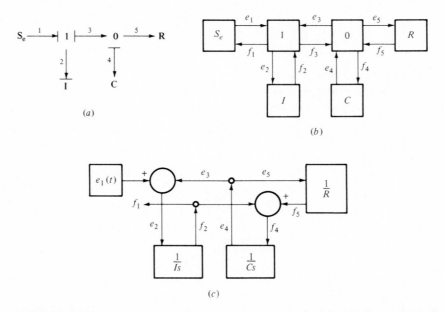

Figure 6-4 Block diagram from a bond graph: (a) bond graph; (b) signal paths in block diagram; (c) complete block diagram.

Table 6-3 Block diagrams for some augmented bond-graph elements

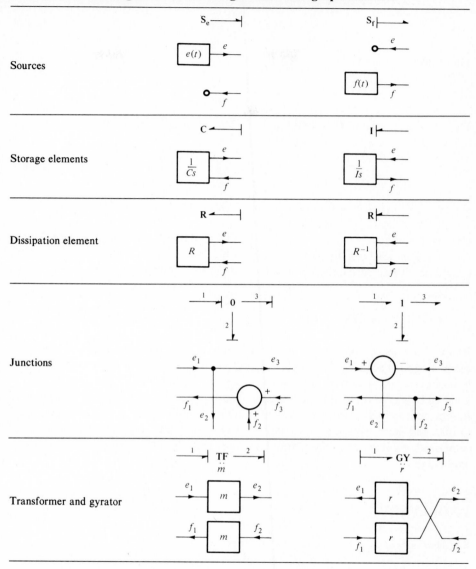

Representation of the I and C blocks requires a little justification. Consider the defining equations of an I element with constant coefficient, namely,

$$f(t) = \frac{1}{I} p(t) \qquad \text{and} \qquad \frac{dp}{dt} = e(t) \qquad (6\text{-}20)$$

The transformed equations are

$$F(s) = \frac{1}{I} P(s) \qquad \text{and} \qquad sP(s) - p(0) = E(s) \qquad (6\text{-}21)$$

Solving for $F(s)$ in terms of $E(s)$ with $p(0) = 0$ gives

$$F(s) = \frac{1}{Is} E(s) \tag{6-22}$$

The transfer function is

$$\frac{F(s)}{E(s)} = \frac{1}{Is} \tag{6-23}$$

This is diagrammed in Fig. 6-4c for the I element. A similar development will show that the C element has a transfer function

$$\frac{E(s)}{F(s)} = \frac{1}{Cs} \tag{6-24}$$

which is also inserted in Fig. 6-4c to complete the block diagram as shown.

A set of useful equivalences is summarized in Table 6-3. Not all augmented element forms are shown. For example, I and C in derivative causality are not shown. Also, various sign configurations are possible for the junctions; nevertheless you should be able to derive the missing block-diagram equivalents with little difficulty.

6-5 CONCLUSIONS

Block diagrams can be drawn directly from any augmented bond graph. In addition, block diagrams are useful for indicating input-output transfer-function relationships. Sometimes these transfer functions are derived from experimental measurements and can be inserted into a system model. Control engineers often use block diagrams to represent amplifier, instrument, and actuator dynamics and control schemes. Bond graphs and state equations are compact ways of representing physical models, but sometimes equivalent block diagrams are useful to show the signal-flow paths more explicitly.

PROBLEMS

6-1 Show that the block diagrams in Fig. P6-1 are equivalent.

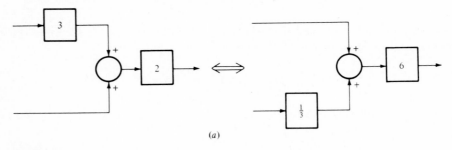

(a)

Figure P6-1

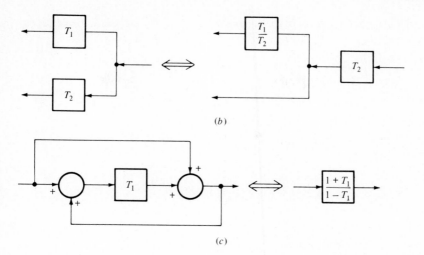

(b)

(c)

6-2 Find the transfer function for each of the systems in Fig. P6-2.

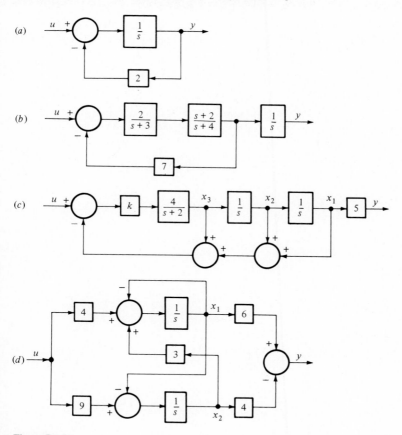

Figure P6-2

6-3 Make a block diagram for each of the bond graphs in Fig. P6-3. Can you reduce the one for Fig. 6-3*c*?

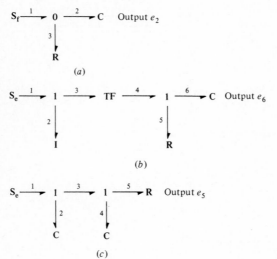

(*a*)

(*b*)

(*c*)   **Figure P6-3**

CHAPTER
SEVEN

COMPUTER SIMULATION OF DYNAMIC SYSTEMS

7-1 INTRODUCTION TO COMPUTER SIMULATION

Computer simulation or (more accurately for our use) the study of the dynamic response of mathematical models using electronic computers has become a standard tool in the kit of the practicing engineer. There are many good reasons for this, among them the following:

1. Problems too large to be solved by hand can be studied.
2. Repeated runs, say for optimizing a design, can be carried out in a reasonable amount of time.
3. Many nonlinear problems can be "solved" only by numerical methods, at which computers excel.
4. Simulation is the most cost-effective way to understand the dynamic behavior of proposed systems.
5. The trend is to make it easier for the engineer to use the computer by improved design of simulation languages.

It is also worth noting that many engineers enjoy sitting at an interactive terminal with a good graphics display and experimenting with system parameters to obtain a desirable type of response.

The cost of computation has shown a continuing decrease, and as programming languages and speed of computation improve, you can expect to have good access to an interactive graphics terminal with an array of useful prestored programs available in almost any engineering organization. You may even get your own small computer or have a dial-up terminal at home. Because of the continuing trends along these lines, we shall conduct our discussion in this chapter in terms of interactive simulation. A later chapter considers numerical analysis for dynamic systems in more detail and at a more basic level.

Three types of electronic computers are in general use for dynamic-system simulation, digital, analog, and hybrid (a combination of digital and analog). A good reference for these various types and their comparative characteristics is Korn and Wait,[1] who point out that the digital computers have proved to be superior in most, but not all, engineering applications. We therefore focus on digital computation. Because analog computers compute in parallel (as opposed to sequentially), they can be superior in speed of computation for many repeated runs of a large-scale model once the model has been properly set up, but even this advantage may disappear as parallel computation is introduced into digital systems.

Another distinction which is slowly disappearing is that between batch and interactive computation. In batch operation an entire problem statement is made at one time and then a complete run is made. This cycle can be repeated indefinitely, but it may require hours for the completion of a cycle. Such a procedure is reasonable for a large problem that is well under control, i.e., already clearly understood, when one just needs more performance data. It is a poor method for gaining insight, making partial runs, using intermediate results to modify subsequent input, and taking many of the steps an engineer must take to understand a system's behavior for design, analysis, or troubleshooting. Batch computing is typically economical of computer resources at the expense of engineers' time, while interactive computing has the inverse characteristics.

How does interactive computing work? You sit at a terminal and begin the description of your problem. First, you enter a set of data, e.g., a bond graph, a circuit graph, or equations. At a certain point you call for some interpretation of, or computation on, the data entered, e.g., verify the data, assign causality, or sort equations. Then you will get a response on that portion of the problem. If all is well, you enter some additional data and continue the interactive process until a stopping point is reached.

From the computer's point of view, if you were its only customer, it would sit idle most of the time, waiting as you enter data or think about some results and decide what to do next. To make better use of its computer resources, an interactive system handles many customers simultaneously, usually giving them short turns so frequently that each is unaware of the existence of the other users. Of course, if too many customers are active simultaneously, or if someone needs a major share of the computer for intensive computation, this can cause interference in the quality of service. You notice such a situation when the computer's response to your commands becomes unusually slow. (Perhaps that is a signal that you should return to your desk and think instead of compute.)

7-2 SIMULATION PROGRAM STRUCTURE

Digital simulation is used for many types of mathematical problems in engineering. We shall focus on programs for continuous-time, lumped-parameter models that lead to coupled first-order ordinary differential equations of the initial-value type. In contrast, some programs specialize in models in which events occur at discrete intervals, synchronous or not; these are called *discrete-time programs*. There are programs in which the *dependent* variables are not continuous, as our usual state variables are, but quantized. And there are even models in which the dependent variables are not deterministic but probabilistic; i.e., one computes, not the exact value at time t, but an expected value or a distribution. The type of program we shall discuss is called Continuous System Simulation Language (CSSL). Efforts at standardization of CSSLs can be followed in the journal *Simulation*.

The typical steps to be taken by a CSSL are shown in Fig. 7-1. The major process blocks include:

1. Reading the initial problem description, in which the circuit, bond-graph, schematic, or block-diagram information is communicated to the computer.
2. Formulating the system and state equations.
3. Performing the numerical integration of the state equations by a suitable method.
4. Storing and displaying the results.

How most simulation languages differ with regard to use by the engineer is in steps 1 and 2. Obviously, a program which accepts circuit descriptions must have a procedure for converting such models into equations or it would not be very useful. On the other hand, a program which accepts state equations directly can be simple in design but shifts the burden of work to the engineer until step 3. Programs also differ

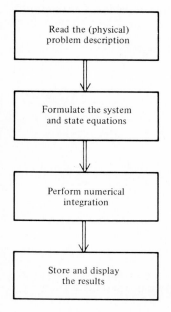

Figure 7-1 Typical simulation flowchart of the CSSL type.

(internally) with respect to their efficiency of computer-resource use, speed of computation, and size and complexity of the problems they can handle. We shall not consider those issues in this chapter. The best guide to a program's characteristics is its user's manual.

Some common simulation programs of various types are SCEPTRE for electronic networks,[2] DRAM for mechanical systems,[3] ENPORT for bond graphs,[4] CSMP for block diagrams, and DARE for state equations. Since new and improved simulation programs are being designed all the time, you would do well to survey the programs available when you need to use the computer.

Example 7-1 Just to see what some of the steps involved in a simulation are, let us consider the mechanical-oscillator example shown in Fig. 7-2a. The state equations can be derived from the bond graph in Fig. 7-2b in terms of p and x, the momentum and spring deflection, respectively

$$\dot{p} = -bm^{-1}p - kx + F(t) \qquad (7\text{-}1a)$$

$$\dot{x} = m^{-1}p \qquad (7\text{-}1b)$$

where $F(t)$ is the input force and m, b, and k are the mass, damping, and stiffness parameters, respectively. Consider a simple numerical case in which the parameters are $m = 1$ kg, $b = 1$ N·s/m, and $k = 9$ N/m. Further, suppose that the force is due to gravity; hence, $F(t) = mg = 9.8$ N. And suppose that the initial conditions are both zero. The equations become

$$\dot{p} = -1p - 9x + 9.8 \qquad p(0) = 0 \qquad (7\text{-}2a)$$

$$\dot{x} = 1p \qquad x(0) = 0 \qquad (7\text{-}2b)$$

Since we know both the initial conditions and the value of the input at time zero, we can compute the derivatives at that time

$$\dot{p}(0) = -(1)(0) - (9)(0) + (9.8) = 9.8 \qquad (7\text{-}3a)$$

$$\dot{x}(0) = (1)(0) = 0 \qquad (7\text{-}3b)$$

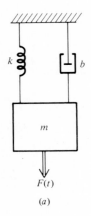

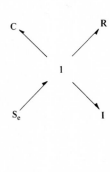

$F(t)$

(a)

(b)

Figure 7-2 An oscillator example: (a) mechanical schematic and (b) bond graph.

With the derivatives known at $t = 0$ we can estimate the value of the state, that is, p and x, at a later time T. A simple estimation rule, which works for any time t including $t = 0$, is Euler's rule

$$p(t + T) = p(t) + \dot{p}(t)(T) \qquad (7\text{-}4a)$$

$$x(t + T) = x(t) + \dot{x}(t)(T) \qquad (7\text{-}4b)$$

Euler's rule is equivalent to using the first two terms of the Taylor series for the state at time t. At time zero, Eqs. (7-4) become, for our example,

$$p(T) = \dot{p}(0)(T) \qquad (7\text{-}5a)$$

$$x(T) = \dot{x}(0)(T) \qquad (7\text{-}5b)$$

How shall we choose T in order to apply Eqs. (7-4)? This question can best be answered in this case by a consideration of the system's *eigenvalues* because they will give us insight into the time response we wish to compute. Since the input force is constant, it has no dynamics which need concern us here. Recall that the eigenvalues are the solution to the characteristic equation

$$\det (sI\text{-}A) = 0 \qquad (7\text{-}6)$$

In our case we get

$$\begin{vmatrix} s + 1 & 9 \\ -1 & s \end{vmatrix} = 0$$

or

$$s^2 + 1s + 9 = 0 \qquad (7\text{-}7)$$

The solutions to Eq. (7-7) are

$$s_1 = -0.5 + 2.958j \qquad s_2 = -0.5 - 2.958j \qquad (7\text{-}8)$$

Thus we see that the system will oscillate with a frequency of $2.958 \approx 3$ rad/s. The period is then $2\pi/3$ (approximately) or $T_p = 2.094$ s (call it 2 s). Now picture trying to draw a reasonable representation of a sine wave by a series of dots. We might want to use at least 20 points to represent a cycle. We could, of course, use more, but too many would waste time and effort without really improving the picture. So a first guess at T for computing the solution might be the period 2 s divided by 20, or $T = 0.1$ s.

Let us compute one period of dynamic response from Eq. (7-4) at fixed intervals of T equal to 0.1 s. The rule is

$$p_{k+1} = p_k + \dot{p}_k T \qquad (7\text{-}9a)$$

$$x_{k+1} = x_k + \dot{x}_k T \qquad (7\text{-}9b)$$

where $k = 0, 1, 2, \ldots , 21$; $P_k = p(t = kT)$, etc., and $p = 0, x = 0$.

The data in the top part of Table 7-1 were obtained from Eqs. (7-2) and (7-9) for a time increment of 0.1 s. They were generated by the program shown in Fig. 7-3, although the results could have been obtained by hand or on a calculator.

Table 7-1 Results of integration

K	T	p	x	$\dot{p}$	$\dot{x}$
			$T = 0.1$ s		
0	0.00	0.000	0.000	9.800	0.000
1	0.10	0.980	0.000	8.820	0.980
2	0.20	1.862	0.098	7.056	1.862
3	0.30	2.568	0.284	4.675	2.568
4	0.40	3.035	0.541	1.896	3.035
5	0.50	3.225	0.844	−1.025	3.225
6	0.60	3.122	1.167	−3.825	3.122
7	0.70	2.740	1.479	−6.252	2.740
8	0.80	2.115	1.753	−8.093	2.115
9	0.90	1.305	1.965	−9.186	1.305
10	1.00	0.387	2.095	−9.443	0.387
11	1.10	−0.558	2.134	−8.846	−0.558
12	1.20	−1.442	2.078	−7.460	−1.442
13	1.30	−2.188	1.934	−5.416	−2.188
14	1.40	−2.730	1.715	−2.905	−2.730
15	1.50	−3.020	1.442	-0.157	−3.020
16	1.60	−3.036	1.140	2.577	−3.036
17	1.70	−2.778	0.836	5.051	−2.778
18	1.80	−2.273	0.558	7.047	−2.273
19	1.90	−1.569	0.331	8.388	−1.569
20	2.00	−0.730	0.174	8.961	−0.730
21	2.10	0.166	0.101	8.722	0.166
			$T = 0.05$ s		
0	0.00	0.000	0.000	9.800	0.000
1	0.05	0.490	0.000	9.310	0.490
2	0.10	0.955	0.024	8.624	0.955
3	0.15	1.387	0.072	7.763	1.387
4	0.20	1.775	0.142	6.751	1.775
5	0.25	2.112	0.230	5.614	2.112
6	0.30	2.393	0.336	4.383	2.393
7	0.35	2.612	0.456	3.087	2.612
8	0.40	2.767	0.586	1.757	2.767
9	0.45	2.854	0.725	0.424	2.854
10	0.50	2.876	0.867	−0.881	2.876
11	0.55	2.832	1.011	−2.131	2.832
12	0.60	2.725	1.153	−3.299	2.725
13	0.65	2.560	1.289	−4.360	2.560
14	0.70	2.342	1.417	−5.294	2.342
15	0.75	2.077	1.534	−6.084	2.077
16	0.80	1.773	1.638	−6.714	1.773
17	0.85	1.437	1.727	−7.176	1.437
18	0.90	1.079	1.798	−7.464	1.079
19	0.95	0.705	1.852	−7.577	0.705
20	1.00	0.327	1.888	−7.515	0.327
21	1.05	−0.049	1.904	−7.286	−0.049
22	1.10	−0.413	1.902	−6.900	−0.413

K	T	p	x	$\dot{p}$	$\dot{x}$
23	1.15	−0.758	1.881	−6.369	−0.758
24	1.20	−1.077	1.843	−5.709	−1.077
25	1.25	−1.362	1.789	−4.939	−1.362
26	1.30	−1.609	1.721	−4.079	−1.609
27	1.35	−1.813	1.640	−3.151	−1.813
28	1.40	−1.971	1.550	−2.177	−1.971
29	1.45	−2.080	1.451	−1.182	−2.080
30	1.50	−2.139	1.347	−0.187	−2.139
31	1.55	−2.148	1.240	0.785	−2.148
32	1.60	−2.109	1.133	1.713	−2.109
33	1.65	−2.023	1.027	2.576	−2.023
34	1.70	−1.894	0.926	3.358	−1.894
35	1.75	−1.727	0.832	4.042	−1.727
36	1.80	−1.524	0.745	4.617	−1.524
37	1.85	−1.294	0.669	5.072	−1.294
38	1.90	−1.040	0.604	5.401	−1.040
39	1.95	−0.770	0.552	5.599	−0.770
40	2.00	−0.490	0.514	5.665	−0.490
41	2.05	−0.207	0.489	5.602	−0.207
42	2.10	0.073	0.479	5.415	0.073

```
C
C
C---    VARIABLES
C        VARM, VARK, VARB = MASS, STIFFNESS, DAMPING
C        P, X, PDOT, XDOT = MOMENTUM, DEFLECTION, ETC.
C        F = LOAD FORCE
C        DELT = TIME INCREMENT
C        TFINAL = FINAL TIME
C
C***    INITIALIZE
C
        F= 9.8
        VARM= 1.0
        VARK= 9.0
        VARB= 1.0
        TFINAL= 2.1
        DELT= 0.05
        KLIM=(TFINAL+.00001*DELT)/DELT
C
        P=0.0
        X=0.0
        K=0
C
        PRINT 1000
1000    FORMAT(2X, 1HK, 4X, 1HT, 8X, 1HP, 9X, 1HX, 8X, 4HPDOT, 6X, 4HXDOT/1X)
C
10      T=K*DELT
        PDOT= (-VARB/VARM)*P -(VARK)*X +F
        XDOT= (1.0/VARM)*P
        PRINT 1100, K, T, P, X, PDOT, XDOT
1100    FORMAT(1X, I2, 2X, F5.2, 4(2X, F8.3))
C
        K=K+1
        IF (K.GT.KLIM) CALL EXIT
C
        P=P+PDOT*DELT
        X=X+XDOT*DELT
        GOTO 10
C
        END
```

Figure 7-3 Program to do Euler integration.

How reliable are these data? Just because the procedure apparently operated successfully is not sufficient reason to accept the results without some verification. If T is too large, the predicted values of p and x at times $t = kT$ will not be close to the true solutions of the differential equations. Let us cut the time increment in half and try again. Again use Eqs. (7-2) and (7-9) in a recursive fashion, this time using 0.05 s as the increment. The results are shown in the bottom part of Table 7-1. What do you think? Shall we repeat the process? Would an increment of 0.2 s be satisfactory? In Sec. 7-5 we shall discuss some accuracy conditions further. Now we turn our attention to another matter.

Suppose the damping parameter is 10.0 N·s/m, changed from 1.0 N·s/m, and all other conditions are unchanged. What effect will this have on our Euler integration procedure? First, we check the new eigenvalues. The characteristic polynomial can be derived from the new state equations, which are

$$\dot{p} = -10p - 9x + 9.8 \qquad (7\text{-}10a)$$

$$\dot{x} = 1\text{p} \qquad (7\text{-}10b)$$

The characteristic equation

$$P(s) = \det(sI\text{-}A) = s^2 + 10s + 9 = 0 \qquad (7\text{-}11)$$

has the solutions $s_1 = -9$ and, $s_2 = -1$. Since there are two real negative roots, we expect two real exponentials in the time response

$$e^{-9t} \quad \text{and} \quad e^{-1t}$$

The associated time constants of the exponentials are $\tfrac{1}{9} = 0.111$ s and 1.000 s. What is an appropriate time increment T, and what is an appropriate time interval to consider; i.e., for how long shall we compute the results?

The smaller time constant (0.111 s) suggests a computing increment of, say, 0.05 s to see several points per time constant and a time interval of about 0.6 s to see most of the exponential decay. On the other hand, the larger time constant (1.000 s) suggests a computing time of about 0.5 s and an interval of about 5.0 s. Now what? In order to preserve accuracy while computing the time response we shall have to use the smaller T value (0.05 s) due to the shorter time constant; but to see most of the slower exponential we shall have to compute for a long interval (5.0 s). This means we should be prepared to compute about 100 points.

It is characteristic of *stiff* systems, i.e., systems with widely separated eigenvalues, that they present us with a dilemma when it comes to numerical integration. Short time constants or high natural frequencies force us to use small time steps, but long time constants or low natural frequencies mean that we must continue the computation for a long time to see the slow parts of the response. Thus we must do a lot of computation for stiff systems. A large spread in eigenvalues can arise in large systems with electrical, mechanical, and hydraulic components all interacting. We shall discuss this point further, but here it is sufficient to alert you to stiff-system problems.

There are many other integration methods besides Euler's, of course. None is simpler in concept or implementation, but many are superior for practical problems where decent accuracy at reasonable speed is important. This matter will be pursued later; here we turn our attention to other types of simulation programs that help us obtain numerical responses.

7-3 THE ENPORT PROGRAM

The ENPORT program is a continuous-system simulation processor that accepts bond-graph models with constant parameters. The tasks performed by the program (Fig. 7-4) should look quite familiar at this point. One advantage of using a simulation program like ENPORT-5 on-line is that you can cycle back to a previous point, make some changes in the data, and continue the process from there. For example, if you make a run and decide to modify a parameter, you need not restate the graph or reassign causality. Of course, the equations must be reformulated and the new equations must be integrated, but the program does this for you.

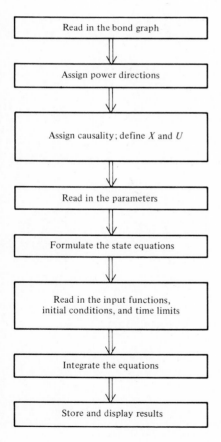

Read in the bond graph

Assign power directions

Assign causality; define X and U

Read in the parameters

Formulate the state equations

Read in the input functions,
initial conditions, and time limits

Integrate the equations

Store and display results

Figure 7-4 The ENPORT-5 main flowchart.

Figure 7-5 The oscillator bond graph with labeled bonds.

```
    COMMAND$
→  NOECHO

    COMMAND$
→  HEADER
    PLEASE GIVE ME A TITLE:
→  EXAMPLE FOR FIG. 7-5,

    COMMAND$
→  GRAPH
→  SE 1 , I 2 , R 3 , C 4 , 1 1 2 3 4

    COMMAND$
→  CAUSALITY

    THERE ARE   2 STATE VARIABLES
           AND   1 INPUT VARIABLES.

    THE STATE VECTOR. . .

       X  1 = P  2
       X  2 = Q  4

    THE INPUT VECTOR. . .

       U  1 = E  1

    COMMAND$
→  PARAMETERS
→  I2= 1.0 , / C4= 9.0 , R3= 1.0 .

    COMMAND$
→  EQUATIONS

    THE A  MATRIX. . .
       1   -0.10000E 01 -0.90000E 01
       2    0.10000E 01  0.00000E 00

    THE B  MATRIX. . .
       1    0.10000E 01
       2    0.00000E 00

    COMMAND$
→  EIGEN
→  VALUES .

    VALUE    REAL       IMAGINARY      NAT FREQ      DAMPING       TIME CONS
       1  -0.50000E 00  0.29580E 01  0.30000E 01  0.16667E 00  0.00000E 00
       2  -0.50000E 00 -0.29580E 01  0.30000E 01  0.16667E 00  0.00000E 00

    COMMAND$

    DO NOT RECOGNIZE ––.

    COMMAND$
→  SOURCE
→  E1 CONSTANT 9.8 .

    COMMAND$
→  INITIAL
→  P2 0.0 , Q4 0.0 .

    COMMAND$
→  LIMITS
→  TIN 0. , TFIN 2.1 , DT 0.1 .

    COMMAND$
→  INTEGRATE
    INTEGRATION COMPLETED.
```

(a)

Figure 7-6 An ENPORT-5 run: (a) example through integration; (b) table of results.

In Fig. 7-5 the oscillator bond graph is shown with labeled bonds. Then it is easy to communicate with the program about bonds, for example 3, elements, for example R_3, and variables, for example E_1. The program uses the following notation:

E, F = effort, flow
P, Q = (generalized) momentum, displacement
X, U = state, input

Now we can read the ENPORT run in Fig. 7-6, which is a complete record of the exchange between a user and the program with user input marked by arrows.

First, we follow the simulation in Fig. 7-6*a*. This takes us from graph input (see GRAPH) through causality to parameters. The mass is specified by the I2 = statement, for example. The state equations are formulated by ENPORT and printed in A,B form, with the state and input vectors defined previously. We then call for eigenvalues and, as expected, we get a complex-conjugate pair. The corresponding natural frequency and damping ratio are also given. This information can guide us later in a choice of time run parameters. Next, the input force and initial conditions are specified, following by the initial time (TIN), final time (TFIN), and the storage interval (DT). The problem is now prepared for integration. The increment DT will

```
        COMMAND$
 —→  DLIMITS
        INITIAL TIME:
 —→  0.
        FINAL TIME:
 —→  2. 1
        TIME INCREMENT:
 —→    1

        COMMAND$
 —→  DVARIABLES
        PLEASE LIST OUTPUT VARIABLES:
 —→  P2 Q4 E2 F4

        COMMAND$
 —→  PRINT

     TABLE OF RESULTS

        TIME            P 2              Q 4              E 2             F 4
     0. 000E  00   0. 00000E  00  0. 00000E  00  0. 98000E  01  0. 00000E  00
     0. 100E  00   0. 91867E  00  0. 47055E-01  0. 84578E  01  0. 91867E  00
     0. 200E  00   0. 16718E  01  0. 17819E  00  0. 65244E  01  0. 16718E  01
     0. 300E  00   0. 22112E  01  0. 37427E  00  0. 42204E  01  0. 22112E  01
     0. 400E  00   0. 25113E  01  0. 61243E  00  0. 17769E  01  0. 25113E  01
     0. 500E  00   0. 25693E  01  0. 86843E  00-0. 58515E  00  0. 25693E  01
     0. 600E  00   0. 24034E  01  0. 11188E  01-0. 26727E  01  0. 24034E  01
     0. 700E  00   0. 20490E  01  0. 13428E  01-0. 43343E  01  0. 20490E  01
     0. 800E  00   0. 15542E  01  0. 15239E  01-0. 54694E  01  0. 15542E  01
     0. 900E  00   0. 97429E  00  0. 16508E  01-0. 60315E  01  0. 97429E  00
     0. 100E  01   0. 36677E  00  0. 17179E  01-0. 60275E  01  0. 36677E  00
     0. 110E  01  -0. 21410E  00  0. 17251E  01-0. 55114E  01-0. 21410E  00
     0. 120E  01  -0. 72150E  00  0. 16775E  01-0. 45759E  01-0. 72150E  00
     0. 130E  01  -0. 11193E  01  0. 15844E  01-0. 33405E  01-0. 11193E  01
     0. 140E  01  -0. 13841E  01  0. 14581E  01-0. 19387E  01-0. 13841E  01
     0. 150E  01  -0. 15060E  01  0. 13124E  01-0. 50547E  00-0. 15060E  01
     0. 160E  01  -0. 14883E  01  0. 11616E  01  0. 83433E  00-0. 14883E  01
     0. 170E  01  -0. 13458E  01  0. 10189E  01  0. 19757E  01-0. 13458E  01
     0. 180E  01  -0. 11024E  01  0. 89576E  00  0. 28405E  01-0. 11024E  01
     0. 190E  01  -0. 78849E  00  0. 80077E  00  0. 33816E  01-0. 78849E  00
     0. 200E  01  -0. 43743E  00  0. 73930E  00  0. 35837E  01-0. 43743E  00
     0. 210E  01  -0. 82582E-01  0. 71340E  00  0. 34619E  01-0. 82582E-01

        COMMAND$

 —→  QUIT
```

(*b*)

be used if a sufficiently accurate solution will result; otherwise a smaller increment will be chosen automatically. We are informed when the integration process is complete.

Turning to the display of results in Fig. 7-6b, we specify display limits as shown in that figure (see DLIMITS), and we choose to look at the results of P2, Q4, E2, and F4 (the x and $\dot{x}$ vectors in this example). These are results for the same problem solved by Euler integration; see Table 7-1 for a comparison. The ENPORT simulation program uses a more accurate integration procedure, which accounts for the difference.

Another feature of a good simulation program is the ability to make a plotted presentation of the data. The data for our run are shown in Fig. 7-7, based on a computer printout. The plot is good for seeing the overall response and estimating peaks, times of zero crossings, etc. The table is good for isolating particular features precisely. Normally in engineering we work back and forth between the various forms of results like plots and tables to build our understanding.

To examine the effect of a change in damping from 1.0 to 10.0 N·s/m we continue the on-line run, as shown in Fig. 7-8. The R3 parameter is modified; the equations are formulated; new eigenvalues are computed; and integration is completed. Again, we print a table of results and then plot them in Fig. 7-9. As predicted by the eigenvalues, the system is now overdamped and nonoscillatory.

There are many advantages to using a CSSL like ENPORT to study dynamic systems: (1) a minimum amount of your time is required for problem preparation; (2) only a few data need be entered; (3) problem conditions can be modified and new results easily obtained; (4) large problems can be treated; (5) the results are presented in convenient forms such as plots and tables. Some programs can conduct multiple runs for design optimization, and some can generate finished report quality graphical output as well. CSSLs have some disadvantages: (1) it must be operating

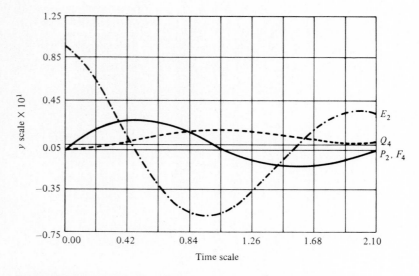

Figure 7-7 Plot of results for oscillator example with $R_3 = 1.0$ N·s/m.

```
      COMMAND$
      PARAMETERS
        R3= 10.

        COMMAND$
      EQUATIONS

        THE A  MATRIX...
          1    -0.10000E 02 -0.90000E 01
          2     0.10000E 01  0.00000E 00

        THE B  MATRIX...
          1     0.10000E 01
          2     0.00000E 00

        COMMAND$
      EIGEN
        VALUES.

        VALUE     REAL         IMAGINARY       NAT FREQ        DAMPING        TIME CONS
          1   -0.90000E 01  0.00000E 00  0.00000E 00  0.00000E 00  0.11111E 00
          2   -0.10000E 01  0.00000E 00  0.00000E 00  0.00000E 00  0.10000E 01

        COMMAND$
      INTEGRATE
        INTEGRATION COMPLETED.

        COMMAND$
      PRINT

        TABLE OF RESULTS

        TIME            P 2            Q 4            E 2            F 4
      0.000E 00   0.00000E 00  0.00000E 00  0.98000E 01  0.00000E 00
      0.100E 00   0.61038E 00  0.35802E-01  0.33740E 01  0.61038E 00
      0.200E 00   0.80045E 00  0.10844E 00  0.81949E 00  0.80045E 00
      0.300E 00   0.82517E 00  0.19053E 00-0.16654E 00  0.82517E 00
      0.400E 00   0.78767E 00  0.27146E 00-0.51987E 00  0.78767E 00
      0.500E 00   0.72939E 00  0.34740E 00-0.62050E 00  0.72939E 00
      0.600E 00   0.66676E 00  0.41721E 00-0.62247E 00  0.66676E 00
      0.700E 00   0.60607E 00  0.48082E 00-0.58804E 00  0.60607E 00
      0.800E 00   0.54951E 00  0.53856E 00-0.54217E 00  0.54951E 00
      0.900E 00   0.49768E 00  0.59088E 00-0.49467E 00  0.49768E 00
      0.100E 01   0.45050E 00  0.63825E 00-0.44926E 00  0.45050E 00
      0.110E 01   0.40771E 00  0.68112E 00-0.40718E 00  0.40771E 00
      0.120E 01   0.36894E 00  0.71992E 00-0.36870E 00  0.36894E 00
      0.130E 01   0.33384E 00  0.75503E 00-0.33373E 00  0.33384E 00
      0.140E 01   0.30208E 00  0.78680E 00-0.30201E 00  0.30208E 00
      0.150E 01   0.27334E 00  0.81555E 00-0.27328E 00  0.27334E 00
      0.160E 01   0.24733E 00  0.84156E 00-0.24728E 00  0.24733E 00
      0.170E 01   0.22379E 00  0.86509E 00-0.22375E 00  0.22379E 00
      0.180E 01   0.20250E 00  0.88639E 00-0.20245E 00  0.20250E 00
      0.190E 01   0.18323E 00  0.90566E 00-0.18319E 00  0.18323E 00
      0.200E 01   0.16579E 00  0.92309E 00-0.16575E 00  0.16579E 00
      0.210E 01   0.15001E 00  0.93887E 00-0.14997E 00  0.15001E 00

      COMMAND$
```

Figure 7-8 Modified ENPORT-5 run.

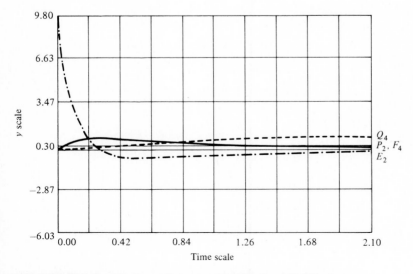

Figure 7-9 Plot of results for oscillator example with $R_3 = 10.0$ N·s/m.

165

on a computer system available to you; (2) you must learn its specific conventions; (3) the size of your problem must be compatible with the capabilities of the program.

The ENPORT program has been used here as an illustration of a high-level simulation program. Next, we examine another practical approach.

7-4 THE HPCG PROGRAM

Review of Fig. 7-1 suggests that if you are willing to formulate the state equations, the computer can provide a numerical solution and display the results. The normal procedure for using such a program is as follows:

1. Derive the state equations.
2. Define the state equations for program use by completing a subroutine, e.g., in FORTRAN.

```
CCCCCCC                     SUBROUTINE DESCRIPTION                     CCCCC
C                                                                          C
C        SUBROUTINE FCT                                                    C
C                                                                          C
CCCCCCCCCCCCCCCCCCCCCCCCCCCCCCCCCCCCCCCCCCCCCCCCCCCCCCCCCCCCCCCCCCCCCCCCCCCCC
CVCCCCCCCC                   VARIABLE IDENTIFICATION                  CCCCCCCC
C                                                                          C
C        X                   THE INDEPENDENT VARIABLE.
C        Y                   VECTOR CONTAINING SOLUTIONS.
C                            Y(1)= MOMENTUM
C                            Y(2)= DEFLECTION
C                                                                          
C        DERY                VECTOR CONTAINING USER DEFINED DERIVATIVES.
C                                                                          C
CCCCCCCCCCCCCCCCCCCCCCCCCCCCCCCCCCCCCCCCCCCCCCCCCCCCCCCCCCCCCCCCCCCCCCCCCCCCC
C                                                                          C
CSCCCCCCC                  ENTRY & STORAGE BLOCK               BLOCK 0000
C                                                                          C
         SUBROUTINE FCT(X, Y, DERY)
C
         REAL*4        DERY  (1),
        +              Y     (1),
        +              PA    (20)
C
C---COMMON FOR FCT PARAMETERS
C
         COMMON/FCTCOM/PA
C                                                                          C
CCCCCCCCCCCCCCCCCCCCCCCCCCCCCCCCCCCCCCCCCCCCCCCCCCCCCCCCCCCCCCCCCCCCCCCCCCCCC
C                                                                          C
CPCCCCCCCC                   PROCESS BLOCK                     BLOCK 0200
C                                                                          C
         IF (X. GT. 0. ) GOTO 20
         M=1. 0
         K=9. 0
         B=1. 0
         F=9. 8
C
C---SET DERIVATIVES
C
20       CONTINUE
         DERY(1) = -(B/M)*Y(1) -K*Y(2) +F
         DERY(2) = -(1. 0/M)*Y(1)
             DERY(3) =0. 0
             DERY(4) =0. 0
             DERY(5) =0. 0
             DERY(6) =0. 0
             DERY(7) =0. 0
             DERY(8) =0. 0
             DERY(9) =0. 0
             DERY(10)=0. 0
C                                                                          C
CCCCCCCCCCCCCCCCCCCCCCCCCCCCCCCCCCCCCCCCCCCCCCCCCCCCCCCCCCCCCCCCCCCCCCCCCCCCC
C
         RETURN
         END
```

Figure 7-10 Listing of FCT subroutine defining the state equations.

```
OK, CO C-HPCG
OK, FTN FCT -64R
0000 ERRORS [<FCT    >FTN-REV18. 2]
OK, LOAD
[LOAD rev 18. 2]
$ MO D64R
$ AU 300
$ CO 200000
$ LI HPCG
$ LO B_FCT
$ LI UTILR
$ LI AGIIR
$ LI TEKR
$ LI
LOAD COMPLETE
$ SA *HPCG
$ QUIT
OK, DELETE B_FCT
```

(a)

```
[HPCG REV 5. 0]

READY, NDIM 2
READY, INIT Y1 0.
READY, INIT Y2 0.
READY, LLIMIT 0.
READY, ULIMIT 2. 10
READY, DELTA . 1
READY, INTE
READY, STATUS

LLIMIT:      0. 0000
ULIMIT:      2. 1000
DELTA :      0. 1000

Y1    =      0. 0000
Y2    =      0. 0000

PA( 1)=      1. 0000
PA( 2)=      1. 0000
PA( 3)=      1. 0000
PA( 4)=      1. 0000
PA( 5)=      1. 0000
PA( 6)=      1. 0000
PA( 7)=      1. 0000
PA( 8)=      1. 0000
PA( 9)=      1. 0000
PA(10)=      1. 0000

READY, FILE RUN1DATA
READY, PLOT TIME Y1
READY, PLOT TIME Y2
READY, DRAW
```

(b)

Figure 7-11 The compile, load and run processes for HPCG. (a) Compile and load; (b) run.

3. Compile the subroutine and load the program package.
4. Run the program to investigate your problem.

Since most engineering-oriented computer centers have good mathematics library packages, this approach is widely applicable. Also, any equations you are able to program can be studied, provided the numerical-integration subroutines available can cope with them.

Figure 7-10 lists the FCT subroutine used by a widely available program called HPCG to define state equations. The subroutine is written in FORTRAN, with a maximum capacity of 10 state variables. We are only using the first two in our example. The program calls Y the dependent-variable set and X the independent variable (time, for us). Once FCT has been compiled and HPCG loaded (Fig. 7-11a), a run such as that shown in Fig. 7-11b can be made. Once integration is complete (INTE), display of the results can occur. Sometimes this is called *postprocessing* of results. The results Y1 and Y2 (momentum and deflection) are plotted in Fig. 7-12. This is not what we expected! The system is responding unstably. The problem could be with

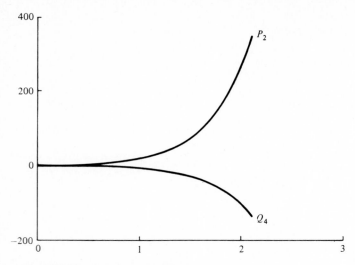

Figure 7-12 Results of the first HPCG run.

the computing increment (DELTA), but in this type of program there typically is an automatic adjustment feature for increment size to maintain numerical stability. Let us check the equations again (FCT). Sure enough, someone entered a negative sign on the P term in the second equation. (See for yourself whether that makes the system unstable.) Once the sign is changed and an HPCG run is repeated, we obtain the plots of Fig. 7-13 for P2 and Q4.

Since we have the problem all set up, let us make one further investigation. The fact that there is damping and a constant load force suggests the possibility of a

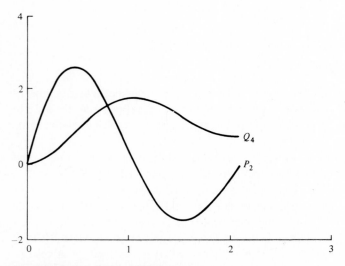

Figure 7-13 Run 2 of the HPCG program.

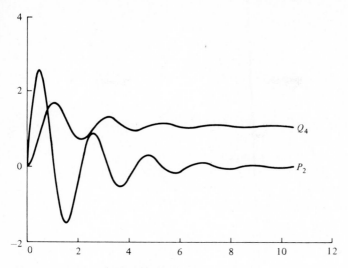

Figure 7-14 Run 3 of the HPCG program.

constant steady state for the system. We shall make a run with the time interval extended to 10.5 s (about five periods). The results are shown in Fig. 7-14, where it appears that P2 → 0 and Q4 → 1. Is this correct? You should always try to find some checks on the correctness of computer solutions.

Simulation programs like HPCG represent a middle ground between the do-it-yourself approach of the Euler program and the we-do-it-all approach of the ENPORT program. Such programs require the ability to code at a level like FOR-TRAN and a familiarity with the language, structure, and behavior of the host program.

7-5 POTENTIAL PROBLEMS

Lest you feel at this point that if you can only find the "right" simulation program all your troubles will disappear, we must discuss some rockier issues lurking below the surface.

Problems with Input and Output

You must be sure to read the user's manual carefully for a given program to see what the available input and output features of a program actually are. How easy are they to use for the problem you want to solve? Do they have a simple level for getting started, and do they allow an experienced user to save time? Are they flexible but also reasonably efficient? Can you control the storage and display of results, or do you get more data than you ever wished to see? What about the size and complexity of problems that can be handled?

Problems with Model Structure

The practical engineering world is filled with systems which do not fit into textbook patterns neatly, if at all. So it is with computer programs too, unless you "build your own" for a specific task, which is normally *very* time-consuming. There are some key features to keep in mind when simulating:

1. Are nonlinear models acceptable? If so, what kind are allowable?
2. Can models with algebraic loops be solved?
3. How are discontinuous functions handled?
4. Are elements that change the model structure, like switches, acceptable?
5. Can pure delays, like conveyor-belt delay times, be treated?
6. Can logic functions and discrete events be combined with continuing time simulation?

This list is merely intended to be suggestive and to increase your awareness of many factors you may not have experienced yet in system dynamics.

Problems with Numerical Integration

Since numerical integration is one of the key contributions of computers in the study of dynamic systems, it deserves a more detailed discussion (see Chap 16). Here, we simply remind you of a few important questions. How does the computation time increment affect the accuracy of the results? Can the method used be unstable numerically? Do stiff systems bring it to its knees?

You can expect to spend many happy hours at a graphics terminal using a responsive simulation program to solve many important dynamics problems in your technical career. Why not get started now?

PROBLEMS

7-1 An electric circuit is shown, together with its bond graph.

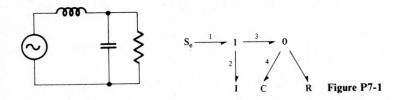

Figure P7-1

(*a*) Derive the state equations in terms of the parameters L, C, and R.

(*b*) Let $L = 1$ H, $C = 1$ F, and R be measured ohms. Evaluate the A matrix from the state equations.

(*c*) For each of the following values of R, discuss the selection of a suitable integration time increment T (eigenvalues are helpful here): (1) $R \to \infty$, (2) $R = 1$, (3) $R = 0.5$, (4) $R \to 0$. You may assume that the input function $V_1(t)$ is constant.

7-2 Reconsider the circuit of Prob. 7-1 with the parameters $L = 1$ mH, $C = 1$ μF, and $R = 1000$ Ω.

(*a*) Select a suitable integration time increment.

(*b*) Would redefining, i.e., scaling, the time unit (now seconds) make the numbers "better"? Show how.

7-3 Investigate the accuracy of Euler integration as the time increment is varied for a linear first-order system goverened by

$$\dot{x}(t) = 1x(t) \qquad x(0) = 10$$

Compare your numerical results with the "exact" solution.

7-4 Repeat Prob. 7-3 for the following system:

$$\dot{x}_1(t) = -1x_2(t) \qquad x_1(0) = 10$$

$$\dot{x}_2(t) = 1x_1(t) \qquad x_2(0) = 0$$

7-5 A mechanical linkage element is shown in Fig. P7-5a. The input force is F_1. We are interested in the reaction force F_2 when point 2 is pinned. A bond-graph model is shown in Fig. P7-5b.

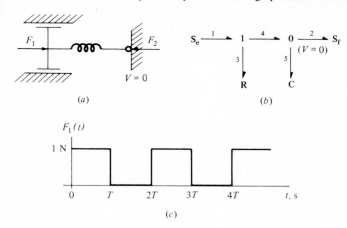

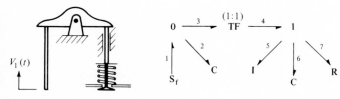

Figure P7-5

(*a*) Write state equations and an output equation for F_2.

(*b*) Derive a single (dynamic) input-output equation relating F_2 to F_1.

(*c*) Let $k/b = 10.0$ s^{-1}. Compute (or solve for) $F_2(t)$ if $F_1(t) = 1$ N and the initial deflection of the spring is zero. What time increment is appropriate?

(*d*) Suppose $F_1(t)$ is a periodic rectangular wave with the form shown in Fig. P7-5c. Find the response $F_2(t)$ to $F_1(t)$ for $T = 0.1, 0.5$, and 0.05 s.

7-6 The pushrod-lifter-valve arrangement common to many internal-combustion engines is shown below, together with a bond-graph model. The parameters are $k_2 = 1000$ N/m, $m_5 = 0.2$ kg, $k_6 = 10$ N/m, and $b_7 = 0.3$ N·s/m.

Figure P7-6

(*a*) Derive state equations for the system. Write an output equation for F, the force in the pushrod (which is also the contact force between the rod and the driving cam).

(*b*) If the input velocity $V_1(t)$ is sinusoidal, as given by $V_1(t) = 2.4 \sin \omega t$ with $\omega = 240$ rad/s, find the response $x_6(t)$, which is proportional to the valve position. Assume a preload in both C_2 and C_6 of 1 N. Does the valve float; i.e., does the contact force F_1 go to zero?

REFERENCES

1. G. A. Korn and J. V. Wait: *Digital Continuous-System Simulation,* Prentice-Hall, Englewood Cliffs, N.J., 1978.
2. J. C. Bowers and S. R. Sedore: *SCEPTRE: A Computer Program for Circuit and System Analysis,* Prentice-Hall, Englewood Cliffs, N.J., 1971.
3. *Users Guide to DRAM,* Mechanical Dynamics, Inc,. Ann Arbor, Mich., 1978.
4. *The ENPORT-5 User's Manual,* A. H. Case Center for Computer-Aided Design, Michigan State University, East Lansing, Mich., 1981.
5. *The HPCG Manual,* A. H. Case Center for Computer-Aided Design, Michigan State University, East Lansing, Mich., 1980.

PART
THREE

MODELING OF ENGINEERING SYSTEMS

In Part Three we expand our horizons to include a variety of system types which can be modeled using the techniques developed in Part Two. They include hydraulic, thermal, and transducer elements, amplifiers, etc. We stress the unity of concept and economy of thought which result by considering power transfers and energy action in a unified modeling approach. Naturally, every particular energy domain has its own special features that can be studied in detail. Here we take only time and space enough to show the main features to be accounted for in the models. You should be impressed with the strong similarities between these system types.

EIGHT

HYDRAULIC AND ACOUSTIC CIRCUITS

8-1 INTRODUCTION

In this chapter an interesting and important class of fluid-flow systems is studied. By *hydraulic* systems we mean systems made of pumps, motors, pipes, pistons, valves, filters, and accumulators of various types in which almost incompressible liquids such as water or hydraulic oil are used. Since such systems can be stiff, high-speed, and capable of generating large forces, they are found in machine tools, earthmoving machinery, power transmissions, and aircraft control-surface servomechanisms. Such systems are usually designed with the fluid velocity low enough to ensure that the static pressure will dominate the dynamic pressure, and for this reason the systems are sometimes called *hydrostatic systems* even though they have lots of interesting dynamics.

A second category to be treated here in a limited way is systems that use compressible gasses as the working fluids. Such *pneumatic systems* are generally more compliant and slower in response than hydraulic systems. Modeling such systems is complicated by many nonlinear effects, and thermodynamic effects must often be taken into account. A useful special case arises when the pressure excursions are relatively small, since then the *acoustic approximation* is useful. The close analogy between hydraulic circuits and acoustic circuits allows us to treat them together.

8-2 VARIABLES AND BASIC ELEMENTS

Table 8-1 shows the fluid variables which play the roles of e, f, p, and q in what follows. A quick glance reveals the sort of problems that always plague system dyna-

175

Table 8-1 Variables in hydraulic and acoustic circuits

Variable	General notation	Fluid circuit	SI unit
Effort	$e(t)$	P, pressure	Pa
Flow	$f(t)$	Q, volume flow rate	m^3/s
Momentum or impulse	$p = \int e\ dt$	P_P, integral of pressure	Pa·s
Displacement	$q = \int\!\!\int f\ dt$	V, volume	m^3
Power	$\mathcal{P}(t) = e(t)f(t)$	$P(t)Q(t)$	W
Energy	$E(p) = \int f\ dp$	$\int Q\ dp_p$, kinetic	J
	$E(q) = \int e\ dq$	$\int P\ dV$, potential	

micists. What seems like obvious notation to the fluid dynamicist conflicts with notation in other fields. For example, P for pressure conflicts with p for momentum and $\mathcal{P}$ for power, Q for volume flow rate conflicts with q for displacement and electric charge, and V for volume conflicts with V for velocity. As if that were not enough, a momentum type of variable, which we call p_P, is never defined at all in fluid mechanics. You will no doubt find that after you have made a bond graph using physical variables it is easier to number the bonds and to switch to the generalized variables e, f, p, and q.

A fluid port is a place where we can define an average pressure P and a volume flow rate Q. The end of a pipe or tube could be a port, and so could a threaded hole in a hydraulic pump. Since pressure is force per unit area and volume flow rate is velocity of flow times area, P,Q is certainly a component of the power at a port. There are other components of power, but we shall neglect them in this chapter. For example, a moving fluid has kinetic energy, which is carried through the port. This component is not important if the dynamic pressure is much less than the static pressure P. If the area of the port is A, the average velocity of the fluid is Q/A, and we must assume

$$P \gg \frac{1}{2}\rho\left(\frac{Q}{A}\right)^2$$

where ρ is the density. Furthermore, the fluid convects internal energy as it leaves the port, and this is a component of power. For the systems we study here, however, this power is not significant to the system dynamics, and so we assume that P and Q can be considered to be effort and flow and the only power we consider is given by the product $P(t)Q(t)$. We now discuss the forms the basic modeling elements take for such fluid systems.

Fluid Resistance

A 1-port fluid resistor will relate effort to flow at a single port. Figure 8-1 shows such a resistor in its most common form, attached to a 1-junction. The fluid resistor (Fig. 8-1 a) relates P_3 to Q_3 by

$$P_3 = P_3(Q_3) \tag{8-1}$$

in general, or

$$P_3 = R_3 Q_3 \qquad (8\text{-}2)$$

in the linear case. The units of the resistance R_3 are $\text{Pa}/(\text{m}^3/\text{s}) = \text{N} \cdot \text{s}/\text{m}^5$. When the resistor is associated with a length of pipe, it is attached to a 1-junction as shown in Fig. 8-1a. Then the three flows are equal

$$Q_1 = Q_2 = Q_3 \qquad (8\text{-}3)$$

and P_3 is a pressure drop

$$P_3 = P_1 - P_2 \qquad (8\text{-}4)$$

Actual resistance laws are fairly easy to measure, at least for steady-flow conditions. Under dynamic conditions, however, steady laminar- or turbulent-flow conditions often have no time to develop fully, so some experimentation with resistance-law parameters may be necessary to match dynamic experimental data. Figure 8-1b to f shows some typical situations where one can make at least a first estimate of the type of resistance law.

Figure 8-1b shows a porous plug in which viscous forces would be expected to predominate. In this case a linear pressure-flow relationship, as in Eq. (8-2), holds for incompressible flow. The resistance R_3 is usually based on experiment. A similar law holds for long thin tubes in which laminar flow develops. In this case, a specific value for the resistance can be given

$$R_3 = \frac{128\mu l}{\pi d^4} \qquad (8\text{-}5)$$

where μ = viscosity, $\text{N} \cdot \text{s}/\text{m}^2 = \text{Pa} \cdot \text{s}$
l = length, m
d = inside diameter, m

and l must be long enough for uniform laminar flow to dominate the pressure losses from other causes.

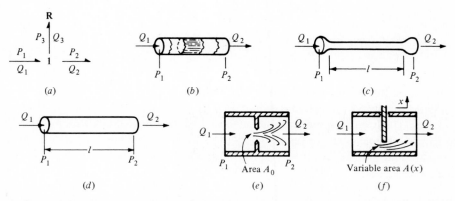

Figure 8-1 Fluid resistors: (a) general symbol, (b) porous plug, (c) capillary tube, (d) long pipe, (e) orifice, (f) valve.

For incompressible flow in long pipes it is useful to compute a Reynolds number Re given by

$$\text{Re} = \frac{4\rho Q}{\pi d\mu} \tag{8-6}$$

where ρ is the density of the fluid. When Re is low, say about 2000, viscous forces predominate and Eqs. (8-2) and (8-5) can be used. At higher Re the flow becomes turbulent, and the pressure-volume-flow relationship for steady flow is nonlinear. The transition to turbulent flow depends on the pipe dimensions l and d, the surface roughness, and fluid properties. For Re greater than about 5000, the flow is likely to be turbulent, and an approximate expression for the general form of Eq. (8-1) is

$$P_3 = a_t Q_3 |Q_3|^{3/4} \tag{8-7}$$

where the absolute-value sign is necessary so that P_3 will be negative for negative Q_3 and a_t is a constant often determined experimentally.[1] While this relationship can be verified for steady flow, it is based on steady turbulent flow and may not hold under oscillatory or other dynamic conditions; therefore for many cases it must be regarded only as a crude approximation.

Figure 8-1e and f shows two important cases in which pressure drops occur in short lengths. The orifice is a constriction of fixed area A_0, and the valve is much like an orifice with a variable area $A(x)$, where x represents a position coordinate. Although Fig. 8-1f looks like a gate valve, many other valve configurations function similarly. It is a standard exercise in fluid mechanics to derive orifice laws using energy, momentum, and continuity, and the main result is that the pressure drop is proportional to the flow rate squared. One form of this law is

$$P_3 = \frac{\rho}{2C_d^2 A_0^2} Q_3 |Q_3| \tag{8-8}$$

where, again, an absolute-value sign has been used to correct the sign on the pressure P_3 for negative Q_3, C_d is the *discharge coefficient,* and A_0 is the orifice area.[2] For a round, sharp-edged orifice, $C_d = 0.62$, but for other shapes and for holes that are not sharp C_d varies. For valves, $A(x)$ plays the role of A_0, and C_d varies also with valve position, so that

$$P_3 = \frac{\rho}{2C_d^2(x)A^2(x)} Q_3 |Q_3| \tag{8-9}$$

which at least follows the general form of valve laws.

Figure 8-2 shows the two most common forms of resistor constitutive laws used in practice. The linear law may occasionally be valid for porous plugs or flow through long narrow channels but is usually thought of as a *linearized* law for perturbations in pressure and flow about steady values. The signed quadratic law often holds for orifices and valves.

Figure 8-3 shows a valve representation of the type shown in Eq. (8-9) using an active bond to indicate the effect of valve position x. The active bond is just like a signal in a block diagram and is shown with a *full* arrow. The active bond implies

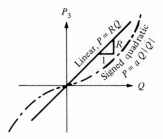

Figure 8-2 Linear and nonlinear fluid-resistance constitutive laws.

that no back effect is considered; i.e., if an effort or flow signal is transmitted in one direction, the complementary signal does not flow in the opposite direction, as in a normal bond. Thus in Fig. 8-3 the $\dot{x}$ signal emanates from a 1-junction on an active bond. All the other bonds on the 1-junction also have $\dot{x}$ as their flows, but the normal bonds also have forces that sum to zero algebraically at the 1-junction. No force is associated with the active bond. For our valve model, this means that we neglect any fluid forces associated with movement of the valve. Friction forces and inertial forces would, of course, be modeled by extra elements attached to the junction.

Because the active bond is just like a block-diagram signal, we can include a piece of block diagram to indicate that $\dot{x}$ must be integrated to x to influence the fluid resistance. Valves are *position-modulated resistors* and are common components of hydraulic systems. They are analogous to electric potentiometers.

Fluid Capacitors

Fluid capacitors provide relationships between pressure and volume variables and store potential energy. Often 1-port capacitors are combined with 0-junctions as shown in Fig. 8-4a. The capacitor itself has the general constitutive law

$$P_3 = P_3(V_3) \tag{8-10}$$

where

$$V_3 = \int^t Q_3 \, dt \tag{8-11}$$

In the linear or linearized case

$$P_3 = \frac{1}{C_3} V_3 \tag{8-12}$$

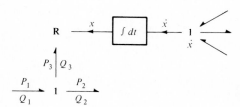

Figure 8-3 Valve representation using an active bond; $P_3 = a(x) \, [Q_3 | Q_3|]$.

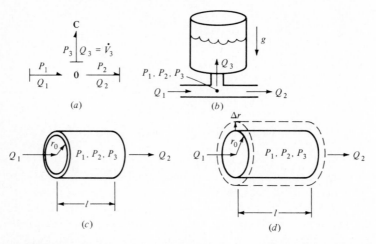

Figure 8-4 Fluid capacitors: (*a*) generalized symbol, (*b*) constant-area water tank, (*c*) rigid pipe, and (*d*) flexible pipe.

where C_3 is the fluid capacitance. The 0-junction sometimes looks like a tee pipe fitting (Fig. 8-4*b*) and in any case has the relations

$$P_1 = P_2 = P_3 \tag{8-13}$$

and
$$Q_3 = Q_1 - Q_2 \tag{8-14}$$

To show how several physical effects lead to linear capacitor relations of the type of Eq. (8-12) we consider first a constant-area tank in a gravity field (Fig. 8-4*b*). The volume of liquid in the tank V_3 is just the integral of the flow Q_3. We relate the pressure at the bottom of the tank P_3 to V_3 by first computing the height h

$$P_3 = \rho g h = \frac{\rho g V_3}{A} \tag{8-15}$$

where A is the tank area. The capacitance C_3 in Eq. (8-12) is then

$$C_3 = \frac{A}{\rho g} = \text{gravity-tank capacitance} \tag{8-16}$$

Next, we consider the rigid pipe length shown in Fig. 8-4*c*. The pressure change when a liquid is compressed is found using the *bulk modulus B*, usually defined as

$$dP = -B \frac{dV}{V} \tag{8-17}$$

where dP is the pressure change when a volume of fluid V *increases* an amount dV. The minus sign merely indicates that pressure decreases when the fluid volume expands. Taking into account Eq. (8-14), we note that V_3 is the amount of *decrease* of volume. For small changes in volume $\Delta V_3 = -dV$, and so we write

$$\Delta P_3 = +\frac{B}{V_0}\Delta V_3 \tag{8-18}$$

where the original volume of fluid is

$$V_0 = \pi r_0^2 l = Al \tag{8-19}$$

Then we have

$$C_3 = \frac{V_0}{B} = \text{liquid-volume capacitance} \tag{8-20}$$

Some typical values of B are 2.18 GPa for water at $20°C$ and 1.52 GPa for well-deaerated hydraulic oil.[3] Obviously we can use Eq. (8-18) only for fairly small excursions away from a reference pressure. The value of B varies with the reference condition of the liquid and in particular with the amount of gas, e.g., air, entrained in the liquid.

In acoustic circuits[4] a gas is compressed in the volume of Fig. 8-4c. The pressure-volume-change law is

$$\Delta P_3 = \frac{\rho_0 c^2}{V_0}\Delta V_3 \tag{8-21}$$

and so

$$C_3 = \frac{V_0}{\rho_0 c^2} = \text{acoustic-volume capacitance} \tag{8-22}$$

where ρ_0 is the reference density and c is the speed of sound. Again, the notation ΔP_3 and ΔV_3 is used to remind us that the relations are valid only for small deviations in pressure and volume. For air at $20°C$ and 1 atm pressure, $\rho_0 = 1.21$ kg/m^3 and $c = 343$ m/s.

If the pipe shown in Fig. 8-4d is elastic, a capacitor relationship can be found even when we neglect fluid compression. In this case ΔV_3 represents the expansion of volume over the length of pipe. Here we make an approximate analysis which will help us decide whether the pipe flexibility or the fluid compressibility is more important in determining the capacitance of a length of pipe. We neglect any stress or strain effects in the longitudinal direction and simply compute the change in volume due to the hoop stress in the pipe.

The pressure P generates a hoop stress in the pipe of

$$\sigma = \frac{r_0}{t_w}P \tag{8-23}$$

where r_0 is the nominal inner radius of the pipe, t_w is the wall thickness, and we assume that $r_0/t_w \gg 1$. The circumferential strain ϵ is then

$$\epsilon = \frac{r_0 P}{E t_w} \tag{8-24}$$

Table 8-2 Some values for E

Material	E, GN/m^2
Steel	195
Aluminum	71
Hard rubber	2.3
Soft rubber	0.005

where E is the elastic modulus. The change in volume due to the increase in radius is

$$\Delta V = \Delta(\pi r^2 l) = \pi l (2r) \Big|\Delta r, \;_{= r0} \tag{8-25}$$

where for small strains

$$\epsilon = \frac{\text{change in circumference}}{\text{nominal circumference}} = \frac{\Delta(2\pi r)}{2\pi r_0} = \frac{\Delta r}{r_0} \tag{8-26}$$

Using Eqs. (8-24) to (8-26), we find

$$\Delta V = \frac{2\pi l r_0^3}{E t_w} P$$

If we consider $V_0 = \pi r_0^2 l$ to be the nominal volume at the nominal pressure $\Delta V = \Delta V_3$ and $P \rightarrow \Delta P$ to be the change in pressure, then

$$\Delta V_3 = V_0 \frac{2r_0}{E t_w} \Delta P \tag{8-27}$$

so that

$$C_3 = V_0 \frac{2r_0}{E t_w} = \text{capacitance due to pipe flexibility} \tag{8-28}$$

Some values for E are given in Table 8-2.

Although the capacitance of a pipe segment may be clearly dominated by either fluid compliance or pipe flexibility, sometimes both effects are important. Then we can combine the compliances as shown in Fig. 8-5, where a C for liquid compliance

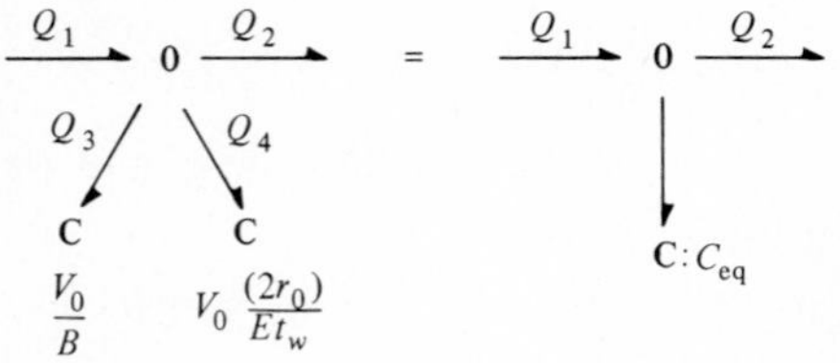

Figure 8-5 Single equivalent compliance for liquid compression and pipe flexibility for a length of pipe.

is shown with a C for pipe flexibility. The bond graph implies that

$$Q_3 + Q_4 = Q_1 - Q_2$$

or

$$\Delta V_3 + \Delta V_4 = \int^t (Q_1 - Q_2)\, dt$$

The two changes in volume can both be related to the single pressure at the 0-junction. Using Eq. (8-18) gives

$$\Delta V_3 = \frac{V_0}{B}\, \Delta P$$

and using Eq. (8-27) gives

$$\Delta V_4 = \frac{V_0(2r_0)}{Et_w}\, \Delta P$$

resulting in

$$\left(\frac{V_0}{B} + \frac{V_0 2r_0}{Et_w}\right)\Delta P = \int^t (Q_1 - Q_2)\, dt$$

What this means is that on a 0-junction compliances add. The equivalent single compliance

$$C_{eq} = V_0\left(\frac{1}{B} + \frac{2r_0}{Et_w}\right) \tag{8-29}$$

is illustrated in Fig. 8-5.

Of course, one could combine acoustic compliance with the pipe-flexibility compliance, but in most practical cases the acoustic compliance is far greater than the pipe compliance; i.e., the pipe appears to be nearly rigid to acoustic oscillations.

Although much of the dynamic modeling of fluid systems can be done using linearized systems, in other cases nonlinear relationships are essential. Consider the compressed-gas accumulator shown in Fig. 8-6. Under dynamic conditions, the gas usually does not have time to exchange much heat with its surroundings. In this case, the isentropic law

$$PV^\gamma = P_0 V_0^\gamma = \text{constant} \tag{8-30}$$

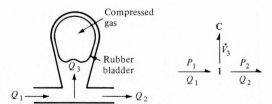

Figure 8-6 Compressed-gas accumulator.

is a good approximation, where γ is the ratio of the specific heat at constant pressure to that at constant volume. For air at atmospheric pressure, $\gamma \approx 1.4$, for example. If P_0 and V_0 are the initial pressure and volume of gas, then

$$V = V_0 - V_3(t) \tag{8-31}$$

for the bond graph of Fig. 8-6 and the nonlinear capacitive law is just

$$P_3 = \frac{P_0 V_0 \gamma}{V_0 - V_3} \tag{8-32}$$

Incidentally, if we linearize Eq. (8-30) by computing

$$\left. \frac{dP}{dV} \right|_0 = -\frac{P_0 \gamma}{V_0}$$

and relate dP to ΔP_3 and dV to $-\Delta V_3$, we arrive at the relation

$$\Delta P_3 = \frac{P_0 \gamma}{V_0} \Delta V_3 \tag{8-33}$$

or

$$C_3 = \frac{V_0}{P_0 \gamma} = \text{compressed-gas compliance} \tag{8-34}$$

This is the same result as Eq. (8-22) for the acoustic compliance since the sound speed can be expressed[4] as

$$c = \sqrt{\frac{\gamma P_0}{\rho_0}} \tag{8-35}$$

Fluid Inertia

The final passive 1-port is the fluid inertia of a pipe segment (Fig. 8-7). Although the inertia element arises due to the mass of the fluid, the inertia coefficient assumes an almost counterintuitive final form when related to P and Q effort and flow variables. The simplest derivation for the fluid inertia involves a constant-area pipe segment. The effort P_3 in Fig. 8-7 is

$$P_3 = P_1 - P_2 \tag{8-36}$$

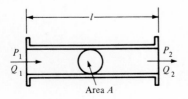

Figure 8-7 Fluid inertia for a pipe segment.

and the momentum variable p_{P_3} is

$$p_{P_3} = \int^t P_3 \, dt = \int^t (P_1 - P_2) \tag{8-37}$$

For a linear inertia element, this momentum is related to the flow Q_3 by an inertia coefficient I_3, just as velocity and momentum are related for a mass

$$I_3 Q_3 = p_{P_3} \tag{8-38}$$

or

$$Q_3 = \frac{1}{I_3} p_{P_3} \tag{8-39}$$

Let us see whether we can deduce the coefficient I_3. Suppose we use $F = ma$ for the pipe. Since the net force on the slug of fluid is $P_1 A - P_2 A$, the fluid velocity is Q_3/A, and the mass of the slug is $\rho A l$, we have

$$\frac{(\rho A l) \dot{Q}_3}{A} = P_1 A - P_2 A \tag{8-40}$$

where $\dot{Q}_3/A$ is the acceleration. Simplifying and using Eq. (8-36) gives

$$\frac{\rho l}{A} \dot{Q}_3 = P_3 \tag{8-41}$$

Then from Eq. (8-37) we find that

$$\frac{\rho l}{A} Q_3 = p_{P_3} \tag{8-42}$$

which, comparing with Eq. (8-38), implies that

$$I_3 = \frac{\rho l}{A} = \text{fluid inertia of pipe segment} \tag{8-43}$$

It is not surprising that the inertia coefficient is proportional to the density ρ and length l, but you may be surprised to find that small-area pipes have more fluid inertia than large-area pipes. This is because we are using pressure as effort, not force. For a given pressure difference, the flow accelerates quickly for large pipes and slowly for small ones.

This derivation, although commonly used, is not rigorous since we used Newton's law on a control volume without considering momentum fluxes at the ends. The results are correct if the two end areas are equal, as a more sophisticated analysis shows. In fact, if s is a centerline length variable running from 0 at the left end to l at the right, and if the area varies as $A(s)$, the inertia is

$$I_3 = \int_0^l \frac{\rho \, ds}{A(s)} \tag{8-44}$$

which shows that places where the area is small contribute most to the inertia effect. When $A(0) = A(l)$, the momentum effects cancel out and the bond graph of Fig.

Table 8-3

Substance	ρ, kg/m^3
Air (1 atm)	1.21
Water	998
Hydraulic oil	900

8-7 represents the inertia effects completely. In nozzles and diffusers the end areas differ, and static pressure is converted into dynamic pressure and vice versa; for that reason another element, the *Bernoulli resistor* needs to be added. We neglect these refinements for now, but they can be found in sec. 9.5 of Ref. 5. Some density values you may find handy are given in Table 8-3.

8-3 MODELING THE FLUID LINE

Consider a length of hose filled with a liquid such as hydraulic oil or water or a duct carrying a gas such as air. We can use the elements discussed so far to make a model of this type of system. Figure 8-8 shows a line of length L broken up conceptually into a number of shorter segments of length l. For each segment we can combine the resistance effect shown in Fig. 8-1 with a capacitance effect shown in Fig. 8-4 and an inertia effect from Fig. 8-7. The result is the bond graph of repeated basic units shown in Fig. 8-8. The C elements represent acoustic compliance, liquid compression, pipe-wall flexibility, or a combination of these, depending upon the application. In each lump, the I and R elements on the 1-junction model pressure drops associated with acceleration and flow resistance, while the C element on a 0-junction allows the

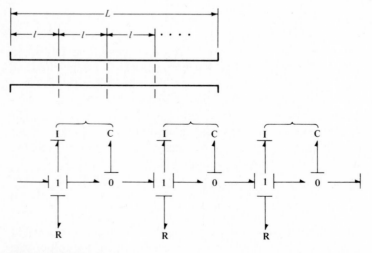

Figure 8-8 Model of a long line.

input and output flows at the ends to differ and some fluid to accumulate in the segment.

In reality, the fluid line is a *distributed-parameter system* that we are modeling with finite-sized lumps. At each lump there is a pressure at the 0-junction and a flow at the 1-junction, but really we can imagine that pressure and flow vary continuously along the length of the line as well as with time. A way to model such a system is with a *partial differential equation,* but our lumped-parameter model will yield a finite set of ordinary differential equations, one for each I and C element in integral causality. If we let $l \rightarrow 0$, the number of lumps will grow without bound and so will the number of state equations.

The question of how many lumps is enough is clearly important, but the answer always depends on the application. The best we can do is to mention a few of the ideas which should go through your mind in setting up a model. In practice, it is normal to vary the amount of detail in models in order to find the simplest one that seems to answer the questions posed about the system.

The simplest model of a line is just a bond which neglects all inertial, capacitive, and resistive effects. This is appropriate if the system dynamics are dominated by other components. For a short but large-diameter line, only a single capacitor may be necessary for a more accurate model. A longer but smaller-diameter line may require a single lump of the type shown in Fig. 8-8. Finally, a long line subjected to high-frequency excitation may require several lumps.

How to guess the number of lumps or, equivalently, the lump length l needed? One way is to compute the natural frequency of a lump, neglecting the resistance effect. Just as the natural frequency of a mass m connected to a spring of spring constant k is $\sqrt{k/m}$, the natural frequency of a linear I element connected to a linear C element is

$$\omega_n = \sqrt{\frac{1}{IC}} \tag{8-45}$$

where I and C are the inertia and capacitance parameters. This natural frequency in radians per second is the *highest* frequency that the lumped-line model can be expected to model; therefore if you can guess something about the highest frequencies of interest, you can adjust l so that ω_n is somewhat higher than this frequency. For example, using Eq. (8-42) for I and Eqs. (8-19) and (8-20) for C, we have

$$\omega_n = \frac{1}{l} \sqrt{\frac{B}{\rho}} \tag{8-46}$$

which shows that if l is small, ω_n is large but so is the number of lumps. Similar expressions apply for the other types of compliance. Note that ω_n is not really a natural frequency of the entire system: it is merely a crude estimate of the highest frequency at which the lumped model could possibly have any validity.

In Fig. 8-8 the causal strokes are consistent with integral causality on all elements. Note that on the left side the first element is an I element; this means that a pressure input to the line leads to integral causality, while at the right side a flow

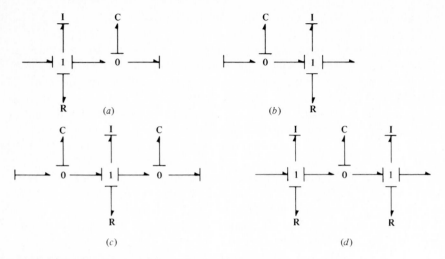

Figure 8-9 Alternative models for a line segment.

input to the C element is required for integral causality. In reality, we can end a model with either a C element or an I element, whichever is most convenient from the point of view of the rest of the system. Figure 8-9 shows several alternatives for line-segment models. In Fig. 8-9a and b the order of the I and C elements has simply been reversed. In Fig. 8-9c half of the total capacitance has been put on each side of the I element. This model accepts flow inputs from both ends conveniently. In Fig. 8-9d the total inertia and resistance have been split in two so that pressure inputs from both ends result in integral causality. As you can imagine, there are many ways to split up the total inertia, capacitance, and resistance of a line into a finite number of parts, and for specific problems one way may be better than another. When the line is just one component of the dynamic system, the approach outlined here is often sufficient even if the number of lumps is quite small.

8-4 PRESSURE AND FLOW SOURCES, PISTON TRANSDUCERS

Effort sources are visualized as supplying constant or time-varying pressures independent of the flow, and flow sources are visualized as delivering a constant or time-varying volume-flow rate independent of the back pressure. As usual, pressure and flow sources do not exist in ideal form as physical components, but there are situations where S_e and S_f are reasonable approximations of the physical situation, either alone or in combination with other elements. Figure 8-10, for example, shows two cases where a constant effort source might be used. The pressure P_0 where water in a reservoir begins its journey to a power turbine is determined by the height of the water. The flow to the turbine hardly affects the pressure at all except by slowly lowering the water level. Thus, we can use S_e to set this pressure. (Of course, the turbine may not see P_0 at its intake due to the dynamics of the line from the reservoir to the turbine inlet.)

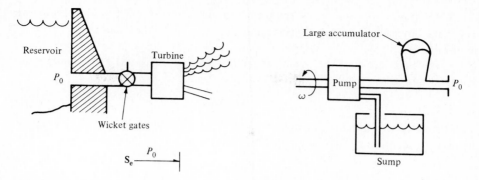

Figure 8-10 Some constant-pressure sources modeled as an effort source.

In a similar way, the gas-pressure accumulator in Fig. 8-10 can be used with a hydraulic pump to keep the oil pressure nearly constant even when the flow exceeds the pump capacity for a time. Of course, we could model the accumulator as a capacitor and model the pump and its associated drive and control system, but for some purposes it may be sufficient to assume that the pressure is essentially constant.

Figure 8-11 shows a highly schematic diagram of a *positive-displacement* pump. The crank enforces a geometric relationship between the crank angle θ and the piston motion x

$$x = x(\theta) \qquad (8\text{-}47)$$

The piston of area A displaces fluid at the volumetric rate of $A\dot{x}$

$$\dot{V} = Q = A\dot{x} = A\frac{dx}{d\theta}\dot{\theta} = A\frac{dx}{d\theta}\omega \qquad (8\text{-}48)$$

The two check valves act like electric diodes and rectify the flow.

During the discharge part of the cycle, the upper check valve opens, and the lower one closes so that Q from Eq. (8-48) passes out of the discharge pipe. During the intake part of the cycle the upper valve is closed and the lower opened so that Q, which has changed sign, is the intake flow and there is no discharge flow.

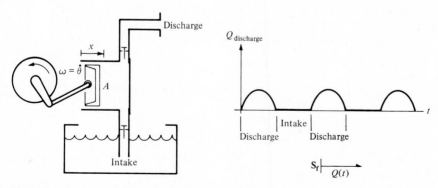

Figure 8-11 A constant-displacement pump modeled as a flow source.

If the pump is driven by a large prime mover such that ω is constant, the pump acts as a flow source with a time-varying flow similar to that shown in Fig. 8-11. This might be the case, for example, if a small pump for accessory drives were driven by a large bulldozer diesel engine. To make Eqs. (8-47) and (8-48) more meaningful, let us assume that

$$x = R \sin \theta \qquad (8\text{-}49)$$

a relationship which cannot be achieved by a crank and connecting rod but can be by other mechanisms. Then the displaced-volume rate is

$$Q = (AR \cos \theta)\omega \qquad (8\text{-}50)$$

When ω is constant and we take $\theta = \omega t$, we have

$$Q = AR\omega \cos \omega t \qquad (8\text{-}51)$$

and the discharge flow will be the positive part of a cosine wave during the discharge phase and zero when Q is negative and the pump is in the suction phase.

Real pumps usually have many cylinders arranged so that the total discharge flow is nearly constant, with just a small ripple in the flow rate. Since each cylinder contributes to the flow, as shown in Fig. 8-11, and each is individually proportional to ω, such a pump will produce a flow proportional to ω. If ω is substantially constant, such a pump is a nearly constant flow source.

Naturally, when pumps discharge to high pressures, much power is involved and ω may not stay constant. All the hydraulic power must be supplied mechanically, which means that there must be a torque on the driver. It is not hard to compute the approximate size of this torque from power considerations.

Suppose we measure the fluid volume which comes out of a multicylinder pump per radian of shaft rotation and call this amount V_0 m^3/rad. Then the output volume V is related to shaft angle θ by

$$V = V_0\theta \qquad (8\text{-}52)$$

Assuming that V_0 is (nearly) constant and not a function of θ, we have

$$Q = \dot{V} = V_0\dot{\theta} = V_0\omega \qquad (8\text{-}53)$$

The hydraulic power is PQ if P is the discharge pressure, and the mechanical power is $\tau\omega$, where τ is the shaft torque. If the pump friction is neglected, as are all dynamic effects, we can equate the mechanical and the hydraulic powers

$$PQ = PV_0\omega = \tau\omega \qquad (8\text{-}54)$$

where we have multiplied Eq. (8-53) by P. Now, however, we see that

$$V_0P = \tau \qquad (8\text{-}55)$$

is complementary to Eq. (8-53). Thus V_0 is the coefficient which relates two flows Q and ω and two efforts P and τ. This is just the definition of an ideal transformer. Figure 8-12 shows the pump with its bond-graph representation and a basic piston

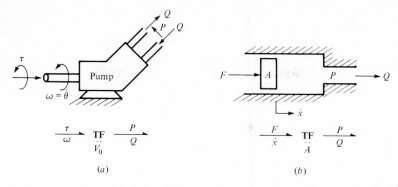

Figure 8-12 Two hydromechanical transformers: (a) $\tau = V_0 P$, $V_0 \omega = Q$; (b) $F = AP$, $A\dot{x} = Q$.

or ram; its equations are more obvious, but it also conserves power instantaneously. The idea of using power conservation on idealized models to derive necessary relationships is generally useful. To model real pumps we shall start with the lossless-transformer model and add loss and dynamic elements to it.

8-5 A PROCEDURE FOR CONSTRUCTING SYSTEM MODELS

Since the fluid systems considered here are similar to electrical systems, our procedure is based upon establishing 0-junctions for each effort (here, pressure) of interest. However, it is often not useful to choose a pressure to be "ground" or zero at the end. Usually we decide in advance whether absolute pressures are used or whether gage pressures are meant when the system is open to the atmosphere. Also, many fluid capacitors do not react to pressure differences as electric capacitors react to voltage differences, and they do not have flows passing through them. Thus the procedure differs slightly from the one used for electric circuits. Here is how to proceed:

1. For each pressure of interest in the system, establish a 0-junction. If you choose to use absolute pressure, include a 0-junction for atmospheric pressure. If you use gage pressure, the atmospheric pressure is zero and no junction is necessary for it. If you do include a junction, it will be eliminated as a ground node is in electric networks.
2. If a compliance is associated directly with a pressure, attach a C element to the appropriate 0-junction. If I, R, or TF elements react to pressure differences, attach them to 1-junctions and connect the 1-junctions between the 0-junctions representing the appropriate two pressures. By using a "through" sign convention, make sure that the difference between the two pressures is applied to the I, R, or TF elements.
3. Simplify the bond graph if there are 2-port 0- or 1-junctions with through sign conventions.

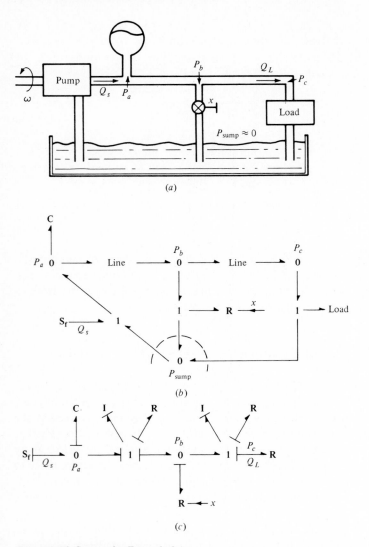

(a)

(b)

(c)

Figure 8-13 System for Example 8-1.

Example 8-1 The system in Fig. 8-13 illustrates the procedure. The pump is considered to be a flow source, which is reasonable for a positive-displacement pump if the speed can be held constant during dynamic conditions in the hydraulic circuit. The flow to the load is controlled by a bypass valve with a position variable x. Power flow to the load can be varied quickly by stroking the bypass valve quickly. An accumulator near the pump outlet will help smooth out fluctuation in the pump flow Q_s and will store energy, which limits the pressure rise if the bypass valve is shut too quickly.

Figure 8-13b shows a preliminary step in the modeling process. Four 0-junctions have been established for P_a, P_b, P_c, and P_{sump}. You can see the need for the P_a and

P_b 0-junctions, since they are tee junctions where there is a single pressure and three flows come together. The 0-junction for P_c can be eliminated if we use "through" sign conventions. If we use gage pressure, so that the atmospheric pressure at the sump is considered zero, we do not really need the 0-junction for P_{sump}, but it will not hurt to put it in.

The pump, bypass valve, and load all work on the pressure difference between P_a, P_b, or P_c and P_{sump}; they are therefore appended to 1-junctions and strung between the appropriate 0-junctions. The accumulator, however, does not care what P_{sump} is; nor does it have a flow returning to the sump. Therefore, we bond the C element directly to its 0-junction. This merely means that the accumulator pressure has only to do with the integral of the net flow into it. The use of the words "line" and "load" means that we want to consider various models for these elements.

Figure 8-13c shows the result of eliminating the P_{sump} 0-junction and simplifying, as well as some specific models for line and load. The lines are assumed to be long enough for inertia and resistance effects to be important but not so long that a multilump model is necessary. The capacitance in the line due to fluid compressibility and pipe flexibility has been judged small compared with the accumulator capacitance. The load may well be a motor with a mechanical load. Since dynamically a motor is just a pump run backward (Fig. 8-12), there may be more dynamics in the load than the equivalent hydraulic resistance we have shown. In any event, you can see that it is easy to replace line and load with other models if necessary.

The causality applied to Fig. 8-13c shows that the system model is third order. The state variables are the volume in the accumulator and the two pressure momenta for the line models. The inputs are the pump flow and the valve position as functions of time. Following the procedure for assigning causality, we found no derivative causality and all bonds had their causality determined as soon as integral causality was imposed. This means that writing equations will be as straightforward as in any other example in Chap. 4.

Example 8-2 To show how changes in the system might affect the model let us assume that it might be much more convenient to mount the accumulator on a long hose than immediately adjacent to the pump outlet. Part of the system is shown in Fig. 8-14, where we now assume that the pressure in the accumulator P_d is not dynamically identical to the discharge pressure P_a. If we include inertia for the hose, the bond graph in Fig. 8-14b results. Note that now if the hose inertia is in integral causality, the inertia for the line is forced into derivative causality. Although we have not studied this situation before, we shall see in Part Four that the equations are somewhat harder to formulate in explicit form when there is derivative causality. This does not mean that the model is not a good one, only that the long hose affects both the dynamics of the system and the formulation procedure. You may be interested to try alternative line models such as those in Fig. 8-9 to see how they might fit into Fig. 8-14. Can you see that Fig. 8-9b would plug into Fig. 8-14 with no necessary derivative causality? This in no way means that the capacity of the line is really important—only that equation formulation is easy with line capacity. Before bond graphs were invented, it was hard to see how submodels interact without getting

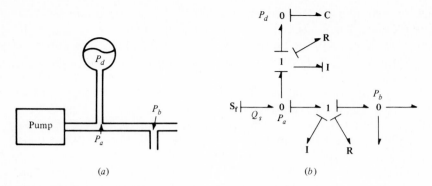

Figure 8-14 Accumulator attached to long hose for Example 8-2.

bogged down in equations. At least now we can study causality before we start trying to organize the system equations, and we can see what algebraic problems will arise.

Judgment based on experience is the best guide to the formulation of dynamic models. We hope you can see how straightforward it is to consider alternative component models, even though you cannot be expected to have an intuitive understanding of important physical effects without a good deal of experience.

PROBLEMS

8-1 Compute the velocity of flow v in a high-pressure hydraulic line such that the dynamic pressure $\rho v^2/2$ will equal the static pressure P when $P = 3000$ lb/in^2 = $3000(6.89 \times 10^3)$ Pa = 2.067×10^7 Pa = 20.67 MPa. (In this chapter, we assume the flow velocity to be much smaller than this value.) Use $\rho = 900$ kg/m^3 for hydraulic oil.

8-2 Use Eq. (8-5) to compute a resistance coefficient for a tube of 1 mm internal diameter and 0.5 m length. Estimate a range of flow rates for which this result is probably valid. The fluid density $\rho = 900$ kg/m^3, and the viscosity $\mu = 0.95$ N·s/m^2.

8-3 Suppose a fluid resistor has a nonlinear law $P = P(Q)$. Consider that there is a steady flow Q_0 upon which a small deviation in flow ΔQ is superimposed. There is also, therefore, a steady pressure P_0 and deviation in pressure ΔP. We can find a *linearized resistance* relating ΔP and ΔQ by expanding $P(Q)$ in a Taylor series at $Q = Q_0$ and keeping only the term linear in ΔQ.

$$P_0 = P(Q_0)$$

$$P_0 + \Delta P = P(Q_0) + \frac{dP}{dQ}\bigg|_{Q_0} \Delta Q + \cdots$$

so that

$$\Delta P \approx \frac{dP}{dQ}\bigg|_{Q_0} \Delta Q$$

Find the linearized resistance for the orifice law, Eq. (8-8), and show on a sketch similar to Fig. 8-2 that this resistance is the local slope of the nonlinear law at $Q = Q_0$.

8-4 A large gate valve is moved by a screw arrangement such that $x = p\theta$ and $\dot{x} = p\dot{\theta}$ the screw can therefore be represented by a transformer. Although the valve can be represented by an active bond, as in Fig. 8-3, the torque required to move the valve is not zero; it is influenced by the wheel inertia, a friction torque due to the stem packing, and a friction force on the slider. Make a bond graph of the valve including these effects.

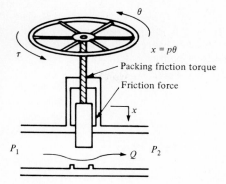

Figure P8-4

8-5 A gravity tank consists of a lower cylindrical portion of height h_1 and area A_1 and an upper cylindrical portion of area A_2. Sketch the capacitive constitutive law for this system, labeling important points.

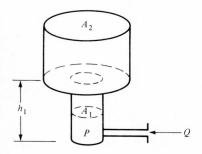

Figure P8-5

8-6 Using the values given in the text for the bulk modulus of oil and the elastic modulus of steel, compute the ratio of wall thickness to radius of steel tubes such that pipe flexibility contributes about as much as fluid compressibility to the capacitance effect. Repeat the calculation for a hose with a modulus about that of hard rubber.

8-7 A Helmholtz oscillator consists of a bottle with a long neck of area A and length l, attached to a body of volume V_0. A bond graph for this system, considering pressure deviations from atmospheric pressure and neglecting resistance, is shown. If I is the inertia coefficient and C is the capacitance, the natural frequency is

$$\omega_n = \sqrt{\frac{1}{IC}} \quad \text{rad/s}$$

just as it would be for an electric or mechanical oscillator. Derive an expression for ω_n for this acoustic oscillator.

Figure P8-7

8-8 A shell-and-tube muffler consists of three volumes V_a, V_b, and V_c connected by tubes and discharging to atmospheric pressure. Assume a time-varying inlet flow $Q_{in}t$ and an acoustic pressure (deviation from constant atmospheric pressure) of zero at the exit. Model each tube purely as an inertia element. Set up a bond-graph model and make equivalent electrical and mechanical devices.

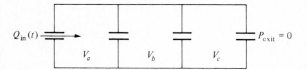

Figure P8-8

8-9 Take a single lump of a fluid line including linearized capacitance, resistance, and inductance, as shown in Fig. 8-9a. Assume an effort input on the left and a flow input on the right, as the causal marks indicate. Number the bonds and write state equations using generalized variables. Give expressions for the reaction flow on the left and the reaction effort on the right in terms of the state variables. What is the physical significance of the state variables and the reaction quantities?

8-10 Make a bond-graph model of this hydromechanical accumulator including the mass of the piston, resistance in the connection, and seal friction.

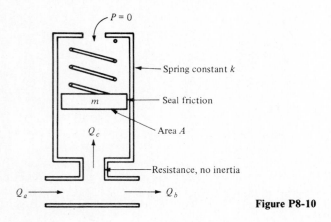

Figure P8-10

8-11 Show that a pipe whose ends differ in height by a distance h must include a pressure source equal to ρgh in its model even if inertia and resistance effects are neglected.

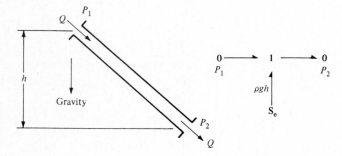

Figure P8-11

8-12 Make a bond-graph model for this hydraulic generating system. Assume that the pipes have inertia and resistance and that the turbine acts like a hydraulic resistor. Use the idea in Prob. 8-11.

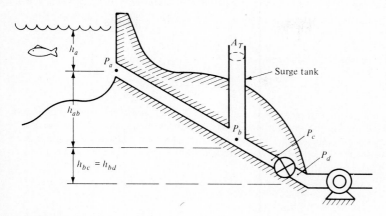

Figure P8-12

8-13 The single-sided ram has a piston of area A_p and a rod of area A_r. Thus

$$Q_a \neq Q_b \quad \text{and} \quad F = P_b A_p - P_a(A_p - A_r)$$

if ram inertia is neglected. Make a bond-graph model for this system.

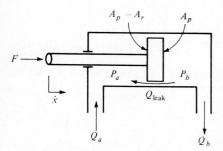

Figure P8-13

8-14 Make a bond-graph model of the pressurized-gas shock absorber shown based on the results of Prob. 8-13.

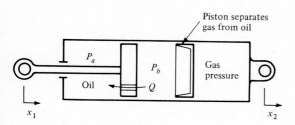

Figure P8-14

8-15 Consider a pipe 2 m long containing air at 20°C. Suppose you wish to make a lumped-parameter model of this fluid line which will be valid to about 100 Hz. Estimate the minimum number of lumps of the type shown in Fig. 8-8 (but without the R elements) which would be necessary by deriving a formula for the natural frequency of a single lump of length l. What is the order of this system model?

8-16 A positive-displacement pump produces a volume of fluid V as a function of input shaft angle θ

$$V = V(\theta)$$

The flow rate Q is thus proportional to the angular rate

$$Q = \frac{dV}{dt} = \frac{dV(\theta)}{d\theta}\frac{d\theta}{dt} = \frac{dV(\theta)}{d\theta}\,\omega$$

In Eq. (8-53) the term V_0 was a constant, but now a term $dV(\theta)/d\theta$ appears; this function of θ may contain a constant plus a varying part that accounts for ripple in the flow rate at constant ω. Using the power-conservation argument of the text, derive the torque-versus-pressure relationship and show that the pump can be represented as a transformer whose modulus varies with θ.

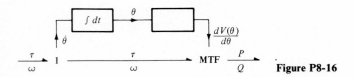

Figure P8-16

8-17 Two cylindrical water tanks of areas A_1 and A_2 are allowed to empty into the atmosphere. The flows are controlled by two identical orifices with pressure-flow constitutive laws

$$\begin{array}{c} P \\ |{\,\rightarrow\,}R \\ Q \end{array} \qquad P = KQ|Q|$$

or

$$\begin{array}{c} P \\ {\rightarrow}|\ R \\ Q \end{array} \qquad Q = (|P|K)^{1/2}\,\mathrm{sgn}\,P$$

which in either causality are valid for either flow direction. The pressure-volume relationships for the tanks are of the form

$$P = \frac{\rho g}{A} V$$

Make a bond graph for the system neglecting inertia in the pipes. Write equations using e, f, p, q variables but using the physical parameters K, ρ, q, A_1, A_2 as given above in the equations.

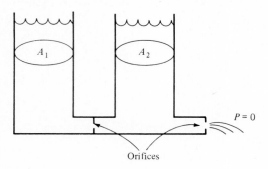

Orifices

Figure P8-17

REFERENCES

1. J. F. Blackburn, G. Reethof, and J. L. Shearer: *Fluid Power Control,* Wiley, New York, 1960.
2. J. L. Shearer, A. T. Murphy, and H. H. Richardson: *Introduction to System Dynamics,* Addison-Wesley, Reading, Mass., 1967.
3. J. Thoma: *Modern Hydraulic Engineering,* Trade and Technical Press, London, 1970.
4. L. E. Kinsler, and A. R. Frey: *Fundamentals of Acoustics,* Wiley, New York, 1962.
5. D. Karnopp and R. Rosenberg: *System Dynamics: A Unified Approach,* Wiley, New York, 1975.
6. E. O. Doebelin: *System Modeling and Response,* Wiley, New York, 1980.

NINE

GENERAL MECHANICAL SYSTEMS

9-1 INTRODUCTION

In Chap. 2 we studied a variety of mechanical systems, and you will be pleased to learn that in principle our methods need little extension in this chapter. The systems in Chap. 3 were quite restricted, however. Masses were allowed to translate in straight-line paths in inertial spaces, and only fixed-axis rotary systems were allowed. In these cases, simple 1-port I elements were all that were needed to model inertial effects. Furthermore, connections between springs, dampers, and force or velocity sources could be represented by 0- and 1-junctions and by simple transformers for pulleys and levers.

As you have seen, bond graphs automatically show striking analogies between mechanical, electrical, and hydraulic systems. Like most analogies, they are not complete. In considering a broader class of mechanical dynamic systems in this chapter we shall find that they have their own peculiarities, not shared by other physical systems.

General mechanical systems have connection laws influenced by *geometry,* but many other physical systems have connection laws based only on *topology.* This means that electric-circuit diagrams (and their bond graphs) yield the same equations even when redrawn in a distorted way, as long as the same elements are connected together. We do not worry about the angles conductors make in a circuit graph as they come together at a node, the angles bonds make at a junction, or the length of a bond.

In a mechanical system, however, when one balances forces on a mass, the angle of the force vectors *is* important and, for deformable systems, the angles generally change as the system moves. Thus, it is not enough to know that several springs, for example, are attached to a mass; we must also know the instantaneous geometric state of the system to know how to resolve the forces. In a similar way, velocity constraints in a system can be expressed only after taking the geometry of the system into account.

Mechanical systems exhibit *geometric nonlinearities* due to these effects which do not appear in electric networks or hydraulic systems. Since bond graphs are themselves topological entities, it may seem strange that geometric relationships can be incorporated into them, but it is possible when the transformer with a constant modulus is generalized to a *displacement-modulated transformer,* where the modulus varies with displacement quantities. This necessary but seemingly minor generalization of a 2-port element actually can cause major inconvenience in equation formulation:

1. In addition to the usual state variables, p's on I's and q's on C's, we may need extra displacements which serve to describe the positions of parts of the system.
2. The geometrical description of mechanical systems can be quite complicated, and usually several choices of variables are possible. Since the ancient science of kinematics has wrestled with these problems for centuries, you should not be surprised to find that some mechanical problems are difficult to visualize and to describe in equation form.
3. We shall find that many mathematical models of mechanical systems impose constraints on the motions of inertial elements that result in differential causality. The algebra necessary to bring the equations into final explicit form can be quite complicated when nonlinear geometric relationships are involved.

In Part Four we shall face these problems, but in this chapter we formulate equations in which derivative causality is not a problem, perform studies of causality to show whether or not a system will have derivative causality, and introduce alternative procedures for avoiding derivative causality. Since the subject of mechanical dynamics is far too complex to treat completely in a single short chapter, we shall be content to show some of the special features of mechanical systems and the basic methods for attacking such systems.

9-2 GEOMETRIC NONLINEARITIES AND DISPLACEMENT-MODULATED TRANSFORMERS

The system of Fig. 9-1 will give us a chance to see how the same mechanical system can sometimes act as a simple linear system like an electric circuit and at other times exhibit the typical complexities of geometry which make mechanics an interesting (or frustrating) science. We shall also have a chance to introduce displacement-modulated transformers, which play a central role in mechanics.

In Fig. 9-1 we consider a point mass attached to a spring in a gravity field. The spring is assumed to be like a rubber band or a coil spring with a freely pivoting end, so that the spring always lines up between the pivot and the mass. For now, we

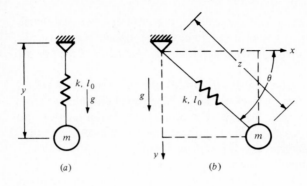

Figure 9-1 Mass-spring system: (*a*) moving in vertical direction only, (*b*) moving in a vertical plane.

assume that the spring stays in tension and does not buckle, as it probably would if it were compressed. The difference between Fig. 9-1 *a* and *b* is that in the former the mass moves only in one dimension (in the vertical direction *y*) and in the latter the mass moves in a vertical plane. Its position is then described by coordinates *x* and *y*.

The spring has a free length l_0 and a spring constant *k*, so the constitutive law relating the force *F* to the coordinate *y* in Fig. 9-1 *a* is

$$F = k(y - l_0) \tag{9-1}$$

That is $F = 0$ if $y = l_0$. [For linear systems, we often express spring deflections as displacements away from equilibrium, for example, $y' = y - l_0$. Then the simpler expression $F = ky'$ can be used in place of Eq. (9-1). In this case, it proves convenient to measure the absolute position of the end of the spring; so we use *y* in Fig. 9-1 *a* or *r* in Fig. 9-1 *b* and just incorporate the free length l_0 in the *C*-element constitutive law.]

Newtonian mass elements are fundamentally linear, but only if velocities and momenta are expressed in an inertial coordinate system. For example, in Fig. 9-1 *a*, if *y* is an inertial coordinate,

$$p_y = m\dot{y} \qquad \dot{p}_y = \Sigma F_y \tag{9-2}$$

and this means that the mass can be represented by an *I* element attached to a 1-junction, as we have done in Chap. 2. However, in Fig. 9-1 *b* $\dot{r}$ is only a relative velocity since the direction of *r* rotates as the mass moves.

If we define $p_r = m\dot{r}$, then

$$\dot{p}_r = m\ddot{r} \neq \Sigma F_r \tag{9-3}$$

In fact, if θ is the angle from the *x* axis to *r*, the correct expression for Eq. (9-3) is

$$\dot{p}_r = m\ddot{r} = \Sigma F_r + mr\dot{\theta}^2 \tag{9-4}$$

in which the mysterious centrifugal force must be added to the sum of real radial forces. Thus in rotating or accelerating coordinate systems the simplicity of $F = m\ddot{x}$ is lost, and we cannot simply use an *I* element and a 1-junction to represent the mass.

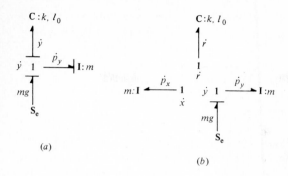

Figure 9-2 Construction of bond graphs for the systems in Fig. 9-1

Therefore in Fig. 9-1*b* we shall be careful to use $\dot{x}$ and $\dot{y}$, absolute-velocity components, to be represented by *I*-element flow variables.

Our general procedure for representing mechanical systems was to establish 1-junctions for each velocity of interest, with the provision that *I* elements should be represented by inertial velocities. This has been done in Fig. 9-2. For Fig. 9-2*a* there really is only one important velocity, $\dot{y}$, which is the velocity of both the mass and spring deflection. The bond graph is easily completed, and we note that the two state variables are y (from the spring) and p_y (from the mass). The state equations are readily written following the causal marks and turn out to be linear

$$\dot{y} = \frac{p_y}{m} \tag{9-5}$$

$$\dot{p}_y = -k(y - l_0) + mg \tag{9-6}$$

For the plane-motion case we need three 1-junctions for $\dot{x}$, $\dot{y}$, and the spring-deflection velocity $\dot{r}$. Much of the bond graph attaches easily, as shown in Fig. 9-2*b*. Now we need some bond-graph structure to relate $\dot{x}$, $\dot{y}$, and $\dot{r}$. Can you see that x, y, and r are not independent? For example, if x and y are known, it is true that

$$r = (x^2 + y^2)^{1/2} \tag{9-7}$$

This *displacement* relation can be differentiated to yield a *velocity* relationship

$$\frac{dr}{dt} = \frac{\partial r}{\partial x}\frac{dx}{dt} + \frac{\partial r}{\partial y}\frac{dy}{dt}$$

$$= \frac{1}{2}(x^2 + y^2)^{-1/2}2x\dot{x} + \frac{1}{2}(x^2 + y^2)^{-1/2}2y\dot{y}$$

$$= [x(x^2 + y^2)^{-1/2}]\dot{x} + [y(x^2 + y^2)^{-1/2}]\dot{y}$$

$$\dot{r} = m_x(x,y)\dot{x} + m_y(x,y)\dot{y} \tag{9-8}$$

A general pattern emerges from this example. Starting with a displacement relation like Eq. (9-7), a velocity constraint can be found by differentiation which will always be linear in the velocities but with coefficients such as m_x and m_y in Eq. (9-8), which are functions of displacements. In our example, the two coefficients are

$$m_x = \frac{x}{(x^2 + y^2)^{1/2}} \tag{9-9}$$

and

$$m_y = \frac{y}{(x^2 + y^2)^{1/2}} \tag{9-10}$$

These coefficients act like transformer moduli except that they are not constant but vary with the position x,y. Since Eq. (9-8) says that $\dot{r}$ is the sum of two scaled velocities, let us see whether we can use a 0-junction to do the velocity summing and two modulated transformers to do the scaling with m_x and m_y. The general idea is shown in Fig. 9-3a.

You should see that at least the velocity relations are correctly represented by this bond graph. The velocity $\dot{r}$ is indeed equal to $m_x\dot{x} + m_y\dot{y}$ because of the 0-junction, and the modulated TFs (MTFs) multiply $\dot{x}$ and $\dot{y}$ by m_x and m_y, respectively. The active bonds and integrators indicate that x and y are needed for m_x and m_y in the MTF constitutive law, but there are no forces associated with the active bonds at the $\dot{x}$ and $\dot{y}$ 1-junctions.

If this bond graph is correct, however, some special force relationships are implied and we should check to see that they are correct. At the 0-junction all bonds carry the same force, F_r, the force in the radial direction. (In our example, this will be provided by the spring.) But then the MTFs imply

$$F_x = m_x F_r \tag{9-11}$$

and

$$F_y = m_y F_r \tag{9-12}$$

If you study the expressions for m_x and m_y, you will discover to your delight that Eqs. (9-11) and (9-12) are merely (but correctly) expressions resolving the radial force into x and y components. Let us see in a more general way why this seemingly fortuitous result is actually inevitable.

If we multiply Eq. (9-8) by F_r, a power expression results

$$F_r\dot{r} = F_r m_x\dot{x} + F_r m_y\dot{y} \tag{9-13}$$

Equations (9-11) and (9-12) must be true because in relating $\dot{r}$ to $\dot{x}$ and $\dot{y}$ and F_r to the x and y components of force F_x and F_y we are simply making transformations that cannot supply or dissipate power. Thus the power relationship

$$F_r\dot{r} = F_x\dot{x} + F_y\dot{y} \tag{9-14}$$

must be equivalent to Eq. (9-13). Comparing terms in these two equations, we conclude that Eqs. (9-11) and (9-12) must hold. Of course there are some "through" sign conventions built into our equations that are reflected in the bond graph of Fig. 9-3a.

The explicit structure shown in Fig. 9-3a is too complex to be used in all bond graphs. The type of MTF that occurs in mechanics is special in that the modulus

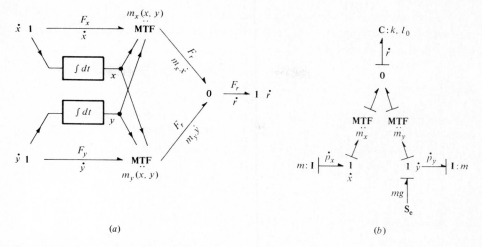

(a) (b)

Figure 9-3 Completing the bond graph of Fig. 9-2b.

(m_x or m_y, for example) is a function of one or more displacements, which are the integrals of the flow variables associated with the external bonds of the MTF structure. Knowing this, we do not have to indicate the active bonds and integrators in block-diagram form as shown in Fig. 9-2b.

A simpler form of notation is shown in Fig. 9-3b, where the structure of Fig. 9-3a has been inserted into the rudimentary graph of Fig. 9-2b. We just note that m_x and m_y are the transformer moduli and we need to remember the functions of Eqs. (9-9) and (9-10) when it comes time to write equations. Just how m_x and m_y enter the equations is clear from the original differentiation of the displacement constraint that resulted in Eq. (9-8).

We must remember that the MTFs require certain displacements that may not otherwise be state variables. This is because the MTFs incorporate information on the geometric state of the system which may not be specified completely by the state variables for the C elements of the system. Thus we shall often be required to write extra state equations for the MTF displacements. These equations are basically due to the active bond-integrator blocks shown in Fig. 9-3a that are implied even when we do not show them, as in Fig. 9-3b.

Let us write state equations for the causally augmented bond graph of Fig. 9-3b. Note that after we applied causal strokes in the standard way, all I and C elements ended up in integral causality and no bonds are left without causal strokes. This means that we should be able to write the equations purely by substitution. (Unfortunately, differential causality and algebraic loops, which occur when some bonds can be given arbitrary causality, are quite common in nonlinear mechanics. This gives a measure of job security to kinematicians.) If you learned your lessons in Chap. 4 well, you will expect the model to be third order with state variables r from the C and p_x and p_y from the I's. The new twist is that the MTFs require x and y, so we must add them to our state-variable list.

The equation for $\dot{r}$ following the causal strokes is determined from x and y, as indicated by Eq. (9-8), but now we continue in the graph to find $\dot{x}$ and $\dot{y}$ from p_x and p_y. The complete result is

$$\dot{r} = \frac{x}{(x^2 + y^2)^{1/2}} \frac{p_x}{m} + \frac{y}{(x^2 + y^2)^{1/2}} \frac{p_y}{m} \tag{9-15}$$

The equation for $\dot{p}_x$ from the graph again follows by substitution. At the $\dot{x}$ 1-junction we see that $\dot{p}_x = -F_x$, as defined in Fig. 9-3. Then we find F_x from the MTF, Eq. (9-11), and finally use the C-element constitutive law to find F_r in terms of r

$$\dot{p}_x = - \frac{x}{(x^2 + y^2)^{1/2}} k(r - l_0) \tag{9-16}$$

The equation for p_y is similar except that the force mg is added

$$\dot{p}_y = - \frac{y}{(x^2 + y^2)^{1/2}} k(r - l_0) + mg \tag{9-17}$$

Now we make x and y state variables to allow their use in m_x and m_y. We look for $\dot{x}$ and $\dot{y}$ on the graph and then follow the causal strokes to write state equations. This time the equations are very easy

$$\dot{x} = \frac{p_x}{m} \tag{9-18}$$

$$\dot{y} = \frac{p_y}{m} \tag{9-19}$$

Now we have five state equations, Eqs. (9-15) to (9-19), for the five state variables r, p_x, p_y, x, and y, and we are ready for simulation or analysis. If you are a dynamicist, you can say that what we have here is a two-degree-of-freedom system which can be represented by two second-order equations, i.e., a fourth-order system. A typical somewhat hidden constraint in this formulation is Eq. (9-7), which constrains three of our state variables, x, y, and r. Thus, we must choose initial values for x, y, and r that are *not* independent but satisfy Eq. (9-7). Subsequent to our initial choice, our state equations will generate $x(t)$, $y(t)$, and $r(t)$ satisfying Eq. (9-7), since this relation is the basis of our MTF structure. This extra order arose in our formulation because r is a state variable from the C element and we made convenient use of x and y as two more. In principle, we could eliminate x or y from m_x and m_y in favor of r using Eq. (9-7). But since this is algebraically awkward, we prefer to use the fifth-order system and to recognize the initial-condition constraint.

We have now completed the formulation of equations for the elementary system shown in Fig. 9-1. In one case we constrained the motion to translation in one dimension, and in the other case we let the system move in two dimensions. The difference between the state equations, Eqs. (9-5) and (9-6), and Eqs. (9-15) to (9-19) is quite dramatic. The general features of mechanical systems which have been illustrated are listed below.

1. Linear *I, R, C,* TF, GY elements result in linear state equations only if the connection laws can be represented with linear junction-structure elements 0, 1, TF, GY. This was the case for the elementary mechanical systems of Chap. 2 and for the system of Fig. 9-1*a*. Generally, however, geometric nonlinearities can arise. They can be conveniently incorporated in bond graphs using displacement-modulated transformers, special MTFs which simultaneously constrain forces and velocities. As the system of Fig. 9-1*b* shows, even simple systems can have complex state equations. Since nothing like these MTFs is found in electrical or hydraulic systems or in many other nominally analogous dynamic systems, they constitute an essential feature of mechanical systems.

2. When differentiated, any kind of displacement constraint yields a velocity relationship which is linear in the velocities but whose coefficients are functions of position. Thus we need only structures consisting of 0's, 1's, and MTFs to handle mechanical systems. Because nonlinear algebra is involved, there are choices in expressing constraints and there may be major differences in the complexity of alternative forms.

3. While there is a minimum order for the mathematical model of a mechanical system, higher-order models may arise depending upon how displacement relations are handled. The state variables for higher-order formulations may not be fully independent.

4. The *I*-element representation of mass points (and the translation of rigid bodies) is possible only under definite conditions on the coordinates describing the motion. For mass points simple *I*-element representation requires *inertial coordinates*.

5. The basic procedure for setting up a bond graph for elementary mechanical systems works in general. Velocities of interest are established by the use of 1-junctions; then 1-port elements are attached where appropriate. The only new feature is that where velocities are constrained we need to insert MTF structures. They can often be derived from differentiated position constraints.

Although it is easy to describe in principle how to set up a bond graph for a general mechanical system, there are many practical pitfalls in writing state equations successfully. We examine some in the next sections.

9-3 TYPICAL FORMULATION DIFFICULTIES

Derivative Causality

In Chap. 4 we noted how derivative causality can arise when using the standard procedure for augmenting a graph, but we deferred detailed discussion of this case until Chap. 14. The idea is that the state variable for an *I* or *C* in derivative causality is *not* independent of the other state variables, and so no state equation for the variable is needed. However, some algebraic manipulations are necessary in the remaining state equations to include the influence of the dependent state variable in the final explicit state equations. When geometric nonlinearities are involved, this algebra can be nasty indeed, and unfortunately many mechanical systems are modeled with rigid constraints that lead to derivative causality.

Example 9-1 The pendulum pivot in Fig. 9-4 is on an oscillator, and so we cannot simply apply the $\tau_0 = J_0\ddot{\theta}$ idea as if the pivot were a fixed point. A more fundamental approach is to use x and z as inertial coordinates for m and X for M. Three 1-junctions are established in Fig. 9-4 and three I elements; the C for the spring and S_e for the gravity force can be applied directly. We now investigate the connections between X, x, and z.

If you have experience in mechanics, you should be able to see that X, x, and z are not independent. For example, if you moved M over a distance X and m over a distance x, the height z would be known. (Of course, x cannot be too much larger than X or m would be ripped off its stick of length l. Also, given a suitable X and x, there are two possible z values, one positive and one negative. This means that we must keep the physical problem in mind in picking values of X, x, and z that fit the constraints; otherwise our mathematics could produce nonsensical results.)

One way to write the displacement constraint is to consider one of the right triangles shown with l as its main diagonal

$$l^2 = z^2 + (x - X)^2 \qquad (9\text{-}20)$$

Since l is constant,

$$\frac{d}{dt} l^2 = 0 = 2z\dot{z} + 2(x - X)(\dot{x} - \dot{X}) \qquad (9\text{-}21)$$

which is one equation constraining the three velocities $\dot{z}$, $\dot{x}$, and $\dot{X}$. To be specific, solve for $\dot{z}$ in terms of $\dot{x}$, and $\dot{X}$

$$\dot{z} = \frac{x - X}{z}\dot{X} + \frac{X - x}{z}\dot{x} \qquad (9\text{-}21a)$$

or

$$\dot{z} = \frac{X - x}{z}(\dot{x} - \dot{X}) \qquad (9\text{-}21b)$$

Now we think of choosing $\dot{X}$ and $\dot{x}$ independently and solving for $\dot{z}$.

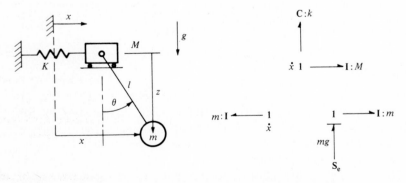

Figure 9-4 System for Example 9-1 and the first step in writing its bond graph.

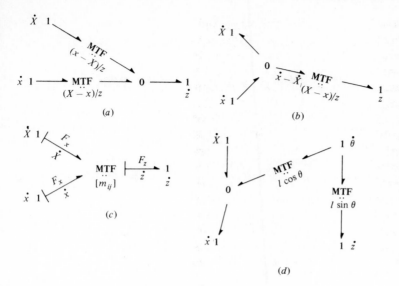

Figure 9-5 MTF junction structures for example 9-1.

Figure 9-5 shows several forms of MTF structures which incorporate the velocity constraint in Eqs. (9-21), (9-21 a), or (9-21 b). Figure 9-5 a is based on Eq. (9-21 a). The 0-junction adds the two flows to get $\dot{z}$, and the two MTFs scale $\dot{X}$ and $\dot{x}$ with modulus functions of X, x, and z. Figure 9-5 b derives from the special form of Eq. (9-21 b). Here the 0-junction first combines $\dot{x}$ and $\dot{X}$ into $\dot{x} - \dot{X}$, and then a single MTF provides the single variable modulus.

Figure 9-5 c represents a further simplification in notation, in which a multiport modulated transformer is defined with a matrix of moduli. We think of Eq. (9-21 a) as being written

$$[\dot{z}] = [m_{11} \quad m_{12}] \begin{bmatrix} \dot{X} \\ \dot{x} \end{bmatrix} \qquad (9\text{-}21\,c)$$

where

$$m_{11} = \frac{x - X}{z} \qquad (9\text{-}22)$$

and

$$m_{12} = \frac{X - x}{z} \qquad (9\text{-}23)$$

The causality shown indicates that $\dot{z}$ is specified as a function of $\dot{x}$ and $\dot{X}$, but it also indicates that the forces F_X and F_x should be functions of F_z. The proper relations are

$$\begin{bmatrix} F_X \\ F_x \end{bmatrix} = \begin{bmatrix} m_{11} \\ m_{12} \end{bmatrix} [F_z] \qquad (9\text{-}24)$$

where the *transpose* of the $[m]$ matrix appears, again because of power conservation. With the "through" sign convention of Fig. 9-5 c we require

$$F_z \dot{z} = F_X \dot{X} + F_x \dot{x} \tag{9-25}$$

but when we multiply Eq. (9-21) by F_z and use the notation of Eqs. (9-22) and (9-23), we have

$$F_z \dot{z} = F_z m_{11} X + F_z m_{12} \dot{x} \tag{9-26}$$

Comparing Eqs. (9-26) and (9-25) shows that Eq. (9-24) must hold. In fact, for any multiport modulated transformer, if the velocities transform according to a matrix $[m]$, the corresponding forces transform[4] according to $[m]^T$, where T stands for transpose. The matrices may be square or nonsquare, as in this example.

Finally, we can introduce the variable θ to help in writing the position constraints

$$z = l \cos \theta \tag{9-27}$$

$$x = X + l \sin \theta \tag{9-28}$$

Then
$$\dot{z} = -l \sin \theta \, \dot{\theta} \tag{9-29}$$

$$\dot{x} = \dot{X} + l \cos \theta \, \dot{\theta} \tag{9-30}$$

which are represented in Fig. 9-5d. This structure is not so elegant, since a superfluous 1-junction and variable θ have been introduced, but if you wanted to include pivot bearing friction, $\dot{\theta}$ might be useful. Any of the structures in Fig. 9-5 can be inserted in Fig. 9-4.

In Fig. 9-6 we have used Fig. 9-5a as the junction structure. Since this form of the junction structure is *explicit*, i.e., broken down into 2-port MTFs, 0's and 1's, it is easy to see how causality will work out. (The multiport MTF short form really is set up for a single causality, as shown in Fig. 9-5c.) The causality in Fig. 9-6 resulted from the attempt to put two I's for $\dot{x}$ and $\dot{X}$ into integral causality. As you can see from the rules of causality, this immediately results in the I element for z being forced into derivative causality. We should expect this, since if $\dot{x}$ and $\dot{X}$ are produced

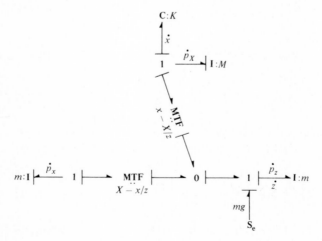

Figure 9-6 Bond graph of Fig. 9-4 with junction structure of Fig. 9-5a.

as functions of p_x and p_X, our fundamental constraint, Eq. (9-21), shows that $\dot{z}$ must be known. A little experimentation should convince you that any two of the three I elements can be given integral causality but the remaining one will always be forced into derivative causality.

We are not yet in a position to write out the state equations for the bond graph of Fig. 9-6 because of the derivative causality, but we do know what the state variables for this causality will be, namely X, p_X, and p_x for the I and C elements in integral causality plus x, X (again), and z because of the way we wrote the MTF moduli. This is a fifth-order system, but again the initial conditions of x, X, and z must obey Eq. (9-20); thus the system is really only fourth order since we can eliminate one of the geometric-displacement variables if we wish.

Unfortunately, explicit state equations are hard to write when derivative causality and MTFs get mixed up. You will see why in Chap. 14. For now, let us examine why derivative causality arises in the first place. The main reason, simply stated, is that inertial elements are coupled by some sort of rigid constraint. This would have happened in the system of Fig. 9-1b if a rigid rod had replaced the spring. The derivative causality in the system of Fig. 9-4 will disappear if we change the rigid rod of length l to a spring of variable length r.

This is done in Fig. 9-7. Now we need 1-junctions for $\dot{X}$, $\dot{x}$, $\dot{z}$, and $\dot{r}$, and we can find a relation between r, X, x, and z. Among the ways this can be expressed is to start with the idea behind Eq. (9-20)

$$r^2 = z^2 + (x - X)^2 \tag{9-31}$$

Then
$$2r\dot{r} = 2z\dot{z} + 2(x - X)(\dot{x} - \dot{X})$$

or
$$\dot{r} = \left(\frac{z}{r}\right)\dot{z} + \left(\frac{x - X}{r}\right)\dot{x} + \left(\frac{X - x}{r}\right)\dot{X} \tag{9-32}$$

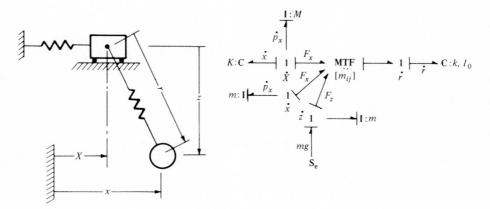

Figure 9-7 Removal of derivative causality in example of Fig. 9-4 by allowing pendulum to be extensible.

In the bond graph of Figure 9-7 this relation has been incorporated into a multiport MTF. The matrix $[m]$ is found from Eq. (9-32)

$$[\dot{r}] = \begin{bmatrix} \dfrac{z}{r} & \dfrac{x - X}{r} & \dfrac{X - x}{r} \end{bmatrix} \begin{bmatrix} \dot{z} \\ \dot{x} \\ \dot{X} \end{bmatrix} \tag{9-32a}$$

which implies

$$\begin{bmatrix} F_z \\ F_x \\ F_X \end{bmatrix} = \begin{bmatrix} \dfrac{z}{r} \\ \dfrac{x - X}{r} \\ \dfrac{X - x}{r} \end{bmatrix} [F_r] \tag{9-33}$$

in which the transpose of the matrix of Eq. (9-32 a) appears and we list the forces in the order corresponding to the order of the velocities.

It is easy to write equations for this system because complete integral causality is possible. The $\dot{r}$ equation is basically Eq. (9-32), but we follow the causality back to the p's

$$\dot{r} = \frac{z}{r}\frac{p_z}{m} + \frac{x - X}{r}\frac{p_x}{m} + \frac{X - x}{r}\frac{p_X}{m} \tag{9-34}$$

Then

$$\dot{X} = \frac{p_X}{M} \tag{9-35}$$

$$\dot{p}_X = -KX - F_x = -KX - \frac{X - x}{r}F_r$$

$$= -KX - \frac{X - x}{r}k(r - l_0) \tag{9-36}$$

where the pendulum-spring law $F_r = k(r - l_0)$ used previously has been substituted. From the inertia elements we find that

$$\dot{p}_x = -F_x = -\frac{x - X}{r}F_r = -\frac{x - X}{r}k(r - l_0) \tag{9-37}$$

$$\dot{p}_z = +mg - F_z = mg - \frac{z}{r}F_r = mg - \frac{z}{r}k(r - l_0) \tag{9-38}$$

These are the five state equations from the I and C elements. The MTF needs X, x, z, and r, of which only X and r already have state equations. Therefore, we also need

$$\dot{x} = \frac{p_x}{m} \tag{9-39}$$

and

$$\dot{z} = \frac{p_z}{m} \tag{9-40}$$

You may find it hard to believe, but these seven equations are much easier to write in explicit form than the five equations of the rigid model. For many purposes, such as computer simulation, seven easy equations are preferable to five hard ones.

Don't expect to let $k \to \infty$ in the seven equations in order to get results for the rigid-rod case. Analytically, this just gets you back into the algebraic thicket we were trying to avoid. Computationally it gets you into numerical problems, but you may be able to pick a fairly high value of k that gives good results without excessive computation time. We shall discuss the problem of stiff systems in more detail when we discuss simulation.

Backward Transformer Structures

Even with simple 2-port transformers you may already have noticed that it is easy to write down constitutive laws in one form only to find that in the final equations an inverted form is required. For example, you may have thought of a lever as producing a force F_1 proportional to F_2 in the form

$$F_1 = \frac{a}{b} F_2$$

but causality may require F_2 in terms of F_1

$$F_2 = \frac{b}{a} F_1$$

This inversion is so simple for 2-port transformers that it is hardly worth worrying about, but for more complex multiport transformers it is much better to put the relationships in the form that will actually be used than to have them in correct but inverted form.

Example 9-2 To illustrate the problem we study a system involving plane motion of a rigid body. Figure 9-8 shows a general representation of a rigid body that has inertial velocities X and Y and angular velocity ω. Recall that the center of mass of a body has some special properties. The acceleration of the center of mass and the angular acceleration in response to moments about the center of mass are uncoupled. Using our form of the equations, we can write

$$p_X = M\dot{X} \qquad p_Y = M\dot{Y} \qquad p_\theta = J_c\dot{\theta} = J_c\omega \qquad (9\text{-}41)$$

where p_X, p_Y = linear momenta
$\qquad p_\theta$ = angular momentum
$\qquad M$ = mass of body
$\qquad J_c$ = centroidal moment of inertia of body

Also

$$\dot{p}_X = \sum_i F_{X_i} \qquad (9\text{-}42)$$

$$\dot{p}_Y = \sum_i F_{Y_i} \qquad (9\text{-}43)$$

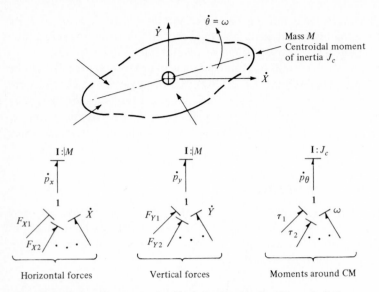

Figure 9-8 Representation of a rigid body in plane motion (Example 9-2).

and
$$p_\theta = \sum_i \tau_i \tag{9-44}$$

where F_{X_i} = X forces
F_{Y_i} = Y forces
τ_i = moments about center of mass

Equations (9-41) to (9-44) are incorporated in the bond graph of Fig. 9-8, which also shows the integral-causality form. The decoupled nature of translation and rotation when the motion is described using the velocities of the center of mass is clear.

Example 9-3 In Fig. 9-9 a typical vibration problem is shown, in which a suspended body is assumed to vibrate in small angular and translational motion in the plane. In this case we neglect X motion. Four velocities are of interest; we need $\dot{Y}$ and ω for the rigid body and $\dot{x}_1$ and $\dot{x}_2$ for the springs and dampers. After establishing four 1-junctions and attaching the 1-ports to them, we arrive at the partial bond graph of Fig. 9-9. (Note that the force of gravity can be considered to pass through the center of mass, exerting no moment about the center of mass.)

Now we come to the crux of the problem. The four velocities $\dot{Y}$, ω, $\dot{x}_1$, and $\dot{x}_2$ can be related in various ways, and each leads to a particular transformer junction structure. We do it the easy way first. If we want integral causality, as shown in Fig. 9-8, it seems that the I elements should determine $\dot{Y}$ and ω, which should then be the inputs to the junction structure and should determine $\dot{x}_1$ and $\dot{x}_2$. Assuming small angles, we can write $\dot{x}_1$ and $\dot{x}_2$ directly in terms of $\dot{Y}$ and ω as

$$\dot{x}_1 = \dot{Y} - a\omega \tag{9-45}$$

and
$$\dot{x}_2 = \dot{Y} + b\omega$$

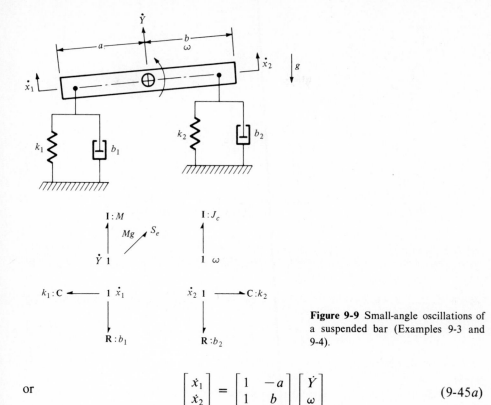

Figure 9-9 Small-angle oscillations of a suspended bar (Examples 9-3 and 9-4).

or

$$\begin{bmatrix} \dot{x}_1 \\ \dot{x}_2 \end{bmatrix} = \begin{bmatrix} 1 & -a \\ 1 & b \end{bmatrix} \begin{bmatrix} \dot{Y} \\ \omega \end{bmatrix} \qquad (9\text{-}45a)$$

Figure 9-10 shows two representations of Eq. (9-45). The first is an explicit structure using 0's and 2-port transformers. (We don't need MTFs because of the small-angle assumption.) Note that we have used simple bonds instead of the equivalent unit transformers and instead of using a transformer modulus of $-a$ the sign-convention arrows have been reversed. Check to see that Eq. (9-45) is really represented by this structure.

The second form uses a 2×2 multiport transformer for which the $[m_{ij}]$ matrix appears in Eq. (9-45a). We have shown the natural causality for this representation; i.e., given $\dot{Y}$ and ω, $\dot{x}_1$ and $\dot{x}_2$ are produced. The combination of this multiport TF

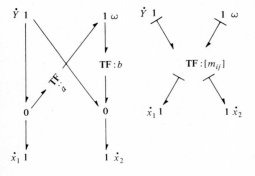

Figure 9-10 Two versions of the transformer junction structure for Eq. (9-45).

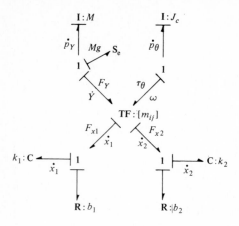

Figure 9-11 Bond graph of Fig. 9-9 completed.

and the remainder of the bond graph is shown in Fig. 9-11 which also shows four efforts, three forces, and a torque that are related by the transpose of the $[m_{ij}]$ matrix of Eq. (9-45 a). The efforts are listed in the same order as the corresponding flows in Eq. (9-45 a). The result is

$$\begin{bmatrix} F_Y \\ \tau_\theta \end{bmatrix} = \begin{bmatrix} 1 & 1 \\ -a & b \end{bmatrix} \begin{bmatrix} F_{x_1} \\ F_{x_2} \end{bmatrix} \tag{9-46}$$

After we write the state equations, it will be quite easy to check that Eq. (9-46) is correct.

The state equations for $\dot{x}_1$ and $\dot{x}_2$ are nearly Eq. (9-45 a), except that the bond graph indicates that we should replace $\dot{Y}$ and ω with functions of p_Y and p_θ

$$\dot{x}_1 = 1\dot{Y} + a\omega = \frac{1 p_Y}{M} - \frac{a p_\theta}{J_c} \tag{9-47}$$

$$\dot{x}_2 = 1\dot{Y} + b\omega = \frac{1 p_Y}{M} + \frac{b p_\theta}{J_c} \tag{9-48}$$

The equations for $\dot{p}_Y$ and $\dot{p}_\theta$ are based on Eq. (9-46) but have some other terms

$$\dot{p}_Y = +Mg - F_Y = Mg - F_{x_1} - F_{x_2}$$

$$\dot{p}_\theta = -\tau_\theta = +aF_{x_1} - bF_{x_2}$$

Continuing the substitution, we must find F_{x_1} and F_{x_2} from the C and R elements

$$\dot{p}_Y = Mg - k_1 x_1 - b_1 \dot{x}_1 - k_2 x_2 - b_2 \dot{x}_2$$

$$\dot{p}_\theta = +ak_1 x_1 + ab_1 \dot{x}_1 - bk_2 x_2 - bb_2 \dot{x}_2$$

All that is now required is to follow the causal marks to express the velocities $\dot{x}_1$ and $\dot{x}_2$ in terms of p_Y and p_θ. Since this has already been done in Eqs. (9-47) and (9-48), we have

$$\dot{p}_Y = Mg - k_1 x_1 - b_1 \left(\frac{p_Y}{M} - \frac{a p_\theta}{J_c} \right) - b k_2 x_2 - b_2 \left(\frac{p_Y}{M} + \frac{b p_\theta}{J_c} \right) \quad (9\text{-}49)$$

$$\dot{p}_\theta = +a k_1 x_1 + a b_1 \left(\frac{p_Y}{M} - \frac{a p_\theta}{J_c} \right) - b k_2 x_2 - b b_2 \left(\frac{p_Y}{M} + \frac{b p_\theta}{J_c} \right) \quad (9\text{-}50)$$

It is easy enough to forget a minus sign in equations as complex as these. For linear systems it often is helpful to arrange the equations in matrix form to see whether patterns develop

$$
\begin{bmatrix} \dot{x}_1 \\ \dot{x}_2 \\ \dot{p}_Y \\ \dot{p}_\theta \end{bmatrix}
=
\begin{bmatrix}
0 & 0 & \dfrac{1}{M} & \dfrac{-a}{J_c} \\[2mm]
0 & 0 & \dfrac{1}{M} & \dfrac{b}{J_c} \\[2mm]
-k_1 & -k_2 & \dfrac{-b_1 + b_2}{M} & \dfrac{ab_1 - bb_2}{J_c} \\[2mm]
ak_1 & -bk_2 & \dfrac{ab_1 - bb_2}{M} & \dfrac{-a^2 b_1 - b^2 b_2}{J_c}
\end{bmatrix}
\begin{bmatrix} x_1 \\ x_2 \\ p_Y \\ p_\theta \end{bmatrix}
\quad (9\text{-}51)
$$

$$
+ \begin{bmatrix} 0 \\ 0 \\ 1 \\ 0 \end{bmatrix} Mg
$$

The pattern you see in the matrix is quite general; elements on the main diagonal are zero or negative, terms symmetrically disposed about the main diagonal due to *I* and *C* elements have opposite sign, and terms symmetrically disposed about the main diagonal having to do with *R* elements may have the same signs.

We have used physical variables here rather than generalized variables and numbered bonds in an attempt to show clearly how the physics of the problem is being handled. The $\dot{x}_1$ and $\dot{x}_2$ equations simply relate velocities at the end of the bar to the linear momentum of the center of mass and the angular momentum using the multiport transformer. The $\dot{p}_Y$ equation simply adds up the net force in the *Y* direction from the springs and dampers and gravity. Again the TF is prominent, but now the transpose of the matrix is used for velocities. The $\dot{p}_\theta$ equation adds up the torques from the springs and dampers. The signs basically take care of themselves if you keep your wits about you.

Example 9-4 Now we return to Fig. 9-9 and devise a different transformer structure to couple the system. Suppose we thought of finding $\dot{Y}$ and ω from x_1 and x_2

$$\dot{Y} = \frac{\dot{x}_1}{2} + \frac{\dot{x}_2}{2} \quad (9\text{-}52)$$

$$\omega = -\frac{\dot{x}_1}{a + b} + \frac{\dot{x}_2}{a + b} \quad (9\text{-}53)$$

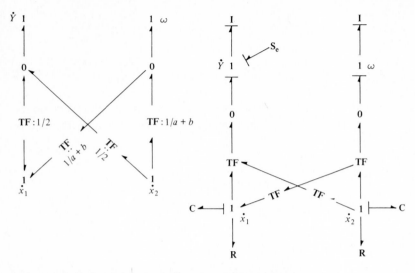

Figure 9-12 Alternative transformer structure for Example 9-4.

This version of the kinematic constraints is represented in Fig. 9-12. Comparison of Figs. 9-12 and 9-10 shows that the 0s and 1s in the explicit structures are interchanged. When this structure is inserted in the partial bond graph of Fig. 9-9, a valid bond graph results. However, this TF structure is backward. As shown in Fig. 9-12, integral causality can be applied to all *I* and *C* elements, but when this is done, the causality does not propagate throughout the graph as it did in Fig. 9-11. As we shall see later, this implies the existence of one or more algebraic loops, but we can already see the problem. The structure set up for Eqs. (9-52) and (9-53) just realizes the inverse of Eqs. (9-45). The algebra we would have to go through to use the (correct) bond graph of Fig. 9-12 is the algebra we would need to get from Eqs. (9-52) and (9-53) to their inverse, Eq. (9-45).

The message is that, if possible, we should set up transformer structures to start from velocities determined by *I* elements and to compute velocities for force-generating elements such as *C*'s and *R*'s. When this is possible, as in Figs. 9-10 and 9-11, the state equations will be easy to formulate.

9-4 SOLUTIONS TO THE DERIVATIVE-CAUSALITY PROBLEM

The problems of geometric nonlinearities and derivative causality have plagued analysts for centuries. Practical problems in mechanics are often much more complicated than the form of Newton's equations would suggest. Among the works of genius suggested are Lagrange's equations,[1,2] explicit *second-order* equations which can be derived for *any* bond-graph model.[2] However, their usefulness is particularly evident when displacement-modulated transformers are present, i.e., when geometric nonlinearities are present in the connection laws. Such systems typically occur in mechanics. On the other hand, converting Lagrange's equations into explicit first-order state equations is often not easy.[3] Again, what is algebraically possible in principle is sometimes impossible in practice.

Computationally, algebraic constraints can be satisfied by a process of iteration, so that it is possible to use implicit equations in which the derivatives of state variables are mixed up in functions with the state variables. What is usually done is to iterate until a set of state-variable derivatives is found which satisfies the equations, take a step of integration to get new state variables, then iterate to find new derivatives, and so on. As you can imagine, this takes extra computer time compared with computations based on explicit state equations.

A general way to avoid geometric or algebraic difficulties is so simple that it is often neglected. Change the model! Often a slightly simpler or even more complex model will be algebraically simpler. Since there is no single correct model for a physical system, you should consider ease of formulation as one of the valid criteria for choosing a model.

Example 9-5 In Fig. 9-13 imagine that the torque τ acting on the rotary inertia and the force F acting on the mass actually come from other system components coupled by the crank-and-slider mechanism. It is not too hard to see that the position of M is related to the crank angle by

$$x(\theta) = R \cos \theta + (l^2 - R^2 \sin^2 \theta)^{1/2} \tag{9-54}$$

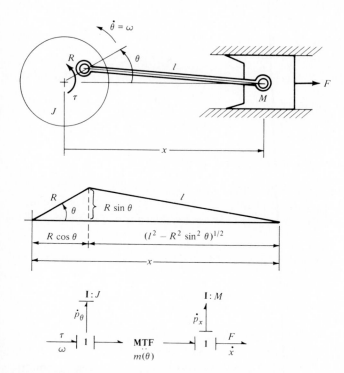

Figure 9-13 A crank-and-slider mechanism (Example 9-5).

Then we can constrain $\dot{x}$ and θ using a 2-port MTF by differentiating Eq. (9-54)

$$\dot{x} = \frac{dx(\theta)}{d\theta} \dot{\theta} = m(\theta)\dot{\theta} \qquad (9\text{-}55)$$

The resulting bond-graph fragment shown in Fig. 9-13 is simple enough, but if the I element for the rotary inertia is assigned integral causality, the one for the slider must be in derivative causality. Even for this elementary system, this fact causes algebraic difficulty.

Figure 9-14 shows one easy way around the problem—eliminating the slider mass. If the slider is a light aluminum piston and the crank is attached to a heavy steel flywheel, this may be perfectly reasonable. (All along, we have been neglecting the connecting-rod mass.) The equation for this part of the system is simplicity itself

$$\dot{p}_\theta = \tau(t) - m(\theta)F(t) \qquad (9\text{-}56)$$

and we also need an equation for θ

$$\dot{\theta} = \frac{p_\theta}{J} \qquad (9\text{-}57)$$

This might be a good model for many purposes; for example, if $F \gg m\ddot{x}$ or if the flywheel were massive and the details of the wrist-pin force were not of interest.

If this way is too simple for you, we can make a more complex system which still is easy to put in explicit form. We eliminate the rigid connecting rod in favor of a relatively stiff but flexible one. The new geometry is shown in Fig. 9-15. Now x and θ are independent and determine r and hence the spring deflection $r - l$, where l is the free length. One way to set up the MTF structure is to differentiate

$$r^2 = R^2 \sin^2 \theta + (x - R \cos \theta)^2 \qquad (9\text{-}58)$$

After some simple manipulations we find

$$\dot{r} = \frac{x - R \cos \theta}{r} \dot{x} + \frac{xR \sin \theta}{r} \theta, \\ = m_1(x,r,\theta)\dot{x} + m_2(x,r,\theta)\dot{\theta} \qquad (9\text{-}59)$$

This relationship is shown in an explicit junction structure in Fig. 9-15. The causality shows that all three elements can be in integral causality and the MTFs are not backward; i.e., the spring-displacement rate is expressed in terms of I-element flows. The equations are

$$\dot{p}_\theta = \tau(t) - m_2(x,r,\theta)k(r - l) \qquad (9\text{-}60)$$

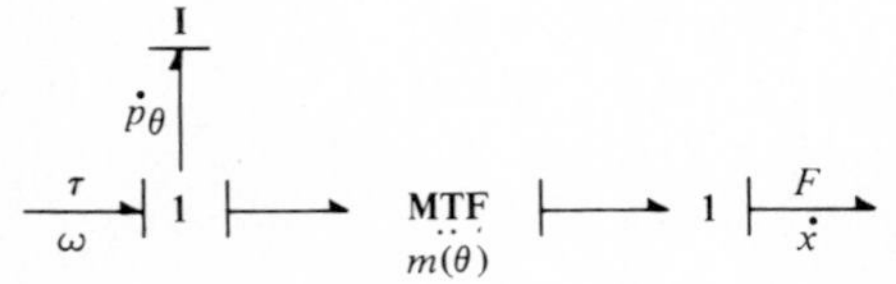

Figure 9-14 Elimination of the mass in the crank-and-slider model.

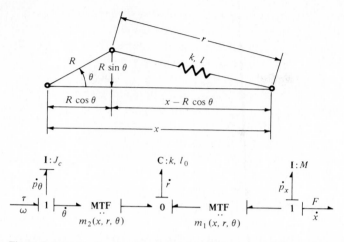

Figure 9-15 Inclusion of connecting-rod flexibility in crank-and-slider mechanism.

$$\dot{p}_x = -F(t) - m_1(x,r,\theta)k(r-l) \tag{9-61}$$

$$\dot{r} = m_1(x,r,\theta)\frac{p_x}{M} + m_2(x,r,\theta)\frac{p_\theta}{J} \tag{9-62}$$

where m_1 and m_2 appear in Eq. (9-59) and we need extra state equations for x and θ

$$\dot{x} = \frac{p_x}{M} \tag{9-63}$$

$$\dot{\theta} = \frac{p_\theta}{J} \tag{9-64}$$

Although the MTFs also require a state equation for r, since r is a state variable for C, Eq. (9-62) is already available.

It is not so clear which of the three models of Figs. 9-13 to 9-15 is the best. Of the two that are easy to formulate, one may be oversimplified and the other overcomplex. Letting k in our last equations be very large will surely cause numerical problems in simulation, for example. We cannot make definitive judgments on models, and the point of this exercise is to open your mind to alternatives.

9-5 CONCLUDING REMARKS

The dynamic equations of mass particles and rigid bodies have been known in modern form for two centuries (since the time of Leonard Euler, 1707–1783), but the problems of modeling mechanical systems and analyzing or computer-simulating the resulting equations are still under study. It is clearly not possible to do more than indicate the basic features of mechanical-system models in a short chapter like this. We have tried to show that although general mechanical systems can be modeled using bond graphs showing analogies to other physical systems, the geometrically influenced connection laws require a unique type of junction structure involving displacement-modulated transformers.

It is quite possible to use a multiport bond-graph approach even for complex three-dimensional rigid-body motion,[3-5] but, as you can imagine from the rather simple cases studied here, the geometrical problems and their associated algebraic descriptions are complex. For this reason, there is no generally superior method of treating mechanical systems, and most good dynamicists are prepared to vary their method of attack to suit the peculiarities of the problem at hand.

Bond-graph methods have been extended[6,7] to many other areas of mechanics such as kinematic mechanisms and normal-mode analyses.

Although there are many ways of formulating equations of motion for mechanical systems, bond-graph methods have some interesting advantages. (1) The causal strokes let one know quickly whether or not explicit first-order state equations will be easy to formulate. (2) A bond-graph representation in mechanics is readily combined with other bond graphs to form a system model. It really does not matter much whether the other models are mechanical, electrical, hydraulic, or from some other energy domain. Special formulation techniques for mechanical systems are often quite awkward to use when the rest of a system is nonmechanical.

We cannot guarantee that a bond-graph approach will always prove to be the most convenient way to attack a mechanical-system problem, but bond graphs are one of the few general-system representations which can handle the most general types of mechanical dynamic systems.

PROBLEMS

9-1 For the systems of Figs. 9-1 to 9-3, add air friction to the suspended mass. Consider the friction to be modeled by a force proportional to the velocity squared (but sign corrected) and in the direction of motion. For Fig. 9-1a use $F_f = c|\dot{y}|\dot{y}$ and for Fig. 9-1b use $F_f = c|\dot{r}|\dot{r}$, where F_f is the damping force in both cases. Add R elements to the bond graphs of Figs. 9-2a and 9-3b and write modified state equations [see Eqs. (9-5), (9-6), and (9-15) to (9-19)]. The completely explicit equations are quite messy, so at least show how the explicit versions would be found by substitution of expressions.

9-2 Use the MTF structure of Fig. 9-5d in Fig. 9-4 so that bearing friction can be included for the pendulum pivot. Show that the derivative-causality problem discussed in the text remains.

9-3 A thin rod with mass m and centroidal moment of inertia $J_c = \frac{1}{2}ml^2$ moves in the x-z plane. The spring force is $F_r = k(r - l_0)$, and gravity force mg acts on the center of mass. Using the relation

$$r^2 = (x - l\cos\theta)^2 + (z - l\sin\theta)^2$$

derive an expression of the form

$$\dot{r} = m_x\dot{x} + m_z\dot{z} + m_\theta\dot{\theta}$$

where m_x, m_z, m_θ are functions of x, z, and θ. Make a bond graph for the system using a multiport MTF and write all necessary state equations.

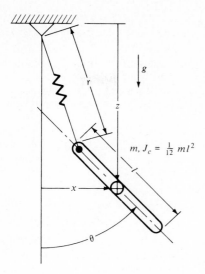

Figure P9-3

9-4 The pitch-and-heave car model is similar to the suspended bar of Fig. 9-9 except that V_1 and V_2 represent vertical velocity inputs at the front and rear wheels. Show how the bond graph of Fig. 9-11 would be modified to represent this case and give modified versions of the state equations (9-47) to (9-50) corresponding to your bond graph.

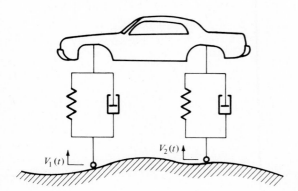

Figure P9-4

9-5 Let $\mathbf{u}_r$ and $\mathbf{u}$ be unit vectors in the r and θ directions, v_r and v_θ velocity components, p_r and p_θ momentum components, and F_r and F_θ force components. Conventionally, the absolute acceleration in polar coordinates is

$$\mathbf{a} = (\ddot{r} - r\dot{\theta}^2)\mathbf{u}_r + (r\ddot{\theta} + 2\dot{r}\dot{\theta})\mathbf{u}_\theta$$

Therefore $\mathbf{F} = m\mathbf{a}$ in component form is

$$F_r = m(\ddot{r} - r\dot{\theta}^2) \qquad F_\theta = m(r\ddot{\theta} + 2\dot{r}\dot{\theta})$$

The claim is that the bond graph with a modulated gyrator of modulus $m\dot{\theta} = p_\theta/r$ represents the same physics in different form. The bond-graph equations are

$$\dot{p}_r = \frac{p_\theta}{r} v_\theta + F_r = \frac{p_\theta}{r} \frac{p_\theta}{m} + F_r$$

$$\dot{p}_\theta = -\frac{p_\theta}{r} v_r + F_\theta = \frac{-p_\theta}{r} \frac{p_r}{m} + F_\theta$$

Using $v_r = \dot{r}$ and $v_\theta = r\dot{\theta}$, show that these momentum equations are equivalent to $\mathbf{F} = m\mathbf{a}$ in this rotating reference frame. (The bond graph is a special case of representations of inertial properties in rotating frames using Euler junction structures.[3,5])

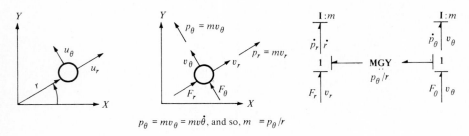

$$p_\theta = mv_\theta = mv\dot{\theta}, \text{ and so, } m = p_\theta/r$$

Figure P9-5

9-6 A mass m oscillates on a massless rod under the influence of a spring of constant k and free length l_0. Friction between the rod and the mass is represented by a radial force $F_r = b_1\dot{r}$. Use the bond graph of Prob. 9-5 to write equations for this system using radial and tangential momenta. Neglect gravity.

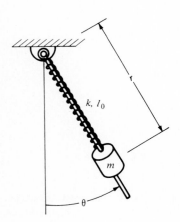

Figure P9-6

9-7 Add pivot friction to the system of Prob. 9-6 by relating $\dot{\theta}$ to v_θ using $\dot{\theta} = (1/r)v_\theta$ and an MTF. Assume a friction torque of the form $\tau = b_2\dot{\theta}$.

9-8 Show that in order to add a gravity force mg to the bond-graph of Prob. 9-7 you could use the relation $y = r \cos \theta$, where y is the vertical height (measured down) of the mass in Prob. 9-6. Show the resulting transformer structure relating $\dot{y}$ to $\dot{r}$ and $\dot{\theta}$ attached to the bond graph developed for Prob. 9-7.

REFERENCES

1. S. H. Crandall, D. C. Karnopp, E. F. Kurtz, and D. C. Pridmore-Brown: *Dynamics of Mechanical and Electromechanical Systems,* McGraw-Hill, New York, 1968.
2. D. Karnopp: Lagrange's Equations for Complex Bond Graph Systems, *Trans. ASME Dyn. Syst. Meas. Control,* vol. 99, no. 4, pp. 300–306, December 1977.
3. D. Karnopp: The Energetic Structure of Multibody Dynamic Systems, *Franklin Inst.,* vol. 306, no. 2, pp. 165–181, August 1978.
4. D. Karnopp: Power-Conserving Transformations: Physical Interpretations and Applications Using Bond Graphs, *Franklin Inst.,* vol. 288, no. 3, pp. 175–201, September 1969.
5. D. Karnopp: Bond Graphs for Vehicle Dynamics, *Veh. Syst. Dyn.,* vol. 5, no. 3, pp. 171–184, October 1976.
6. R. R. Allen and S. Dubowski: Mechanisms as Components of Dynamic Systems: A Bond Graph Approach, *Trans. ASME J. Eng. Ind.,* vol. 99, no. 1, pp. 104–111, 1977.
7. D. L. Margolis: Bond Graphs, Normal Modes and Vehicular Structures, *Veh. Syst. Dyn.,* vol. 7, no. 1, pp. 49–63, 1978.

ELECTRONIC SYSTEMS

10-1 INTRODUCTION

Although previous study of electric networks forms the basis of this chapter on electronic-system models, the increasing use of electronic systems in the control of virtually all other types of engineering systems justifies a separate discussion of how electronic subsystems can be integrated with our modeling treatment of other types of physical systems.

The physical nature of electronic systems is often clearly secondary to their information-processing capabilities. In such cases, signal interactions represented by active bonds or block diagrams are appropriate. From an overall system point of view, an electronic component can be conveniently characterized as a "black box" with a certain input-output behavior. Of course the black-box characterization fails if the component is overloaded or not properly matched to the rest of the system, since then real physical limitations appear.

From another point of view, the components of electronic systems are physical elements themselves and can be usefully characterized from the point of view of power and energy. This aspect of electronics is important for designers of black boxes, who need to understand how individual electronic elements can be combined into a useful dynamic system.

The subject of electronics is far too vast to cover in a short chapter; nevertheless we shall be able to indicate how electronic subsystems and components can be described in the same way as systems of other types. In succeeding chapters we shall also discuss components that link electronic systems with all other types. We call these linking systems *transducers, amplifiers,* and *instruments.*

10-2 BLACK BOXES AND CONTROLLED SOURCES

Many electronic components are carefully designed to operate in a way untypical of physical systems in general. These are the black boxes that respond to some input signal and produce an output signal according to some scheme but with no effect from the output back to the input. This one-way flow of information greatly facilitates the design of complex information-processing systems, since what happens at each stage in a chain of such components can be studied independently, without regard for the final system in which the component will be used. From our point of view, we can say that such components operate only in a single causality. In a sound system, for example, the phonograph needle and cartridge together send a signal to an amplifier, which in turn sends a signal to the loudspeakers. Do what you will to the speakers, the phonograph needle will never know about it.

Furthermore, electronic subsystems may consist of hundreds of elements or complex integrated circuits but still be easy to characterize from an input-output point of view. The designer of the component has to look inside the black box, but the user of the component usually does not want or need to.

We illustrate a black-box approach by studying a useful device called an *operational amplifier*. Figure 10-1 shows conventional symbols for operational amplifiers. Depending upon their output power levels, such amplifiers may be constructed on an integrated-circuit chip or out of discrete components; in any case, the purpose of the several internal components is to make the input-output relationships very simple.

Under normal loading conditions the differential amplifier of Fig. 10-1a behaves according to the law

$$e_3 = G(e_2 - e_1) \qquad (10\text{-}1)$$

where the gain G is a large number, say 10,000 or so. Since this type of amplifier works for constant values of e_2 and e_1, it is called a dc amplifier; but it also works when e_2 and e_1 are sinusoidal at quite high frequencies. As you should expect, there is always some high-frequency limit where Eq. (10-1) begins to break down.

Another important feature of the amplifier is that i_1 and i_2 are very small currents compared with the maximum current values which i_3 can take on. Thus, when the amplifier is used correctly, we can often completely ignore i_1 and i_2 in comparison with other currents in the system

$$i_1 \approx 0 \qquad i_2 \approx 0 \qquad (10\text{-}2)$$

This is clearly the case when e_1 and e_2 are the output voltages of other similar amplifiers, and this is the secret to making analog computers and active filters using operational amplifiers.

Under the assumptions of Eqs. (10-1) and (10-2) the input power is negligible and the output power is finite; so either there is a hidden energy source in our amplifier or we have not paid attention to the necessary power supply.

When adding feedback elements to an operational amplifier, it is necessary to use negative feedback for stability, and so the inverting amplifier shown in Fig. 10-

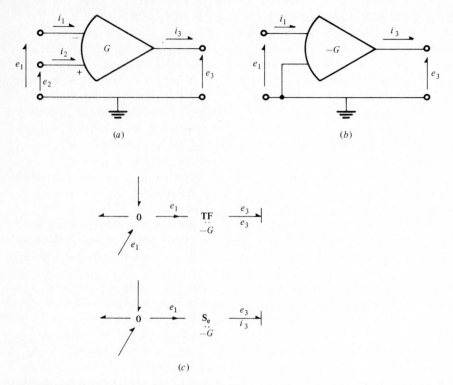

Figure 10-1 Models of operational amplifiers: (a) differential amplifier, (b) inverting amplifier, and (c) activated-TF and controlled-source models.

$1b$ is useful. An inverting amplifier is formed from a differential amplifier, for example, simply by grounding the positive input. We can then characterize the amplifier with the negative gain $-G$

$$e_3 = -Ge_1 \qquad (10\text{-}3)$$

$$i_1 \approx 0 \qquad (10\text{-}4)$$

Figure 10-1c shows two ways of modeling the inverting amplifier. If we regard Eq. (10-3) as half of a transformer constitutive law with modulus $-G$, by activating the input bond we effectively enforce Eq. (10-2). Now the transformer is unusual in two respects: (1) normal transformers always conserve power, but because of the activated bond this one does not; (2) normal transformers have two possible causal patterns, but this one has only one. The e_1 signal must come into the TF, so e_2 must be the output. Although the i_3 signal is supplied to the TF, it has no effect on i_1 since we have suppressed any information about i_1. However, i_3 can be used to calculate the output power. At the 0-junction where i_1 is determined, the algebraic sum of currents vanishes as always, but on bond 1 we assume that i_1 equals zero.

Perhaps a clearer way of representing the inverting amplifier is also shown in Fig. 10-1c, where an effort source S_e is controlled by the e_1 signal. The use of S_e

clearly indicates that there is a source of power. We can even indicate $-G$ as a parameter to help remind us of Eq. (10-3). The use of a source dictates the necessary output causality, and the active bond stemming from a 0-junction clearly carries the common-effort signal (according to the rule that all bonds on a 0-junction have the same effort); it is therefore redundant to indicate a causal mark on the active bond.

It is quite possible that e_3 could be related to e_1 by a nonlinear function or a complicated transfer function instead of a simple gain. If this were the case, one could simply treat the active bond exactly like the signal in a block diagram and insert the appropriate block-diagram boxes between the 0-junction and S_e. This combines the advantages of block diagrams for representing functional relationships with bond graphs for representing physical systems.

We are dealing with a voltage amplifier with a high-impedance input (which means that $i_1 \approx 0$) and a low-impedance output (i_3 is almost arbitrary), but there are four basic types. The four controlled-source models are listed in Fig. 10-2. Other

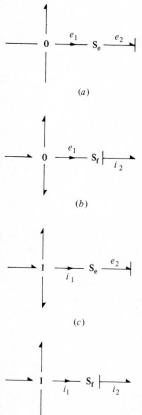

(a)

(b)

(c)

(d)

Figure 10-2 Controlled-source models of electronic amplifiers: (a) voltage-controlled voltage source, (b) voltage-controlled current source, (c) current-controlled voltage source, and (d) current-controlled current source.

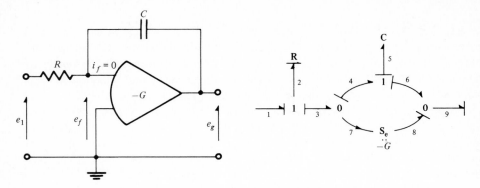

Figure 10-3 Integrator circuit made from an operational amplifier.

uses for these controlled sources in modeling individual electronic devices will appear in Sec. 10-3.

Now we explore the use of a controlled-source model with normal circuit elements. In Fig. 10-3 an inverting operational amplifier is combined with passive resistance and capacitance elements. The result is an integrator circuit, as we shall soon see. The idea behind this circuit is as follows. Although it is not particularly hard to make high-gain (or high-power-gain) active devices, it is hard to make them strictly linear and time-invariant. Often the many nonlinear components involved change as they age or vary with operating temperature, making it hard to create a really stable electronic substystem. If a high-gain amplifier is put into a system so that there is negative feedback influenced only by stable linear passive elements, the designer can make a system almost completely independent of the amplifier's characteristics as long as its gain remains high.

The bond graph of Fig. 10-3 is constructed simply by inserting the controlled-source model of Fig. 10-1 into the bond graph of the rest of the circuit, which can be constructed according to the rules already studied. The causality is fixed partly by the source and partly by our knowledge that this will be used as a voltage-input device; i.e., even though the circuit graph does not indicate a voltage source supplying e_1, we know that this is how such devices are used. Thus we choose the causality on bond 1 as shown. Finally, we pick integral causality on the C; then all remaining bonds must have the causality shown.

Normally the causal assignment just described would lead us to believe that the state equations could be written just by following the causal marks until either a state variable or an input variable is encountered. However, the controlled source introduces a minor difficulty, an algebraic loop, which will be treated more extensively in Chap. 14. For present purposes, it is sufficient to note that the controlling voltage e_7 is determined by e_4, which is partly determined by e_6, e_8, and (through the source) e_7 itself. Thus, if we simply write equations in the normal way, we find ourselves going around and around a loop involving bonds 7, 8, 6, and 4. We break this loop by expressing e_7 as an algebraic function of itself and then by eliminating e_7 from the final state equation.

The single state equation for q_5 is

$$\dot{q}_5 = f_4 = f_3 = f_2 = \frac{E_1 - e_3}{R} = \frac{E_1}{R} - \frac{e_4}{R} = \frac{E_1}{R} - \frac{e_5 + e_6}{R} = \frac{E_1}{R} - \frac{q_5}{RC} - \frac{e_8}{R}$$

$$= \frac{E_1}{R} - \frac{q_5}{RC} - \frac{-Ge_7}{R}$$

$$\dot{q}_5 = \frac{E_1}{R} - \frac{q_5}{RC} + \frac{Ge_7}{R} \tag{10-5}$$

Now the equation involves an input E_1, a state variable q_5, and the algebraic variable e_7, which we proceed to eliminate. (Note that when we wrote $f_4 = f_3$, we used the activation of bond 7 to eliminate f_7 from consideration at the 0-junction.) Writing e_7 in terms of itself gives

$$e_7 = e_4 = e_5 + e_6 = \frac{q_5}{C} + e_8 \, e_7 = \frac{q_5}{C} - Ge_7$$

Then, solving for e_7, we have

$$e_7 = \frac{1}{1 + G} \frac{q_5}{C} \tag{10-6}$$

Substituting Eq. (10-6) into Eq. (10-5) gives an explicit equation

$$\dot{q}_5 = \frac{E_1}{R} - \frac{q_5}{RC} + \frac{G}{R} \frac{1}{1 + G} \frac{q_5}{C} = \frac{-1}{1 + G} \frac{q_5}{RC} + \frac{E_1(t)}{R} \tag{10-7}$$

The output equation for e_9 is readily found by following the causal strokes

$$e_9 = e_8 = -Ge_7$$

which becomes, using Eq. (10-6),

$$e_9 = \frac{-G}{1 + G} \frac{q_5}{C} \tag{10-8}$$

These results are not particularly interesting until we use the assumption that even when G is not constant it is always very large, $G \gg 1$. In this case, Eq. (10-7) yields

$$\dot{q}_5 \approx \frac{E_1(t)}{R} \tag{10-9}$$

or

$$q_5 \approx \int^t \frac{E_1(t)}{R} \, dt \tag{10-10}$$

Also, Eq. (10-8) means that

$$e_9 \approx -\frac{q_5}{C}$$

and after using Eq. (10-10)

$$e_9 = \frac{-1}{RC} \int^t E_1(t) \, dt \tag{10-11}$$

This means that e_9 is proportional to the time integral of $E_1(t)$, that G has disappeared, and that the relationship deals only with the passive R and C elements. Such elements can have very constant R and C values, particularly if their temperatures are well controlled. Our results would hardly be affected if the amplifier had been nonlinear, as long as its equivalent gain was large over its operating range.

10-3 DEVICE MODELS: NODIC RESISTIVE MULTIPORTS

We now take a brief but instructive glimpse into the individual electronic devices used to make components such as operational amplifiers. We shall not descend to the basic physical level of electrons and holes but remain at the more descriptive level of circuit-equivalent models of the type designers use to understand how the device functions.

Although at high enough frequencies all devices begin to exhibit dynamic effects, at low to moderate frequencies the devices statically relate various currents and voltages, which means that they are resistive in nature. They are in fact multiport extensions of the 1-port electric resistors we have already studied. They certainly do not supply net power (in fact some electric power always gets lost as heat), but they are able to switch power around and can amplify when they are given a source of power.

Although multiport R elements occur in all physical domains, the electronic devices are special in that they are restricted to be nodic. This means that a device with n terminals, although it does have n currents, still cannot accumulate charge, so that an algebraic sum of the currents must vanish just as it does for a 0-junction. Also, the device does not react to the absolute voltages at its terminals; it would be unaffected by an equal increase in voltage at all terminals.

Example 10-1 We illustrate this first for a familiar two-terminal device, the diode shown in Fig. 10-4a. We can define two terminal voltages e_1 and e_2 and two currents

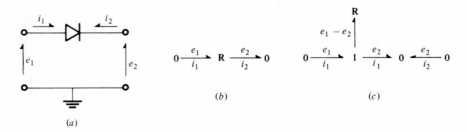

Figure 10-4 A two-terminal electronic device: (a) circuit-graph representation, (b) 2-port resistive multiport, (c) 1-port representation after nodicity conditions are applied.

i_1 and i_2. (We shall explain the odd-looking direction for i_2 shortly.) There is nothing wrong with thinking of this device as a 2-port version of an R element, as in Fig. 10-4b. Since we expect that some power will be dissipated or at least none will be supplied,

$$e_1 i_1 + e_2 i_2 \geq 0 \qquad (10\text{-}12)$$

We might imagine expressing the constitutive laws in inverse resistance form; i.e., given the voltages e_1 and e_2, certain currents i_1 and i_2 would flow. In symbolic equation form

$$i_1 = F_1(e_1, e_2) \qquad (10\text{-}13)$$

$$i_2 = F_2(e_1, e_2) \qquad (10\text{-}14)$$

For a general multiport resistor we would stop here. But for this nodic resistor there are two more restrictions. First, since no charge accumulates,

$$i_1 + i_2 = 0 \qquad (10\text{-}15)$$

which means that

$$F_1(e_1, e_2) = -F_2(e_1, e_2) \qquad (10\text{-}16)$$

Furthermore, it should not make any difference to the diode when we add an arbitrary voltage E to both e_1 and e_2. This means that

$$F_1(e_1, e_2) = F_1(e_1 + E, e_2 + E) \qquad (10\text{-}17)$$

This would certainly be the case if i_1 really depended upon the difference between e_1 and e_2 alone, since $e_1 - e_2 = (e_1 + E) - (e_2 + E)$. A model in which both Eqs. (10-17) and (10-16) are satisfied appears in Fig. 10-4c.

This model is just about what you would have expected. The 2-port R is broken down into a 1-port R and some junction elements. Now i_1 is just a function of $e_1 - e_2$, and the extra 0-junction enforces Eq. (10-15). The sign convention on the 1-junction ensures that the 1-port R sees $e_1 - e_2$ instead of $e_1 + e_2$.

Normally, we would not define both i_1 and i_2 pointing into the diode and the extra 0-junction in Fig. 10-4c would not be required; we leave it in because it plays an essential role for multiterminal devices.

Let us see whether we can argue that Eq. (10-17) really requires the bond graph of Fig. 10-4c. Figure 10-5 shows the e_1-e_2 plane. Any point in this plane corresponds to a value of i_1, according to Eq. (10-13). Now suppose we change to other coordinates to describe points in the plane, namely, $e_1 + e_2$ and $e_1 - e_2$ (Fig. 10-5). Clearly, any points such as A or B can equally well be described by values of e_1 and e_2 or by values of $e_1 + e_2$ and $e_1 - e_2$.

The only difference between points A and B is that we have performed the operation indicated in Eq. (10-17). Then if i_1 is considered to be some new function of the sum and difference voltages, we must get the same i_1 value for points such as A and B. As you can see, the $e_1 - e_2$ value does not change, but the $e_1 + e_2$ value certainly does. Thus, we conclude that i_1 must be a function of $e_1 - e_2$ only and not $e_1 + e_2$. This comes down to saying that our 2-port nodic R can be decomposed into

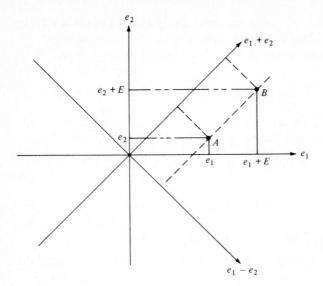

Figure 10-5 Sum and difference voltage coordinates for a nodic two-terminal device.

a 1-port R and a junction structure, as shown in Fig. 10-5; but since we would have looked at the diode as a 1-port anyway, what is the point?

For a 3-terminal element such as the transistor of Fig. 10-6 if we consider the collector, base, and emitter voltages e_C, e_B, e_E and the corresponding currents i_C, i_B, and i_E, the 3-port R representation of Fig. 10-6b leads us to conclude that no net power should be produced since

$$e_B i_B + e_C i_C - e_E i_E \geq 0 \tag{10-18}$$

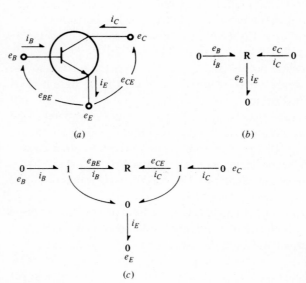

Figure 10-6 Reduction of a nodic 3-port to an embedded 2-port model.

after taking the sign conventions into account. Furthermore, in inverse resistance causality the constitutive laws can be expressed as

$$i_B = F_1(e_B, e_C, e_E) \qquad i_C = F_2(e_B, e_C, e_E) \qquad i_E = F_3(e_B, e_C, e_E) \quad (10\text{-}19)$$

The nodicity conditions now put restrictions on the element that are not obvious in Eqs. (10-19). First, there is the matter of the current sum

$$i_B + i_C - i_E = 0 \qquad (10\text{-}20)$$

and then the voltage relativity

$$F_1(e_B, e_C, e_E) = F_1(e_B + E, e_C + E, e_E + E)$$

$$F_2(e_B, e_C, e_E) = F_2(e_B + E, e_C + E, e_E + E) \qquad (10\text{-}21)$$

$$F_3(e_B, e_C, e_E) = F_3(e_B + E, e_C + E, e_E + E)$$

The current condition means that the three laws of Eq. (10-19) are not independent. For example, if we know i_B and i_C from F_1 and F_2, then i_E (or F_3) is just their sum. The second condition says that instead of expressing the constitutive laws in terms of absolute voltages e_B. e_C, e_E we should use relative voltages. For example, if we use

$$e_{BE} = e_B - e_C \qquad e_{CE} = e_C - e_E \qquad (10\text{-}22)$$

we can express the constitutive laws in the form

$$i_B = F_1' (e_{BE}, e_{CE}) \qquad i_C = F_2' (e_{BE}, e_{CE}) \qquad i_E = F_1' + F_2' \quad (10\text{-}23)$$

and adding E to all voltages will not change e_{BE} and e_{CE}. These reduced laws are built into the bond graph of Fig. 10-6c. The first two equations in Eq. (10-23) are the equations of the embedded 2-port R. The junction structure elegantly constrains i_E according to the last equation and also relates the two relative voltages to the absolute ones.

Actually, there are three ways of expressing relative voltages. We picked the one in which the voltages are relative to e_E, and that is why the emitter node is treated differently from the others in Fig. 10-7. We could just as well have used the base or collector voltage. Figure 10-7 shows a common connection pattern for the transistor, in which the emitter is grounded. Then e_E is zero, and we can simplify the bond

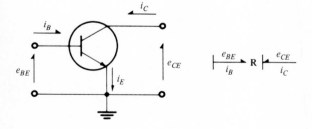

Figure 10-7 A 2-port transistor representation for the grounded-emitter configuration.

graph of Fig. 10-6c, so that only the 2-port R is evident. If we intended to ground e_B or e_C instead, one of the other two realizations would be the appropriate one.

To sum up, a three-terminal electronic device is a·3-port R element at low frequencies, but it can be reduced to a 2-port R and some junction elements by the nodicity conditions. There are three basic forms of the embedded 2-port, depending upon which terminal voltage is selected to be the reference voltage. Perhaps you can see that an n-terminal resistance device can be reduced to an $(n - 1)$-port R with junction elements by continuing our pattern of thought.

The 2-port R of Fig. 10-7 can have its constitutive laws written in four ways. In Eq. (10-23) we thought of the conductance form, i.e., currents were expressed in terms of voltages, but there is a resistance form in which voltages are expressed in terms of currents, as well as two mixed forms. The causality of Fig. 10-7 shows a common mixed form, in which i_C and e_{BE} are expressed as functions of e_{CE} and i_B

$$i_C = F_3(e_{CE}, i_B) \tag{10-24}$$

$$e_{BC} = F_4(e_{CE}, i_B) \tag{10-25}$$

This is an interesting way to present the grounded-emitter constitutive laws since when e_{CE} is positive, e_{BE} is often small enough to be neglected. Thus, $i_B e_{BE}$, which is the base power, is also very small. However, the base current has a large effect on i_C, so that changes in i_B have an amplified effect on i_C and the power $i_C e_{CE}$.

Figure 10-8 is a plot of Eq. (10-24) for a power transistor. You can think of the transistor as a nonlinear resistor relating i_C and e_{CE} which is strongly influenced by i_B. When i_B is small, 10 mA for example, the resistance is high and little collector current flows. The transistor is said to be *blocked*. When $i_B = 700$ mA, the resistance is low and up to 8 or 9 A of collector current can flow when e_{CE} is just a few volts. The transistor is said to be *conducting* in this mode. Power transistors are usually operated in a switching mode between blocked and conducting states so that little power is dissipated in the transistor itself. (Either i_C is small in the blocked state, or e_{CE} is small in the conducting state, so that $i_C e_{CE}$ is small in either case.)

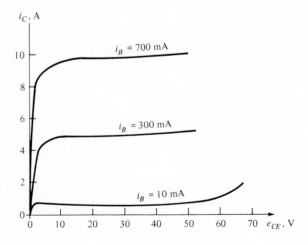

Figure 10-8 Plot of the constitutive law for a typical grounded-emitter power transistor [Eq. 10-24].

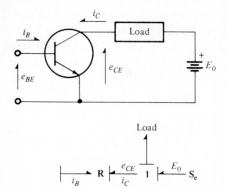

Figure 10-9 Power transistor used in switching mode.

Signal-power-level transistors can operate in the middle of the part of the i_C-e_{CE} plane in Fig. 10-8, where if e_{CE} is set by a voltage source, i_B can be used to control i_C roughly as a current-controlled current source. If we hold e_{CE} constant at 15 V, i_B = 300 mA corresponds to i_C = 5 A for a current gain

$$\frac{i_C}{i_B} = \frac{5 \times 10^3}{300} = 16.6$$

For the transistor shown, this gain is not very constant, but for others the gain is constant over a usefully wide range.

Figure 10-9 shows a possible connection for a power transistor in a switching mode. The control current i_B is switched rapidly between a (low) blocking value and a (high) conducting value. During the blocking phase, i_C is nearly zero even when e_{CE} takes on quite large values. During the conducting phase e_{CE} is small even for fairly large currents, so that the voltage source is effectively applied directly to the load. By varying the proportion of time that i_B spends in the blocking and conducting modes the average load current can be continuously varied. Of course, in the design of the system we have to make sure that the transistor is protected from excessive voltages and currents. If we use 700 mA for the conducting phase, for the transistor of Fig. 10-8 the load and voltage source should not allow more than 8 or 9 A of current to flow; otherwise the transistor might dissipate too much power and burn out, for example.

10-4 CIRCUIT-EQUIVALENT MODELS

In many signal-processing systems, electronic devices are used as nearly linear elements that relate small changes in voltages and currents from steady-state values. There is a steady loss of real power, which in absolute terms may be small but which the device must be designed to handle without harm. Assuming that the power-drain problem has been taken care of, the operation of such devices is best seen by considering the incremental voltages and currents. We show here how circuit-equivalent models can be made for the incremental variables. These models help the designer understand how to make useful systems from basic electronic elements.

Let us return to the 2-port characterization of the grounded-emitter transistor in mixed causality represented by Eqs. (10-24) and (10-25). If the device is biased in such a way that the variables i_C, e_{BE}, e_{CE}, and i_B satisfy these equations, we can write

$$i_C = \bar{i}_C + \Delta i_C \qquad e_{BE} = \bar{e}_{BE} + \Delta e_{BE} \qquad (10\text{-}26)$$

$$e_{CE} = \bar{e}_{CE} + \Delta e_{CE} \qquad i_B = \bar{i}_B + \Delta i_B$$

where Δ means a small change in variables away from the nominal values, which are indicated with overbars. Using a Taylor-series expansion of Eqs. (10-24) and (10-25) and stopping after linear terms in the incremental variables, we have

$$i_C \approx \bar{i}_C + \frac{\partial i_C}{\partial e_{CE}} \Delta e_{CE} + \frac{\partial i_C}{\partial i_B} \Delta i_B \qquad (10\text{-}27)$$

$$e_{BE} \approx \bar{e}_{BE} + \frac{\partial e_{BE}}{\partial e_{CE}} \Delta e_{CE} + \frac{\partial e_{BE}}{\partial i_B} \Delta i_B \qquad (10\text{-}28)$$

where each partial derivative must be evaluated at the nominal point e_{CE}, i_B. Using, then, the first two of Eqs. (10-26) gives the relationship for the incremental variables

$$
\begin{bmatrix} \Delta i_C \\ \Delta e_{BE} \end{bmatrix} =
\begin{bmatrix} \dfrac{\partial i_C}{\partial e_{CE}} & \dfrac{\partial i_C}{\partial i_B} \\[2mm] \dfrac{\partial e_{BE}}{\partial e_{CE}} & \dfrac{\partial e_{BE}}{\partial i_B} \end{bmatrix}
\begin{bmatrix} \Delta e_{CE} \\ \Delta i_B \end{bmatrix} \qquad (10\text{-}29)
$$

in which the matrix of partial derivatives is really a matrix of constants for a given operating point. It is useful to give the partial derivatives the following special symbols when we wish to construct equivalent circuits

$$\frac{\partial i_C}{\partial e_{CE}} = \frac{1}{r_0} \qquad \Omega^{-1}$$

$$\frac{\partial i_C}{\partial i_B} = \beta = \text{current gain, dimensionless}$$

$$\frac{\partial e_{BE}}{\partial e_{CE}} = \mu = \text{voltage gain, dimensionless}$$

Then
$$\frac{\partial e_{BE}}{\partial i e_B} = r_n \qquad \Omega$$

$$
\begin{bmatrix} \Delta i_C \\ \Delta e_{BE} \end{bmatrix} =
\begin{bmatrix} 1/r_0 & \beta \\ \mu & r_n \end{bmatrix}
\begin{bmatrix} \Delta e_{CE} \\ \Delta i_B \end{bmatrix} \qquad (10\text{-}30)
$$

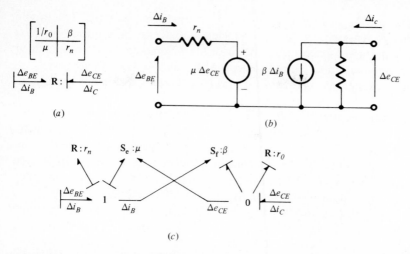

(a)

(b)

(c)

Figure 10-10 Incremental model for grounded-emitter transistor and circuit equivalent.

The actual values of r_0, β, μ, and r_n can be found from experiment by making small changes in e_{CE} and i_B away from a nominal point and recording the corresponding changes in i_C and μ e_{BE}.

The bond-graph representation of the incremental model of the transistor shown in Fig. 10-10a is similar to the nonlinear representation of Fig. 10-6, except that the constitutive laws for the incremental variables can be summed up in a single matrix, as in Eq. (10-30), rather than the nonlinear functions of Eqs. (10-24) and (10-25). It is also sometimes useful to construct circuit-equivalent models as shown in Fig. 10-10b and c, which have the same equations but use two controlled sources of the sort shown in Fig. 10-2. In this example, a voltage-controlled voltage source and a current-controlled current source are needed.

An even simpler model is shown in Fig. 10-11, where we neglect both μ and r_n. Then we see that the transistor is almost a current-controlled current source all by itself. In fact, from the curves of Fig. 10-8 we see that there are large regions in

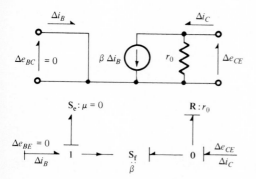

Figure 10-11 Special case of transistor incremental model when $\mu = r_n = 0$.

which the slope $\partial i_C / \partial e_{CE}$ is small, so that $1/r_0$ is small or r_0 is large. This would mean that the R representing the r_0 coefficient could also be removed in these regions. Then our crudest model of the transistor for incremental variables is nothing more than a current-controlled current source. Although this model is not very accurate, it does give insight into the practical workings of transistors in one useful connection pattern.

The simple current amplifier shown in Fig. 10-12 may give some insight into the usefulness of incremental models. In Fig. 10-12a when the input current i_{in} is zero, there will be constant values for $\bar{i}_B$, $\bar{e}_{BE}$, $\bar{i}_C$, $\bar{i}_{CE}$ which will be somewhere on the non-linear transistor curves. Fig. 10-12b shows an incremental model of the simplest type with r_n and μ both neglected. The two voltage sources are not present because they are constant and thus do not contribute to changes in voltage as currents change. The load resistance remains since a change in i_C will cause a change in e_{CE}, according to the resistance law, with slope R_L. The input resistor, which can be used to set $\bar{i}_B$, also remains but in fact is shorted out because we assumed $\mu = 0$. Finally, we have the current source with gain β and r_0 from the transistor model.

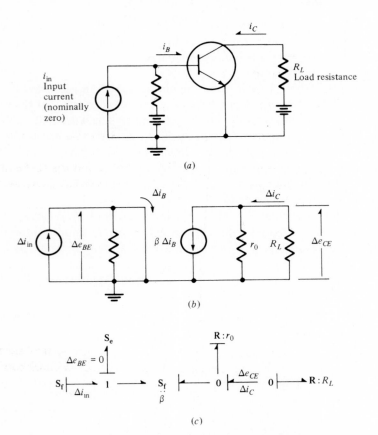

Figure 10-12 A simple transistor amplifier: (a) circuit diagram, (b) incremental circuit model, (c) bond-graph incremental model.

The bond graph of Fig. 10-12c shows how the device works. The input current source acts on a zero voltage source so that the input current can be varied with (almost) no power. This current drives the transistor current source through an active bond, and the load resistance is just in parallel with the resistance r_0 from the transistor. If $r_0 \gg R_L$, most of the current will flow through R_L and we can control large changes in load current with a small input current. Furthermore, not only is Δe_{BE} almost zero but e_{BE} is also small, so that the true power $\Delta i_{in} e_{BE}$ is small. Thus, there is both a current gain β and a power gain in the device.

10-5 CONCLUSION

This brief introduction to electronic systems should leave you with two impressions: (1) Electronic devices are physical systems that can be treated like any other physical systems on an energy and power basis. Many basic electronic devices are nothing more than multiport R elements with significantly nonlinear constitutive laws. The interesting laws allow switching and amplification. (2) Electronic systems are conveniently modeled using the concept of controlled sources. Sometimes a complex system can be designed to act in a rather simple way from input to output, as reflected in a controlled-source model. At other times controlled sources appear because an incremental model is being studied and the efforts and flows are not really power variables but deviations in power variables. Figure 10-12 illustrates how two sources of power (the effort sources) disappear in the incremental model, and the active bond makes it appear as if power came from nowhere to drive the load.

The incremental circuit models are not actual circuits but equivalent circuits, and the incremental-variable bond graphs are really pseudo bond graphs because incremental variables do not measure total electric power. The contribution these bond graphs make toward understanding is considerable, however, even though they do not indicate power relationships completely.

PROBLEMS

10-1 Suppose that the inverting amplifier of Fig. 10-1b has the transfer function

$$\frac{e_3}{e_1} = \frac{-G}{\tau s + 1}$$

rather than just the negative gain of the text, where τ is a (short) time constant. Write a state equation for the amplifier. At what frequency would the simple gain model begin to break down?

10-2 Equation (10-7) for the integrator system was based on an amplifier gain of $-G$. Imagine this gain to be replaced with $-G/(\tau s + 1)$ as a transfer function. Let $E_1(t)$ be zero in Eq. (10-7); rewrite the equation in the Laplace domain as

$$sq_5 = \frac{-1}{1 + G} \frac{q^5}{RC}$$

Substituting the transfer function for G, derive a second-order equation for the free response of the system. Is the system still stable?

10-3 Insert a piece of block diagram into the controlled-source bond graph of Fig. 10-1c to include the transfer function of Prob. 10-1.

10-4 An easy way to compute the behavior of a feedback operational amplifier is just to assume that e_0 is finite and $e = (-1/G)e_0$ is almost zero. Thus the input to the amplifier must be almost at ground and beyond that i, the input current, is also almost zero compared with other currents in the system. Using these ideas, demonstrate that for the system shown

$$e_0 = \frac{-R_1}{R_2}$$

(This works only for inverting amplifiers, since they are stable under feedback of this type.)

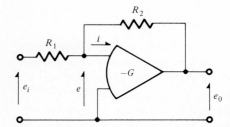

Figure P10-4

10-5 Find e_0 in terms of e_1 and e_2.

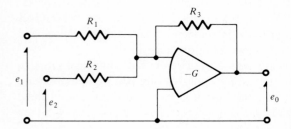

Figure P10-5

10-6 Consider the four-terminal nodic electronic device illustrated and show that it can be represented as a 3-port R connected to a junction structure.

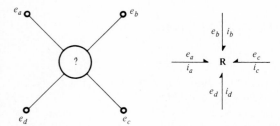

Figure P10-6

10-7 What form would the constitutive laws of the 2-port of Fig. 10-7 take in general in the causality shown?

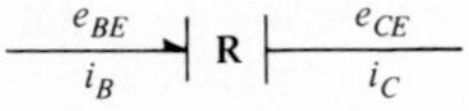

$$\underset{i_B}{\overset{e_{BE}}{\longrightarrow}} \boxed{R} \underset{i_C}{\overset{e_{CE}}{\longmapsto}}$$

Figure P10-7

10-8 The claim is that there is an analogy between the transistor and the gate valve. What variables are analogous?

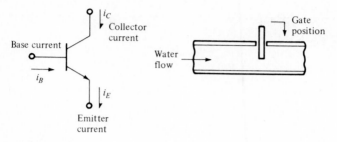

Figure P10-8

10-9 Suppose R is a resistance heater to be controlled by the transistor of Fig. 10-7 by switching the control current i_B between 10 and 700 mA. The manufacturer states that the transistor can dissipate 20 W with the cooling system you plan to use. Estimate the largest V and the smallest R you can control with this transistor. How much heat power is flowing to R in the "on" and the "off" states?

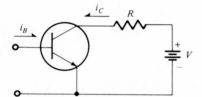

Figure P10-9

10-10 In Fig. 10-12 β is the current gain between Δi_C and Δi_{in}; suppose we were really interested in the gain between the voltage Δe_{CE} and Δi_{in}. This voltage would appear across R_L. What would this gain be?

ELEVEN

THERMAL SYSTEMS

11-1 INTRODUCTION

In previous chapters we modeled energetic systems as if they were isothermal, i.e., as if energy dissipated caused no changes in temperature and hence changes in constitutive laws. Here we show how to model systems in which heat-energy and temperature effects are important. We shall be dealing with the science of thermodynamics (more specifically, irreversible thermodynamics, since we shall consider changes of state which occur at finite rates) and with the applied field of heat transfer. Since these fields are wide-ranging and of fundamental importance in virtually all physical systems, we shall consider how the modeling process for thermal effects fits into our general scheme.

Two features of thermal systems set them apart from other types: (1) There is no need for an I element. One cannot make a lumped-parameter thermal oscillator as one can in any system containing both I and C elements, and, considered as distributed systems, thermal systems are diffusive rather than wavelike. (2) Although there are effort and flow variables that fit our general power and energy pattern, engineering practice has firmly established the use of other effort and flow variables that do not quite fit the general pattern. This gives us an opportunity to explore the useful concept of a *pseudo bond graph,* which retains most of the advantages of true bond graphs without some of their energy and power interpretations. Pseudo bond graphs have proved particularly useful for modeling open systems in which several types of quantities pass into and out of control volumes. We discuss briefly how true bond graphs are established for thermal systems and then show how heat-transfer systems can be modeled using pseudo bond graphs.

11-2 TRUE BOND GRAPHS FOR THERMAL SYSTEMS

In Fig. 11-1 an amount of a pure substance is confined in a chamber of volume V. The substance has entropy S and is at pressure P and thermodynamic temperature T. The substance can exchange thermal energy with its environment if it is heated or cooled, and it can exchange mechanical energy as it changes volume. Let U stand for the internal energy of the substance.

For the extensive quantities U, S, and V, it is convenient to work with specific quantities u, s, and v on a per unit mass basis. For mass m, then, $U = mu$, $S = ms$, $V = mv$. If you remember your thermodynamics, you should recall that in the absence of motion and electromagnetic and surface-tension forces a pure substance has only two independent properties and *equations of state* relate other properties to any two independent ones. From our point of view, the most fundamental properties are specific entropy s and specific volume v, and a basic equation of state relates the specific internal energy u to s and v

$$u = u(s,v) \tag{11-1}$$

Then the well-known Gibbs equation relates changes in u to changes in s and v

$$du = T\,ds - P\,dv \tag{11-2}$$

This energy expression can be changed into a power expression if we imagine that the changes du, ds, and dv occur in time dt (we must now assume that changes occur slowly enough to ensure that the confined substance has a substantially uniform pressure and temperature throughout). Then

$$\frac{du}{dt} = T\frac{ds}{dt} - P\frac{dv}{dt} \tag{11-3}$$

on a unit mass basis or

$$\frac{dU}{dt} = T\frac{dS}{dt} - P\frac{dV}{dt} \tag{11-4}$$

for the mass m in Fig. 11-1.

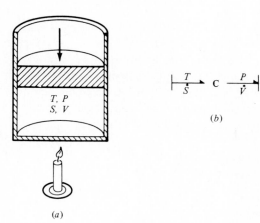

$$\vdash \frac{T}{\dot{S}} \rightarrow \ \mathrm{C} \ \frac{P}{\dot{V}} \dashv$$

(b)

(a)

Figure 11-1 True-bond-graph representation of a fixed quantity of a pure substance.

Table 11-1 Thermal bond-graph variables

Variable	General notation	Notation	SI units
Effort	$e(t)$	T, thermodynamic temperature	K
Flow	$f(t)$	$\dot{S}$, entropy flow rate	$J/K \cdot s$
Momentum or impulse	Not needed		
Displacement	$q(t) = \int^t f \, dt$	S, entropy	J/K
Power	$P = e(t)f(t)$	$P = T(t)\dot{S}(t)$	W
Energy	$E(q) = \int e \, dq$	$\int T \, dS$	J

The product $P \, dV/dt$ reminds us of pressure times volume flow rate, the effort-flow product used previously in hydraulic and acoustic systems. Similarly, we can regard temperature T and entropy flow rate $\dot{S}$ as effort and flow.

Table 11-1 shows e, f, and q variables for true thermal bond graphs based on the Gibbs equations. This table is analogous to earlier ones except that no momentum (p) variable is necessary and only one form of energy (the internal energy) exists. Let us see now what form of element represents the pure substance.

Considering Eqs. (11-1) and (11-2), we note that

$$du = \frac{\partial u}{\partial s} \, ds + \frac{\partial u}{\partial v} \, dv \tag{11-5}$$

so that

$$T = T(s,v) = \frac{\partial u}{\partial s} \tag{11-6}$$

and

$$P = P(s,v) = -\frac{\partial u}{\partial v} \tag{11-7}$$

This means that the two effort variables T and P are functions of the two displacement variables s and v (or S and V since $s = S/m$ and $v = V/m$). Thus the pure element is described by two C-element relations. Now, however, two e's are related to two q's, so that we have a 2-port version of a C element shown in Fig. 11-1. Such multiport elements are discussed in detail later, but for now we need only note that T and $\dot{S}$ form an effort-flow pair on a thermal bond and that state equations such as those in Eqs. (11-6) and (11-7) are just the constitutive relations for a 2-port C element.

Finally, it is useful to note that because both $T(s,v)$ and $P(s,v)$ can be derived from an energy expression $u(s,v)$, there is a reciprocal relation

$$\frac{\partial T}{\partial v} = \frac{\partial^2 u}{\partial v \, \partial s} = \frac{\partial(-P)}{\partial s} \tag{11-8}$$

which must hold if the conservation of energy implied by the Gibbs equation is to be maintained. As we shall see later, this type of reciprocity based on energy principles is very common.

Although bond graphs based on thermodynamic temperature and entropy flow rate are theoretically satisfying, in practice it is far more common to use other temperature scales and heat-energy flow rates. The resulting models are no longer analogous to other models we have studied *on an energy basis,* although one can operate on the resulting pseudo bond graphs in the standard way. The main disadvantage to pseudo bond graphs comes when they must be coupled to true bond graphs for a system model, since special elements have to be invented for the coupling.

11-3 PSEUDO BOND GRAPHS FOR HEAT TRANSFER

Since it is given to few of us to feel comfortable with the concept of entropy, it is common to substitute heat (energy) flow $\dot{Q}$ for entropy flow rate $\dot{S}$ as a flow variable. Also, it does not seem sensible to use a thermodynamic temperature scale when dealing with ordinary temperatures, so the Celsius or Fahrenheit scale is often used. Although temperature and heat flow rate are commonly presented as analogous to voltage and current in electrical systems, Table 11-2 shows that the analogy is more mathematical than physical.

Table 11-2 Thermal pseudo-bond-graph variables

	Notation		
Variable	General	Thermal	SI units
Effort	$e(t)$	T, temperature†	K‡
Flow	$f(t)$	$\dot{Q}$, heat flow rate	J/s
Momentum or impulse	Not needed		
Displacement	$q(t) = \displaystyle\int^t f\,dt$	Q, heat energy	J
Power	$P = e(t)f(t)$	$\dot{Q}$, no correspondence	W
Energy	$E(q) = \displaystyle\int e\,dq$	Q, no correspondence	J

† Common symbols for temperature are t and θ, but since t conflicts with our symbol for time and θ for our symbol for angle, we use T for temperature without implying that thermodynamic temperature is meant.

‡ For temperature differences 1 Celsius degree is the same as 1 kelvin, and K is used. Temperatures on the Celsius scale are found by subtracting 273.15 from temperature on the kelvin (absolute) scale.

The product of effort and flow $T\dot{Q}$ has no particular significance since $\dot{Q}$ alone is power flow. Further, since the displacement Q is energy, the normal expressions for energy are not physically meaningful either. On the other hand, useful bond-graph models can be made using the variables in Table 11-2, and the resulting equations do indeed resemble equations from electric RC circuits.

In Fig. 11-2 basic elements for heat-conduction system models are shown. In Fig. 11-2a a relatively thin slab of material allows the transmission of heat $\dot{Q}_2$ in response to a temperature difference

$$T_2 = T_1 - T_3 \tag{11-9}$$

The 1-junction implies not only Eq. (11-9) but also the fact that no heat energy is considered to accumulate in the slab, so that the heat flow in one side $\dot{Q}_1$ is matched by the flow out the other side $\dot{Q}_3$

$$\dot{Q}_1 = \dot{Q}_2 = \dot{Q}_3 \tag{11-10}$$

The *thermal resistance* R_2 relates T_2 to $\dot{Q}_2$.

When the temperature difference T_2 is not large, a linear resistance law is often accurate enough. Two common forms of the constitutive law involve a *heat-transfer coefficient h* or a *thermal conductivity k*

$$\dot{Q}_2 = hAT_2 = hA(T_1 - T_3) \tag{11-11}$$

or

$$\dot{Q}_2 = \frac{kA}{\Delta l}T_2 = \frac{kA}{\Delta l}(T_1 - T_3) \tag{11-12}$$

where A is the surface area and $\dot{Q}_2$ is a power quantity, so the heat-transfer coefficient h has SE units of $W/m^2 \cdot K$. Often h refers not to a slab of material but to an interface, e.g., the surface of a solid in contact with a gas or liquid.

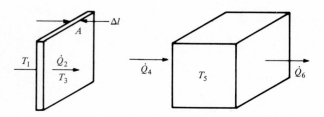

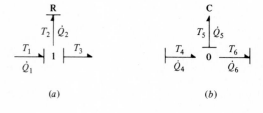

(a) (b)

Figure 11-2 Pseudo-bond-graph elements: (a) thermal resistance and (b) thermal capacitance.

The thermal resistance of a slab of thickness Δl and area A is sometimes given in terms of the conductivity $k\text{W}/\text{m}\cdot\text{K}$ in Eq. (11-12). If we define a thermal resistance R by

$$R\dot{Q}_2 = T_2 \tag{11-13}$$

two versions of R corresponding to Eqs. (11-11) and (11-12) are

$$R = \frac{1}{hA} \quad \text{K/W} \tag{11-14}$$

and
$$R = \frac{\Delta l}{kA} \quad \text{K/W} \tag{11-15}$$

If the same flow of heat passes through a slab and an interface, the resistances of Eqs. (11-14) and (11-15) can be added to get an overall resistance coefficient.

The thermal capacitance of Fig. 11-2b represents an amount of material at a common temperature T_5, which is determined by how much energy has been stored in it. In the bond graph we have shown only two heat flows impinging on the 0-junction, so that

$$\dot{Q}_5 = \dot{Q}_4 - \dot{Q}_6 \tag{11-16}$$

represents the rate of energy storage; but in other cases more bonds could be attached to the 0-junction to model other heat flows. The C element which relates T_5 to the stored energy Q_5 can be related to the true bond-graph element in Fig. 11-1. If we neglect any energy associated with change in volume (either because the volume really cannot change or because the $P\ dV$ work is small compared with the heat energy involved), we shall find that the energy varies with the entropy from Eq. (11-1). Further, the temperature varies with the entropy also, according to Eq. (11-6) — but this would mean that in principle we could eliminate the entropy from the two relations and express temperature directly in terms of energy.

Often heat capacity is defined directly in terms of *specific heat c*

$$c = \frac{\partial u}{\partial T} \tag{11-17}$$

Strictly speaking, this specific heat is for constant volume, but for solids and liquids the expansion work done is so small that the specific heats at constant volume and constant pressure are almost identical. In using Eq. (11-17) for a gas keep in mind that this implies that the gas really is constrained to constant volume.

Although c is not strictly constant but depends upon the temperature, for modest changes in temperature we can replace Eq. (11-17) with a version involving finite changes in u and T

$$\Delta T = \frac{\Delta u}{c} \tag{11-18a}$$

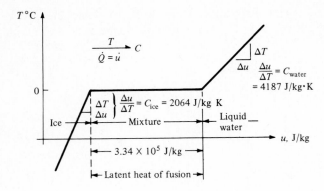

Figure 11-3 Thermal-capacity constitutive law for 1 kg of water.

Then, for an amount of material of mass m the change in internal energy ΔU would be $m\,\Delta u$, or

$$\Delta T = \frac{m\,\Delta u}{mc} = \frac{\Delta U}{mc} \tag{11-18b}$$

In Fig. 11-2b we identify the integral of $\dot{Q}_5$ with a change in U and a change in T_5 with ΔT. Thus

$$\Delta T = T_5 - T_{50} - \frac{1}{mc}\int_{t_0}^{t}\dot{Q}_5\,dt \tag{11-18c}$$

$$= \frac{1}{mc}\left[Q_5(t) - Q_5(t_0)\right]$$

When we compare this relationship with the general expression for a linear C element

$$\Delta e = \frac{\Delta q}{C} \tag{11-19}$$

it is clear that the thermal capacitance C is nothing but

$$C = mc \tag{11-20}$$

which shows that the specific heat is perhaps better referred to as thermal capacitance per unit mass.

When a substance undergoes a phase change, the thermal-capacitor constitutive relationship is dramatically nonlinear. Figure 11-3 shows how temperature is related to energy for a unit mass of water as it passes from solid ice to liquid water. At 0°C, the ice begins to melt and the temperature remains constant until an energy of about 334 kJ/kg has been added. At this point, all the ice has melted and the water warms up with a specific heat different from that for the solid ice. The energy required to complete the phase change is often called the *latent heat of fusion,* and at 100°C there is another flat spot in the curve, where the *latent heat of vaporization* must be added to the water to convert the liquid water into steam. Note that, as is typical for

C elements, the zero for the q variable (internal energy, in this case) can be chosen quite arbitrarily. If the water were initially at $-10\,^\circ$C, we might consider u to be zero at this point. Then the curve shown in Fig. 11-3 would pass through the point $T = -10\,^\circ$C, $u = 0$ J/kg, and $\dot{Q}$ when integrated would give instantaneous u values.

11-4 EXAMPLE SYSTEMS

Example 11-1 Figure 11-4 shows a basic thermal system consisting of a mass of material which we assume has a nearly uniform temperature T and thermal capacity C. It is surrounded by insulation with thermal resistance R. We neglect any capacitance in the insulation, but in R we include the resistance associated with the two interfaces between the insulation and the block of material and the insulation and the surrounding atmosphere at temperature T_0. We assume that the atmospheric temperature remains at temperature T_0 even if heat flows to and from the atmosphere. A constant-effort source is used for the atmosphere.

An electric-resistance heater is embedded in the block. From an electrical point of view, the power dissipated is $e(t)i(t)$, and from a thermal point of view this power becomes the source flow $\dot{Q}_s$. We use a flow source for the heater and show an active bond, indicating that $\dot{Q}_s$ is instantaneously equal to the electric power ei.

The pseudo bond graph is assembled in Fig. 11-4 using the elements shown in Fig. 11-2. The system should remind you of an electric RC circuit. A minor complication is the choice of zero point for the temperature. Since the Kelvin, Celsius, and Fahrenheit scales all pick different points for zero, it is not so obvious that we can pick our own point for zero. Otherwise, we would probably pick T_0 to be our zero, just as we pick a certain voltage point to be the ground node at zero voltage.

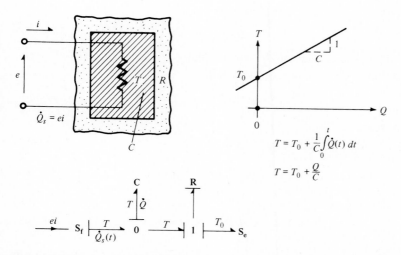

$$T = T_0 + \frac{1}{C}\int_0^t \dot{Q}(t)\,dt$$

$$T = T_0 + \frac{Q}{C}$$

Figure 11-4 System for Example 11-1.

One way to handle this problem is to let T_0 be defined on any scale and to use T_0 in the definition of the capacitor law. If Q is the energy delivered to the thermal capacitor, we might define $Q = 0$ to be the state when the temperature is T_0. Then, instead of our common form $e = Q/C$ or $T = Q/C$, we have

$$T = T_0 + \frac{Q}{C} \qquad (11\text{-}21)$$

as shown in Fig. 11-4.

The system is so simple that there is no need to number the bonds in order to follow the causal marks and to write the single state equation

$$\dot{Q} = \dot{Q}_s(t) - \frac{1}{R}(T - T_0) = \dot{Q}_s(t) - \frac{1}{R}\left(T_0 + \frac{Q}{C} - T_0\right) \qquad (11\text{-}22)$$

$$\dot{Q} = \frac{1}{RC}Q + \dot{Q}_s(t)$$

in which Eq. (11-21) was used. It now appears that by use of the constitutive law of Eq. (11-21) the atmospheric temperature has disappeared altogether from the state equation; but it reappears if we want to express the temperature in terms of the state variable Q. In fact, Eq. (11-21) is the output equation for T.

Since Eq. (11-22) is a linear first-order equation, it is easy to sketch the response of the system to simple inputs. Suppose, for instance, at $t = 0$,

$$T = T_0 \qquad Q = 0 \qquad (11\text{-}23)$$

and for $t > 0$, a constant input power is supplied

$$ei = \dot{Q}_s = \text{constant} \qquad t > 0 \qquad (11\text{-}24)$$

We can see that heat energy will flow into the material until the right-hand side of Eq. (11-22) vanishes. This will happen when

$$Q_{ss} = RC\dot{Q}_s \qquad (11\text{-}25)$$

This yields the steady-state value of the stored energy Q. Using Eqs. (11-21) and (11-25), we find the steady-state temperature to be

$$T_{ss} = T_0 + R\dot{Q}_s \qquad (11\text{-}26)$$

Thus the time histories of Q and T are

$$Q(t) = Q_{ss}(1 - e^{-t/\tau}) \qquad t \geq 0 \qquad (11\text{-}27)$$

and

$$T(t) = T_0 + (T_{ss} - T_0)(1 - e^{-t/\tau}) \qquad t \geq 0 \qquad (11\text{-}28)$$

where the time constant τ is

$$\tau = RC \qquad (11\text{-}29)$$

The time responses are sketched in Fig. 11-5.

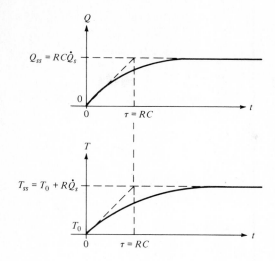

Figure 11-5 Time responses for Example 11-1.

With this formulation, T_0 could be the ambient temperature on any scale (20 °C, for example). A somewhat cleaner formulation results if we consider T to be the temperature deviation from the ambient. This is equivalent to setting $T_0 = 0$ and removing the effort source in Fig. 11-4. You can see the simplification in Eq. (11-21) or by just setting $T_0 = 0$ in the other equations.

Example 11-2 In Fig. 11-6 a typical example of the use of a lumped-parameter model to describe the dynamics of a distributed-parameter heat-transfer system is shown. The cooling fin is designed to conduct heat away from a hot fluid at temperature T_a along the fin and ultimately to the surrounding air at temperature T_f. The temperature in the metallic fin really varies continuously both in time and in distance along the fin. For the model, however, the fin is imagined to be broken up into four lumps with temperatures T_b, T_c, T_d, and T_e. The arrows show heat flow from the hot fluid to the first lump, flows between lumps, and flows from the lumps to the atmosphere.

The bond graph can be constructed by establishing a 0-junction for each temperature. If the temperatures are constant or known functions of time, effort sources such as those for T_a and T_f can be appended. If a thermal capacitance is associated with some temperatures, C elements can be attached to the appropriate junctions. Any time the heat transfer has to do with temperature differences, an R element can be inserted on a 1-junction between 0-junctions.

If you add causal marks to this graph, you will find that the four C-elements can all be in integral causality and that all bonds will have their causality determined at the end of the assignment process. This means that writing state equations would be straightforward. You probably would want to number the bonds and use generalized variables for a graph of this complexity.

The choice of the zero for temperature is with us in this problem as it was in the

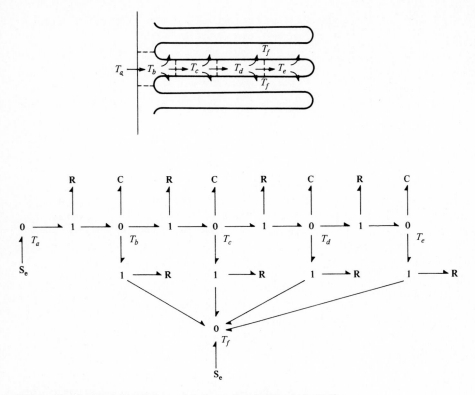

Figure 11-6 Lumped-parameter model of a cooling fin (Example 11-2).

last. If you chose to measure all temperatures as so many degrees above or below T_f, T_f would be zero and all the bonds attached to the T_f 0-junction could be removed. (This is similar to choosing a ground node in electric circuits. We have to remember that T_a is now relative to the ambient temperature T_f.)

Figure 11-7 shows how the bond graph simplifies when we measure each temperature as a deviation from T_f. The reason for choosing T_f as the zero point on our temperature scale is clear from the bond graph of Fig. 11-6; five bonds have T_f as their effort, and if $T_f = 0$, all these bonds and the S_e element can be eliminated.

Figure 11-7 Simplified bond graph for a cooling fin using a temperature scale in which $T_f = 0$.

The equations for the simplified bond graph are quite easy to write. For example,

$$
\begin{aligned}
\dot{Q}_4 &= f_3 - f_5 - f_6 = f_2 - \frac{e_5}{R_5} - f_7 \\
&= \frac{e_2}{R_2} - \frac{e_4}{R_5} - \frac{e_7}{R_7} = \frac{E_1(t) - e_3}{R_2} - \frac{q_4}{C_4 R_5} - \frac{e_6 - e_8}{R_7} \\
&= \frac{E_1(t) - e_4}{R_2} - \frac{q_4}{C_4 R_5} - \frac{e_4 - e_9}{R_7} \\
&= \frac{E_1(t)}{R_2} - \frac{q_4}{R_2 C_4} - \frac{q_4}{R_5 C_4} - \frac{q_4}{R_7 C_4} + \frac{q_9}{R_7 C_9}
\end{aligned}
\tag{11-30}
$$

The remaining equations are written similarly. You probably agree that numbering the bonds and using generalized variables is helpful for complex systems. A more vexing question is how to break up the fin into lumps logically. A certain amount of experience and physical intuition is helpful here. Once you decide on the lumps, it is not hard to estimate the capacitances from specific heats and the mass of the lump using Eq. (11-20). If you know the conductivity of the material, it is not so hard to estimate R_7, R_{12}, and R_{17}, using Eq. (11-15) and making a guess about equivalent lengths and areas of material between the centers of neighboring lumps. The remaining resistances basically come from Eq. (11-14), but estimating the appropriate areas and heat-transfer coefficients may not be so easy. Ultimately, the model parameters may have to be tuned up to match some experimental data.

In conduction heat transfer it is not hard to set up lumped-parameter models, just as we have done here. However, deciding on the number and location of the lumps and estimating resistance and capacitance parameters requires skill and some experience.

PROBLEMS

11-1 Two blocks of metal at temperatures $T_1(t)$ and $T_2(t)$ are separated from the rest of the world by perfect insulation but can exchange heat energy Q through a thermal resistance. If the heat power is

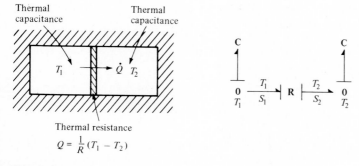

$$Q = \frac{1}{R}(T_1 - T_2)$$

Figure P11-1

$T_1 \dot{S}_1$ coming out of the first block and $T_2 \dot{S}_2$ going into the second, and if both powers are equal to $\dot{Q}$

$$T_1 \dot{S}_1 = T_2 \dot{S}_2 = \dot{Q} = \frac{1}{R}(T_1 - T_2)$$

write the consitutive laws for the 2-port thermal resistor in the causality shown, i.e.,

$$\dot{S}_1 = \dot{S}_1(T_1, T_2) \qquad \dot{S}_2 = \dot{S}_2(T)_1, T_2)$$

Also show that because heat flows from hot to cold, the system entropy rate $\dot{S}_2 - \dot{S}_1$ is positive no matter which way heat flows. Is this an illustration of the second law of thermodynamics for an isolated system?

11-2 A slab of hot steel is immersed in an oil bath to be quenched. Model the steel as one lump with temperature T and assume a single thermal resistance R. Make two pseudo bond graphs and write equations using either physical variables or generalized variables under the following conditions:

  (*a*) The oil bath has so much oil that $T_b \approx$ constant.
  (*b*) The oil bath is modeled as a thermal capacitance.
  Use the relation $t = T_0 + Q/C$ for the steel so that $T = T_0$ when $Q = 0$.

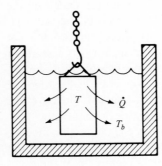

Figure P11-2

11-3 Model a unit length of the metal pipe with a plaster covering, using one lump for the pipe and one for the covering. Assume that the hot fluid remains at temperature T_f and the surrounding air remains at T_0. The thermal resistances R_1, R_2, and R_3 contain the resistance for the three interfaces plus appropriate contributions for the conductivity of the pipe and covering. Make a pseudo bond graph for the system and write state equations assuming linear capacitances and resistances. For capacitances, assume that zero energy corresponds to T_0; that is,

$$T_i = T_0 + \frac{Q_i}{C_i}$$

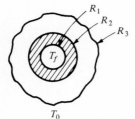

Figure P11-3

11-4 Suppose the system of Fig. 11-4 was designed to melt ice with the constitutive law sketched in Fig. 11-3. Assuming that T_0 is below the freezing point and T_{ss} is above it in the responses of Fig. 11-5, do the sketches over to show roughly how Q and T would vary with time.

11-5 The internal energy u of a perfect gas can be written

$$u(s,v) = c_v T_0 \left[e^{s/c_v} \left(\frac{v}{v_0} \right)^{-R/c_v} - 1 \right]$$

in the form of Eq. (11-1), where a reference state with subscripts 0 has temperature T_0, pressure p_0, specific volume v_0 and at this state $u = s = 0$. Using Eqs. (11-6) and (11-7), compute T and P and show that $P_v = RT$. [c_v, the specific heat at constant volume, is $c_v = (\partial u / \partial T)_v$.]

11-6 Consider the system shown in Fig. 11-1.

(a) Assuming that the piston does not move, make a simple model of the heating of the confined gas using a pseudo bond graph.

(b) Assuming that the piston can do work on a force F so that the volume changes, make a true-bond-graph model using the 2-port C of Fig. 11-1 and a 2-port R as in Prob. 11-1. Represent a piston of area A with a TF.

Note that the true bond graph can be coupled easily to a mechanical system using an $F\dot{x}$ port, while the pseudo bond graph needs some special elements.

11-7 Two pieces of metal exchange heat by radiation (Fig. P11-7a). The heat flow is assumed to be proportional to the difference of the fourth powers of their absolute temperatures. Show that the pseudo bond graph (Fig. P11-7b) will not work even when T_1 and T_2 are constrained to be absolute temperatures and that a special 2-port resistor must be defined in the manner required for true bond graphs as in Prob. 11-1.

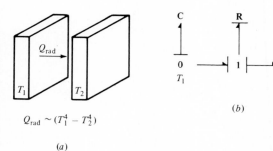

(b)

$Q_{rad} \sim (T_1^4 - T_2^4)$

(a)

Figure P11-7

TRANSDUCER SYSTEMS

12-1 INTRODUCTION

The word "transducer" has almost as many meanings as the word "system," but here we mean a device which couples subsystems in two different energy domains. (The Latin roots of transduce mean "lead across," and we mean that energy or information is led across energy-domain boundaries.) From an energy point of view, there are three distinct classes of transducers.

Signal Transducers

These are essentially zero-power-efficiency devices in which signal-level information is extracted from one system and signal- or power-level effects are produced in another. A strain gage, for example, may exert negligible forces on a mechanical structure, but with proper amplifying devices it can drive an electrical system in response to forces in the mechanical system. This important class of transducers will be discussed with other similar devices in Chap. 13.

Energy-Storing Transducers

Electrostatic loudspeakers and electric relays and solenoids are examples of transducers which store two forms of energy. Since these systems are best studied using general multiport elements, we defer discussion of these types until Chap. 17.

Power Transducers

The transducers we study in this chapter are converters of power. Typically they are very efficient. They store energy or dissipate power only incidentally; in fact, designers strive to minimize energy storage and dissipation in designing them. The simplest models of such transducers are transformers and gyrators.

Transducers are valuable in engineering because they permit the creation of multi-energy-domain systems. The increasing trend to computer control of devices of all sorts has increased the need to understand the transducer sensors and actuators required to make our modern systems perform well. Fortunately, such multiple-domain systems can be modeled effectively using bond graphs.

12-2 TRANSDUCERS OF THE TRANSFORMER TYPE

Figure 12-1 shows schematic diagrams of two hydraulic cylinders or rams which interconvert hydrostatic and mechanical power. In Fig. 12-1a the motion of the piston causes the same volume flow rate out of one port and into the other, so that we can speak of a single Q, measured in cubic meters per second. Similarly, since the pressures on each side of the piston act on the same area A, the force on the rod is the difference in pressure times the area. For this reason, we can speak of a single pressure P, which is actually a pressure difference.

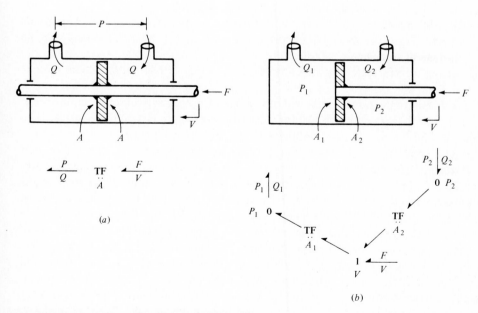

Figure 12-1 Two transformer models of hydraulic rams.

The constitutive laws of the ideal ram are

$$AP = F \tag{12-1}$$

$$Q = AV \tag{12-2}$$

and we see that power is instantaneously conserved

$$PQ = FV \tag{12-3}$$

Of course, for Eq. (12-3) to hold, the variables must be expressed in consistent units, such as those of the SI.

A more common hydraulic-cylinder configuration is shown in Fig. 12-1b. The piston-rod cross-sectional area is often a good fraction of the piston area, so that two distinct areas, A_1 and A_2, are relevant to the two ports. Now

$$Q_1 = A_1 V \tag{12-4}$$

$$Q_2 = A_2 V \tag{12-5}$$

and

$$F = A_1 P_1 - A_2 P_2 \tag{12-6}$$

again modeling the ideal case. If we multiply Eq. (12-6) by V and use Eqs. (12-4) and (12-5), the power-conservation equation

$$FV = P_1 Q_1 - P_2 Q_2 \tag{12-7}$$

results. The bond graph for this system could be constructed according to our rules:

1. Establish a 1-junction for the single mechanical velocity V.
2. Establish two 0-junctions for the two hydraulic pressures P_1 and P_2. Port bonds 1 and 2 and the mechanical port can be attached to these junctions and the proper sign conventions indicated.
3. Using transformers to realize the relations of Eqs. (12-4) and (12-5) or (12-6).

Using bond-graph transformers will always result in a power-conservation law; so if you start with flow laws, you will build in a force-pressure law, and conversely. The sign conventions must be compatible with the schematic diagram.

The transformer models in Fig. 12-1 capture the essence of hydraulic-piston transducers and may well be satisfactory for many design purposes, but they are incomplete. In Fig. 12-2 several extra elements have been added to the basic transformer model of Fig. 12-1b in order to model nonideal effects. Two power-dissipation effects are modeled with R elements. The R attached to the 1-junction models the friction of the piston seal and the rod seals. This may be important at low levels of pressure and force. The R between the two 0-junctions models leakage past the piston, which may be important at high pressure levels. The hydraulic-compliance and mechanical-inertia elements do store some energy and affect the system response at high frequencies of operation or under fast transient conditions. Since designers try to minimize mass, compliance, friction, and leakage effects, the idealized model of Fig. 12-1b may be accurate enough under normal conditions.

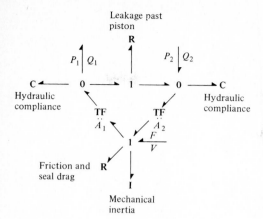

Figure 12-2 Nonideal effects added to the model of Fig. 12-1 b.

The two bond graphs of Figs. 12-1 b and 12-2 show an example of an important modeling concept. We first try to find a fundamental model which represents the essential physics of the situation. Then we consider various nonideal effects that may be important under certain conditions. Extra elements are then added to the basic model until we have the simplest model capable of answering our questions about the system. As you can imagine, many models of a ram can be constructed at various levels of sophistication. Only in the context of the particular system can one decide which model is best, but simplicity is always an important virtue in a model. Only enough complexity to model important effects should be tolerated.

The hydraulic cylinder is an important transducer, both in its own right and as the basic building block for other devices, e.g., the positive-displacement pumps and motors shown in Fig. 12-3. These devices contain several pistons and various means of converting the rotary motion of a shaft into reciprocating motions of the pistons. Porting arrangements allow the several pistons (typically seven or nine) to take in a fairly steady flow of fluid at one port and to discharge it at another pressure at the other port when the shaft turns at constant speed. Because the pistons and the mechanical linkage in the pump can all be described by transformer structures, so can the pump itself.

The simplest model of such a pump or motor (Fig. 12-3 b) can be developed as follows. Suppose we turn the pump shaft through an angle θ and measure the volume V of hydraulic fluid that flows through the machine. The volume is basically proportional to θ, say

$$V = T\theta \qquad (12\text{-}8)$$

Differentiating this expression gives

$$Q = \frac{dV}{dt} = T\frac{d\theta}{dt} = T\omega \qquad (12\text{-}9)$$

This is half of a transformer constitutive law, and the other half must be

$$TP = \tau \qquad (12\text{-}10)$$

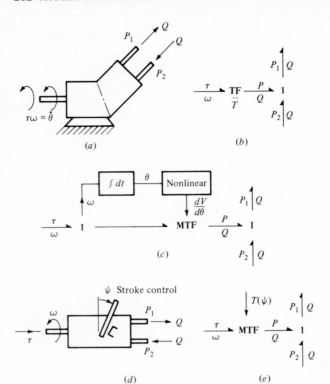

Figure 12-3 Several models for hydraulic pumps.

for the pressure-torque relation; otherwise, power conservation

$$PQ = \tau\omega \tag{12-11}$$

would not hold. The transformer modulus T has the units $N \cdot m/Pa = N \cdot m/(N/m^2)$ or m^3/rad. The simplest transformer model is shown in Fig. 12-3b.

In fact, the volume discharge assumed in Eq. (12-8) is not exactly correct. There is a little ripple in the flow because each piston's contribution to the output flow varies with θ and the summation of the flows from several pistons evens out the total flow only partially. If we replace Eq. (12-8) with

$$V(\theta) = T\theta + a \sin n\theta \tag{12-12}$$

where a is usually quite small and n is the number of pistons, then

$$Q = \frac{dV(\theta)}{d\theta} \frac{d\theta}{dt}$$

or

$$Q = (T + an \cos n\theta)\,\omega \tag{12-13}$$

Then we must also have

$$(T + an \cos n\theta)\,P = \tau \tag{12-14}$$

indicating that the ripple effect shows up in the torque relation as well. This model can be represented by a displacement-modulated transformer much like those used in generalized mechanics (see Fig. 12-3c). (This should come as no surprise; the internal linkage in the pump between the shaft and the pistons could be represented by an MTF.)

Quite another use of an MTF is for a pump or motor which contains a stroke control. Such a control varies the internal linkage so that T in Eq. (12-8) can be varied by the position of some mechanical element. In the schematic diagram of Fig. 12-3d we have shown a lever angle ψ as the stroke control. Since the stroke-control effort has little to do with the power variables τ, ω, P, or Q, we idealize the effect by having the value of T in Eq. (12-8) be a function of ψ. Because ψ is communicated to the pump model by an active bond, ψ can be changed without affecting the basic power conservation $PQ = \tau\omega$ although the constitutive laws do change

$$Q = T(\psi)\omega \tag{12-15}$$

$$T(\psi)P = \tau \tag{12-16}$$

Of course, it really does take some force to move a stroke control, but this force has mainly to do with friction and we can make a model for this force and then communicate the instantaneous value of ψ from this model to the MTF shown in Fig. 12-3e on an active bond.

Any of the models shown in Fig. 12-3 can be supplemented by extra elements to account for nonideal behavior, as we did for the cylinder in Fig. 12-2.

For devices being used as pumps we are typically putting shaft power τ and ω in and adding it to the low-power, i.e., low-pressure, fluid stream. Hydraulic motors use the high-pressure fluid as input power and split it into shaft power, i.e., work, and low-pressure fluid power. The basic structure of the models is remarkably similar in many cases because the type of transducers we are studying are reversible to the direction of power flow.

12-3 TRANSDUCERS OF THE GYRATOR TYPE

The basis of a variety of electromechanical transducers is shown in Fig. 12-4. A length of conductor l moves with velocity V through a magnetic field with flux density B. A voltage e is induced in the conductor and a magnetic force F_m appears. We apply force $F = -F_m$ to the conductor so that FV is mechanical power supplied to the wire and ei is electric power, which flows to whatever completes the electric circuit.

An application of Faraday's law of magnetic induction yields the relation

$$e = BlV \tag{12-17}$$

while the Lorentz force law gives the relation

$$Bli = F \tag{12-18}$$

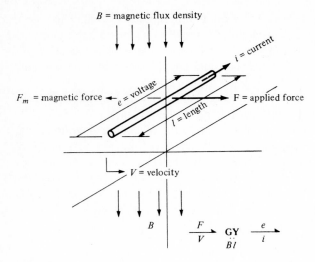

Figure 12-4 Gyrator model of a current-carrying conductor moving in a magnetic field.

These two laws are the constitutive laws of the gyrator shown in Fig. 12-4, and together they yield a power law

$$ei = FV \tag{12-19}$$

which indicates that this is an idealized power-conversion device.

The parameters B and l and variables i, V, F, and e can be thought of as components of vectors. In the sketch we have shown the vectors at right angles to each other so that the scalar relations can clearly be seen. When the vectors are not at right angles, vector versions of the laws can be given, such as

$$\mathbf{F}_m = i\mathbf{l} \times \mathbf{B}$$

but for our purposes it is easier to resolve these relations into components. Thus we might consider B to be that part of the $\mathbf{B}$ vector perpendicular to the vector length $\mathbf{l}$ in Eq. (12-18). Furthermore, in many devices the conductors will be curved to some extent and B will not be uniform; therefore although a gyrator can represent the effect, the gyrator parameter will represent the integration of effects on differential lengths of conductors. In such cases, gyrator parameters can be determined experimentally.

In SI units magnetic flux is measured in webers (equivalent to volt-seconds) and the flux density is in teslas (equivalent to webers per square meter). The gyrator parameter Bl is measured in tesla-meters, equivalent to webers per meter or volt-seconds per meter (Appendix A). With some thought you can confirm that this gyrator parameter makes both Eqs. (12-17) and (12-18) have consistent units.

A model of a simple single-phase alternator is easy to construct based on the gyrator model of Fig. 12-4. A schematic diagram of such a device is shown in Fig. 12-5. Wires wrapped around a rotor or armature are carried through a gap of substantially constant flux density B. The speed of the conductor is

$$V = r\omega \tag{12-20}$$

where r is the radius of the armature and ω is the angular velocity. Since the velocity component perpendicular to the flux is $V \sin \theta$, it is this component which must be used in Eq. (12-17) for the voltage equation

$$e = Blr\omega \sin \theta = (Blr \sin \theta)\omega \qquad (12\text{-}21)$$

where l is the effective length of all conductors cutting the flux lines. Wires may be wound many times around the armature, and the total voltage generated is the sum of all the voltages generated by each wrap. Note that voltage induced on one side of the rotor adds to that on the other because although the direction of flux cutting is opposite on two ends of the rotor diameter, the direction of current flow is also opposite on the two ends. The electric circuit can be completed with slip rings as shown.

Equation (12-21) suggests a modulated gyrator model with a gyrator parameter $Blr \sin \theta$. Does the torque-current relationship implied by the gyrator in Fig. 12-5 make sense? Assuming the "through" sign convention, Eq. (12-21) implies a companion relation

$$(Blr \sin \theta)i = \tau \qquad (12\text{-}22)$$

since we insist that ei equal $\tau\omega$. This equation makes sense because, according to Eq. (12-18), Bli should be a force, perpendicular to **B**. Then the torque would be obtained by multiplying by the moment arm $r \sin \theta$. The result is basically Eq. (12-22). By using power conservation we avoid some sign-convention problems which arise if we derive Eqs. (12-21) and (12-22) separately.

The entire device is modeled by a special kind of modulated gyrator, as shown in Fig. 12-5. The modulation has to do with the displacement θ, which is the integral of the flow ω. Why this should be so is understandable if you think of the device as a combination of the basic gyrator of Fig. 12-4 with a displacement-modulated transformer which relates ω to the flux-cutting velocity component V_{cut}, as shown in Fig.

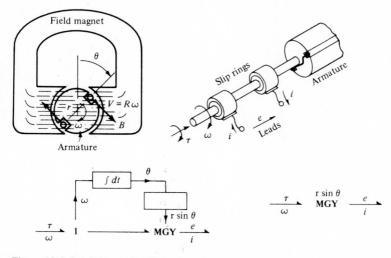

Figure 12-5 Primitive single-phase alternator.

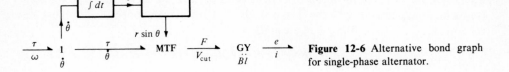

$$\frac{\tau}{\omega} \longrightarrow 1 \underset{\dot{\theta}}{\overset{\tau}{\vphantom{x}}} \underset{\dot{\theta}}{\overset{\tau}{\longrightarrow}} \text{MTF} \underset{V_{\text{cut}}}{\overset{F}{\longrightarrow}} \underset{Bl}{\overset{\text{GY}}{\vphantom{x}}} \overset{e}{\underset{i}{\longrightarrow}}$$

Figure 12-6 Alternative bond graph for single-phase alternator.

12-6. The MTF is due to the mechanical part of the system and is typical of the type of element so common in Chap. 9.

Real alternators have more complicated configurations than the one shown schematically in Fig. 12-5 and may have sets of windings to generate three-phase power, but the basic modulated-gyrator model idea still applies. Naturally, we can add elements to the basic model to represent electric resistance, mechanical friction, and windage and dynamic effects such as inertia and inductance.

A dc machine consists of several windings much like alternator windings but with a commutator that switches the windings into and out of the external circuit as the armature rotates. As shown schematically in Fig. 12-7, a coil is connected for only a small fraction of a revolution. During the time any coil is connected, the sin θ terms in Eqs. (12-21) and (12-22) are nearly unity. The equations therefore are approximately

$$e = Blr\omega \qquad (12\text{-}23)$$

and

$$Blri = \tau \qquad (12\text{-}24)$$

which can be represented, as in Fig. 12-7, by a simple constant gyrator with parameter

$$T = Blr \qquad (12\text{-}25)$$

There is some ripple in the curves for e versus ω and i versus τ due to the finite number of coils in a dc machine, as there is in a hydraulic motor with a finite number of pistons, but Eqs. (12-23) and (12-24) are accurate enough for many design purposes. You might compare them with Eqs. (12-9) and (12-10). It is perhaps better to use T than Br for the gyrator parameter, since we want an overall parameter

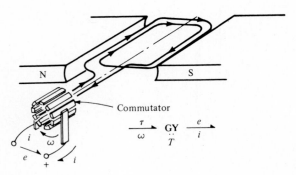

Figure 12-7 Primitive dc motor or generator.

which averages over all current paths, B values in the gap, and radii. The units of this T are $\mathrm{V/(rad/s)} = \mathrm{N \cdot m/A}$.

In a separately excited machine, B is not constant (as it would be with a permanent magnet) but varies as a function of the field current. The basic model in this case is a modulated gyrator with T a function of the field current i_f

$$e = T(i_f)\omega \tag{12-26}$$

$$T(i_f)i = \tau \tag{12-27}$$

The modulated gyrator appears in Fig. 12-8, but we have added circuit elements for the armature and field resistances R_a and R_f, the armature and field inductances L_a and L_f, and the mechanical rotary inertia and friction. The hybrid schematic diagram in Fig. 12-8 does a pretty good job on the circuit components but is less satisfactory in representing the essential transduction process and the mechanical part of the system. The bond graph is considerably more explicit.

The model of Fig. 12-8 is an excellent example of the use of an active bond. Although the signal i_f enters the MGY constitutive laws (12-26) and (12-27), no power is associated with the active bond. Although we still have power conservation in the form $ei = \tau\omega$ for the two normal bonds, there is no back effect from MGY to complement the i_f signal. This is because currents in the armature circuit induce only B-field components essentially normal to those generated by the field windings. Thus, what happens in the armature windings has almost no effect on the field circuit, at least to a good first approximation.

Although the ideal model of the motor or generator (which is a motor run in reverse) is perfectly efficient at transducing energy, real devices contain losses and extra dynamic effects. Thus in Fig. 12-8, since $e_a \neq e$, $\tau \neq \tau'$, and $e_f i_f \neq 0$, viewed from the three external ports, our transducer model will show nonideal behavior. You can almost get at the ideal MGY characteristics experimentally by doing two tests. In an open-circuit test, if you measure e_a when $i_a \approx 0$, then $e_a \approx e$ because the effects of L_a and R_a will be absent. Knowing ω and $e = e_a$, you could find $T(i_f)$.

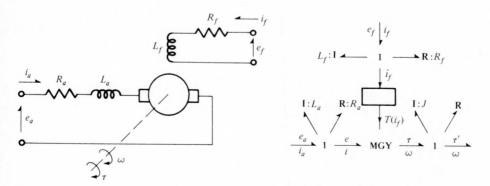

Figure 12-8 A separately excited dc machine modeled as a modulated gyrator with loss and dynamic elements added.

Also, by blocking the rotor so that $\omega = 0$, $\tau' \approx \tau$ because the inertia and most of the friction effects will vanish. Then you could measure τ and i to find $T(i_f)$. You should be suspicious if these two computations of T do not agree reasonably well.

12-4 CONCLUSIONS

In this chapter we have concentrated on efficient power-transducer models based on TF, GY, MTF, and MGY. Many quite different devices have some basic similarities when power relationships are considered. Furthermore, this chapter has been an object lesson on the modeling process. We have shown how idealized models that obey power conservation can be supplemented with elements to model nonideal behavior. We have also tried to dispel the notion that there is a *single* model of a device. There are many models, and the choice of model should always be made with a specific application in mind.

PROBLEMS

12-1 A mechanical damper is made from a cylinder connected to a valve as shown. The valve can be adjusted by turning the stem an angle θ. The pressure drop across the valve is $P = K(\theta)Q|Q|$. Using the bond graph of Fig. 12-1 a, compute the damper law $F = F(V)$.

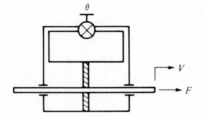

Figure P12-1

12-2 The accelerator consists of a large accumulator (pressure source) connected to a ram blocked initially by a quick-opening valve. Make two bond graphs for this system assuming (a) zero pressure on area A_1 after the valve opens and (b) some resistive loss for the lines from the accumulator to the ram port and from the ram port to the valve.

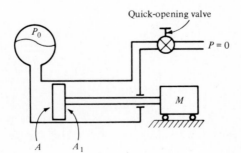

Figure P12-2

12-3 Although hydraulic systems are quite stiff, they are not infinitely so. The hydraulic servo is tested by blocking both ports of the ram and by trying to predict what the natural frequency of the piston and load mass M will be when they are oscillating against the two hydraulic capacitances formed by the volumes of the ram. At the position shown the left chamber has volume V_1 and the right V_2. The areas are A_1 and A_2, and the fluid has bulk modulus B. The fluid capacitance of a volume V is therefore V/B.

(a) Write state equations for the bond-graph model for p, q_1, and q_2 in terms of the physical parameters.

(b) Combine the equations into a single second-order equation for p and deduce the natural frequency.

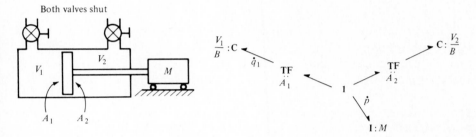

Figure P12-3

12-4 Modify the bond graph of Fig. 12-3b to include mechanical friction and leakage past the piston.

12-5 This bond graph has been developed from that of Fig. 12-3e. Explain the physical effects represented by I_1, R_2, R_3, and R_4.

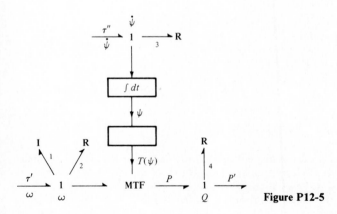

Figure P12-5

12-6 The variable-displacement pump is driven by a large engine, and so $\omega_0 \approx$ constant. The fixed-displacement motor accelerates the rotary inertia J and does work against the friction torque $B\omega$. Since the stroke-control angle ψ can be moved rapidly, a relief valve has been provided to prevent damage from excessive pressure. Its law can be expressed in the causal form $Q = F(P)$, or $-|$ R, but not in the other causality, since when $Q = 0$ many pressures are possible. Show that by including a capacitor for the line, the relief valve can be characterized in its preferred causality. Write state equations for your model.

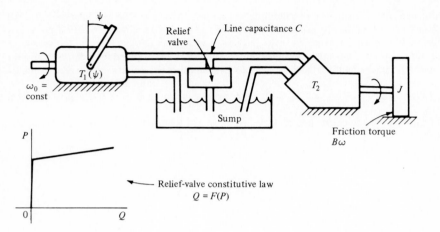

Figure P12-6

12-7 The voice-coil transducer relates the linear mechanical variables F and V to the electrical variables e and i. It is shown in cross section and consists of a magnet which produces a field of density B in an annular air gap. A lightweight cylinder, radius R, is wrapped with N turns of wire and inserted in the gap. Note that the velocity V, the direction of the current, and the direction of the field are almost at right angles with each other, as in Fig. 12-4. Electrodynamic loudspeakers use voice coils like this. Make a bond-graph representation of this device including the resistance of the wire as well as an I element which models self-inductance of the coil. Write the constitutive laws for the basic gyrator involved.

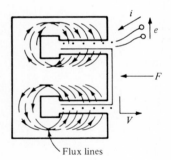

Figure P12-7

12-8 A voice-coil transducer of the type shown in Prob. 12-7 is used to move the spool in a hydraulic valve. The load seen by the voice coil is modeled by a mass M and a viscous damping force $F = BV$, where V is the velocity of the mass. The voice coil has electric resistance R and transduction coefficient T' (the gyrator modulus). At the frequencies of interest the inductance of the coil is not important. Make a bond graph of the system and write state equations in both physical and generalized notation.

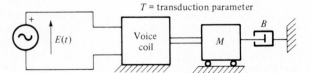

Figure P12-8

12-9 Write a bond graph and state equations for a permanent-magnet dc motor (transduction coefficient T) driven by a voltage source and driving a large rotary inertia of moment of inertia J. Include armature resistance R_a and inductance L_a but no mechanical friction.

12-10 Model the system shown, in which a dc motor is driven by a voltage source and in turn drives a positive-displacement pump pumping fluid into a gravity tank. Include armature resistance and inductance, inertia of the flywheel, rotary friction, and resistance in the long fluid line to the tank. Using generalized variables and bond numbers, write state equations assuming linear models everywhere.

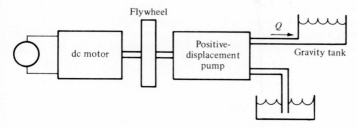

Figure P12-10

12-11 The separately excited dc motor drives a load modeled as a viscous torque and a moment of inertia. Make a bond-graph model of the system similar to that of Fig. 12-8. The transduction coefficient is assumed to be proportional to the field current, as indicated. E_a and E_f are imposed as voltage sources. Use the momentum variables λ_a, λ_f, and p_ω defined in Fig. P12-11 and write state equations for this system.

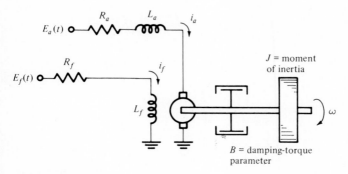

Figure P12-11

$$i_a = \frac{\lambda_a}{L_a} \qquad i_f = \frac{\lambda_f}{L_f} \qquad \omega = \frac{p_\omega}{J}$$

where λ_a = flux-linkage variable in armature
λ_f = flux-linkage variable in field
p_ω = angular momentum
and the transduction coefficient (gyrator parameter) is

$$T(i_f) \approx Ki_f = K\frac{\lambda_f}{L_f}$$

12-12 For the system of Prob. 12-11 show that:
  (a) The state equations are not linear in general.
  (b) If E_f = constant, the system *will* be linear after a time.
  (c) If E_a = constant, the system will not be linear.

12-13 Suppose the armature circuit of Prob. 12-11 is driven by a constant-current source, so that $i_a = I_a$. Make a bond graph eliminating R_a and L_a since the current is constant regardless of the voltages they produce. Show that by controlling the field through E_f (at generally low power) the torque of the motor can be controlled in a linear manner.

12-14 Consider the single-phase alternator of Fig. 12-5 with a drive at constant angular speed ω_0; θ is defined by $\theta = \omega_0 t$. Suppose an external resistance R_e is attached to the output leads so that the total resistance is $R_e + R_a$, where R_a is the armature resistance. Neglect any inductance. Compute the torque required of the drive neglecting any mechanical friction. Note that decreasing R_e increases power dissipation.

THIRTEEN

SIGNALS, GENERALIZED AMPLIFIERS, AND INSTRUMENTS

13-1 INTRODUCTION

Previous chapters have discussed specific applications of activated bonds, e.g., MTFs for general mechanics and MGYs for separately excited dc motors. In these cases the moduli varied according to signals which had neither back effects nor complementary power variables. We also discussed electronic devices modeled by controlled sources where the controlling signal was essentially an active bond on which an effort or flow variable appeared. The effect of its complementary power variable was neglected.

In this chapter we generalize our considerations. First we show how transfer functions, frequency-response functions, and general block diagrams can be converted into state-equation form to fit into our general-system description, a technique that is often useful when a part of a system is known from the results of experiments such as frequency-response tests. Next we discuss some devices which amplify signal levels or power levels and which often can be modeled efficiently using active bonds. Since many such devices are not usually called amplifiers, this discussion will expand your notion of amplification. Finally we discuss instruments and a type of transducer which has almost zero power efficiency. Again, active bonds and transfer functions are useful in this context.

13-2 STATE EQUATIONS FROM TRANSFER FUNCTIONS

Throughout this text we have regarded a set of state equations as the fundamental mathematical form for our system models. For linear models we have used Laplace transforms and have expressed input-output relationships as transfer functions or frequency-response functions, deriving them starting from state equations. Now we reverse the procedure.

Suppose we have an electronic amplifier or other device which is conveniently characterized in an input-output form. Suppose that the input is communicated by an active bond and that the output acts on an effort or flow source, so that back effects are unimportant at both the input and the output port. Assume further that we can approximate the transfer function of the device by a ratio of polynomials in the Laplace variable s and can approximate the frequency response by substituting $s = j\omega$ in the transfer function. It may well be that our amplifier is used in its linear range but the rest of the system contains nonlinearities and can conveniently be represented using state equations. How can the two descriptions be combined?

If the rest of the system were linear, we could convert from state equations to transfer functions and then combine the transfer functions. Generally, however, we must go back to state equations for the entire system, since the transfer-function concept does not apply to nonlinear systems. It is the latter approach that we take now.

Consider, for example, the third-order transfer function

$$\frac{Y(s)}{U(s)} = \frac{g}{s^3 + a_2s^2 + a_1s + a_0} \tag{13-1}$$

where g, a_0, a_1, and a_2 are constants. Writing this out as

$$s^3Y + a_2s^2Y + a_1sY + a_0Y = gU \tag{13-2}$$

we recognize the Laplace transform of

$$\dddot{y} + a_2\ddot{y} + a_1\dot{y} + a_0y = gu \tag{13-3}$$

where $Y(s)$ is the Laplace transform of $y(t)$, $U(s)$ is the transform of $u(t)$, and we assume *zero initial conditions*. It is easy to make a block diagram for Eq. (13-3) by solving for the highest derivative $\dddot{y}$

$$\dddot{y} = -a_2\ddot{y} - a_1\dot{y} - a_0y + gu \tag{13-4}$$

and using integrator blocks to find $\ddot{y}$, $\dot{y}$, and y. Then by feeding back through appropriate gains Eq. (13-4) can be satisfied (see Fig. 13-1).

The integrator blocks can be indicated in the time domain (Fig. 13-1 a) or in the frequency domain (Fig. 13-1 b). (Recall that integration corresponds to dividing by s and differentiation to multiplying by s, so that the two block diagrams have the same meaning.) You may see block diagrams using s in the blocks but what appear to be functions of time such as y or $\dot{y}$ as the signals. This hybrid but rather sloppy usage need not confuse those skilled in the art.

We want to make block diagrams for transfer functions or differential equations using only integrators (instead of differentiators) for two reasons: (1) We can make physical integrators quite easily, as we did in Chap. 10 out of a feedback operational amplifier, but we have severe problems with noise in the signals when we try to make

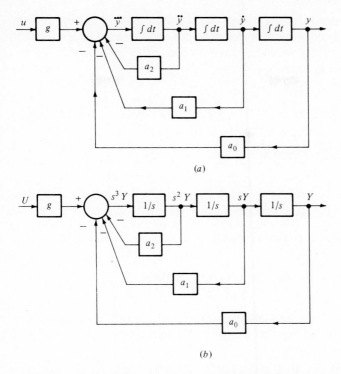

Figure 13-1 Block diagrams for the transfer function of Eq. (13-1): (*a*) time domain; (*b*) frequency domain.

a differentiator. Thus block diagrams like those in Fig. 13-1 can be simulated on analog computers or realized with special electronic devices using operational amplifiers. (2) State-space equations can easily be written from integrator block diagrams.

Figure 13-2 shows a different labeling of the integrator block diagram. The output of every integrator is always a state variable. In Fig. 13-2 we have named these state variables x_1, x_2, and x_3. The input to each integrator is just the derivative of

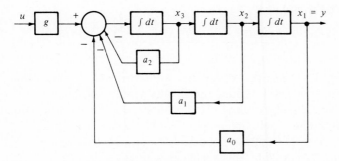

Figure 13-2 State variables for example transfer function.

the output (read the blocks in reverse). The state equations can be written directly from the block diagram by equating integrator inputs to functions of integrator outputs by reading the block diagram. Note that $\dot{x}_1$, which is the input to the x_1 integrator, is now also x_2, and so on; i.e.,

$$
\begin{aligned}
\dot{x}_1 &= x_2 \\
\dot{x}_2 &= x_3 \\
\dot{x}_3 &= -a_0 x_1 - a_1 x_2 - a_2 x_3 + gu
\end{aligned}
\tag{13-5a}
$$

which can be written in matrix form as

$$
\begin{bmatrix} \dot{x}_1 \\ \dot{x}_2 \\ \dot{x}_3 \end{bmatrix}
=
\begin{bmatrix} 0 & 1 & 0 \\ 0 & 0 & 1 \\ -a_0 & -a_1 & -a_3 \end{bmatrix}
\begin{bmatrix} x_1 \\ x_2 \\ x_3 \end{bmatrix}
+
\begin{bmatrix} 0 \\ 0 \\ g \end{bmatrix} u
\tag{13-5b}
$$

The original output y is just the same as x_1, and so $y = x_1$ or, in matrix form,

$$
y = \begin{bmatrix} 1 & 0 & 0 \end{bmatrix} \begin{bmatrix} x_1 \\ x_2 \\ x_3 \end{bmatrix} + [0][u]
\tag{13-6}
$$

Now Eqs. (13-5) and (13-6) can be combined with state equations for the remainder of the system, and the transfer function will be built into a state-equation model. If this seems too easy, it can get harder.

If instead of a simple gain g in the transfer function of Eq. (13-1) we have a more complex numerator, the technique must be extended. Let

$$
\frac{Y(s)}{U(s)} = \frac{b_3 s^3 + b_2 s^2 + b_1 s + b_0}{s^3 + a_2 s^2 + a_1 s^1 + a_0}
\tag{13-7}
$$

One way (among several) to proceed is to consider initially

$$
\frac{Y'(s)}{U(s)} = \frac{1}{s^3 + a_2 s^2 + a_1 s + a_0}
\tag{13-8}
$$

A diagram for this transfer function is just like those in Figs. 13-1 and 13-2 but with $g = 1$. Now, we can make the transfer function of Eq. (13-7) for $Y(s)$ out of the one for $Y'(s)$ in Eq. (13-8) by adding

$$
\frac{Y(s)}{U(s)} = \frac{b_0 Y'(s) + b_1 s Y'(s) + b_2 s^2 Y'(s) + b_3 s^3 Y'(s)}{U(s)}
\tag{13-9}
$$

The result is shown in Fig. 13-3. The part of the diagram from U to Y' is the same as in Fig. 13-1 except that no block is needed for g. The symbols for $\dot{y}'$, $\ddot{y}'$, and $\dddot{y}'$ are labeled ahead of successive integrators as sY', $s^2 Y'$, and $s^3 Y'$. Finally, using Eq. (13-9), we add up signals to find the desired output variable $Y(s)$.

As in Fig. 13-2, each integrator output is a state variable, labeled again x_1, x_2, x_3. We read the state equations directly from the block diagram as before

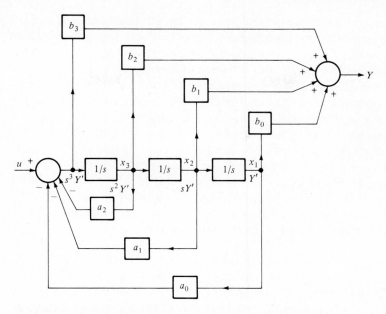

Figure 13-3 Integrator block diagram for example transfer function.

$$\dot{x}_1 = x_2$$
$$\dot{x}_2 = x_3$$
$$\dot{x}_3 = -a_0 x_1 - a_1 x_2 - a_2 x_3 + u \qquad (13\text{-}10\,a)$$

The output variable in terms of the state variables is

$$y = b_0 x_1 + b_1 x_2 + b_2 x_3 + b_3(-a_0 x_1 - a_1 x_2 - a_2 x_3 + u)$$
$$= (b_0 - a_0 b_3) x_1 + (b_1 - a_1 b_3) x_2 + (b_2 - a_2 b_3) x_3 + b_3 u \qquad (13\text{-}11\,a)$$

In matrix form, these equations take on a special pattern

$$\begin{bmatrix} \dot{x}_1 \\ \dot{x}_2 \\ \dot{x}_3 \end{bmatrix} = \begin{bmatrix} 0 & 1 & 0 \\ 0 & 0 & 1 \\ -a_0 & -a_1 & -a_2 \end{bmatrix} \begin{bmatrix} x_1 \\ x_2 \\ x_3 \end{bmatrix} + \begin{bmatrix} 0 \\ 0 \\ 1 \end{bmatrix} u \qquad (13\text{-}10b)$$

$$[y] = [b_0 - a_0 b_3 \quad b_1 - a_1 b_3 \quad b_2 - a_2 b_3] \begin{bmatrix} x_1 \\ x_2 \\ x_3 \end{bmatrix} + [b_3] u \qquad (13\text{-}11b)$$

In cases typically encountered in practice, the numerator order is lower than the denominator order; i.e., $b_3 = 0$. Then Eq. (13-11b) takes on a simpler form. The extension of this procedure to higher-order transfer functions should be quite obvious. There are a number of other standard ways to make integrator block diagrams for a transfer function and thus to derive state and output equations. In fact, an infinite number of linear-state-equation sets exist, each set corresponding to a given transfer function. From our present point of view, all these state- and output-equation sets are equivalent, so that it suffices to know *one* way to make up a set of state equations.

Although the transfer-function idea is limited to linear systems, block diagrams are not. If you can make a nonlinear block diagram of a system using only integrators for the dynamic elements, you can easily convert to state equations much as we just did. The output of every integrator is a state variable, and the input is its derivative. The state equations are written by following the signal representing the state-variable derivative back through the diagram until either a state variable or an input is encountered. This is what we have done with our linear-transfer-function block diagrams, but the concept works just as well when nonlinear blocks exist. In fact, this is exactly what one does to write equations from an augmented bond graph. The augmented bond graph can be viewed as a compact and nonredundant block diagram, the p and q energy variables being integrator outputs.

Example 13-1 A typical example of a mixture of a transfer-function and physical-system model is shown in Fig. 13-4. The audio amplifier is designed as a voltage amplifier and the electrodynamic speaker is a (gyrator) transducer. The amplifier will be modeled as a voltage-controlled voltage source, but we wish to include an approximation to the measured-frequency-response curve sketched in Fig. 13-4b. Ideally, the amplifier should have a constant gain g at all frequencies, but an ac amplifier cannot pass very low frequencies. Furthermore, there is an inevitable falloff in performance at very high frequencies for all real devices. Let us assume that any problems at low frequencies occur below any frequencies of interest in the sound system but that the amplifier shows a minor resonance followed by a reduction in gain at high frequencies in the audio range, as sketched in Fig. 13-4b.

The transfer function (Fig. 13-4c) is an approximation to the measured frequency response. The magnitude of the ratio of e_{out} to e_{in} when both signals are sinus-

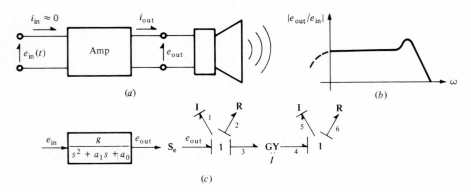

Figure 13-4 Model of a sound system; (b) is a log-log frequency-response plot.

oids with frequency ω can be found by substituting $s = j\omega$ in the transfer function

$$\frac{e_{out}}{e_{in}} = \frac{g}{(j\omega)^2 + a_1 j\omega + a_0} \tag{13-12}$$

From this expression we see that the low-frequency gain is g/a_0, the natural or break frequency is $a_0^{1/2}$, and a_1 can adjust the damping or height of the resonance peak. At high frequencies the frequency response falls off as $1/\omega^2$. While this simple transfer function certainly will not match an experimental curve for an amplifier perfectly, it may suffice to tell us whether the amplifier or the speaker is the limiting device in the high-frequency performance of the sound system.

The voice-coil inductance is I_1, its resistance is R_2, and its transduction coefficient is T in the bond graph of Fig. 13-4c. Then I_5 and R_6 are the mass of the voice coil, speaker cone and some of the air moved by the cone, and the mechanical and radiation resistance of the speaker. Modeling the interaction of the speaker with the air is actually quite complicated, but for this example we shall assume that I_5 and R_6 are constants.

The state and output equations derived from the transfer function are much like Eqs. (13-5) and (13-6), except that they are second order. The role of the input u is played by e_{in} and the role of the output y is played by e_{out}. Even without making the block diagram we can write the equations

$$\dot{x}_1 = x_2 \tag{13-13}$$
$$\dot{x}_2 = -a_0 x_1 - a_1 x_2 + g e_{in}(t)$$
$$e_{out} = x_1 \tag{13-14}$$

For the rest of the bond graph, we note that e_{out} from the transfer function is also the voltage from the source. (This is assuming unit gain in the controlled source.) Using Eq. (13-14), we can simply use x_1 for the voltage-source effort. The entire equation set then is

$$\dot{x}_1 = x_2$$
$$\dot{x}_2 = -a_0 x_1 - a_1 x_2 + g e_{in}(t)$$
$$\dot{p}_1 = x_1 - \frac{R_3}{I_3} p_3 - T \frac{p_5}{I_5} \tag{13-15}$$
$$\dot{p}_5 = T \frac{p_1}{I_1} - \frac{R_6}{I_6} p_6$$

As you can see, x_1 and x_2 are artificial state variables which realize the transfer function fitted to the measured frequency response of the amplifier. They have no particular physical significance, nor are they easily measurable; nevertheless they fit together with the physical state variables p_1 and p_5 to form a unified mathematical model which can be simulated on a computer or otherwise analyzed. Although this entire system is linear, it would be easy to include nonlinear terms in the p_1 and p_5 state equations if the loudspeaker model contained significant nonlinearities.

13-3 AMPLIFYING DEVICES

Amplifiers generally make something big out of something small, but a lever with a large lever ratio is not an amplifier in the sense used here. It is a transformer, and while it may amplify the motion of one end compared with that of the other, it will diminish the force in the same lever ratio. The transformer conserves power, so that what it gains in volocity it must lose in force.

The amplifiers we discuss here amplify power, as shown in Fig. 13-5. A control signal, ideally an active bond without power, controls the flow of power on a real bond. If the control signal is a voltage, it may be that the power-bond voltage is an amplified version of the control signal, as it could be for an electric transformer; but in addition to voltage amplification there is also power amplification.

The apparent violation of the first law of thermodynamics is easily resolved if we recognize that amplifiers need power supplies. A typical pattern is shown in Fig. 13-5 b. This introduces a new type of element, the modulator, which is the essential part of most amplifiers. In this section we introduce several modulators which can be used to create amplifiers.

Some modulating elements have already been studied. Modulated gyrators and transformers can be used to amplify. For example, a separately excited dc motor with its armature connected to a constant power supply can be controlled through the field port with relatively little control power. There is an active bond in the basic model for the field effect which allows this. Similarly, a variable-displacement hydraulic pump can be controlled with little power by the stroke control even when the pump is transducing a large amount of power from a source to a load. Again, an active bond is evident in the modulated transformer of the model of the pump.

Another class of modulator is shown in Fig. 13-6. The examples all are basically variable resistors, in which the change of resistance at a true port is accomplished ideally on an active bond. The transistor, for example, has been studied previously. We saw that in some regions i_B changes the relation between i_C and e_{CE} drastically. Not only are changes in i_B smaller than changes in i_C, but e_{BE} is also much smaller than e_{CE} so that the power gain is large and the simplified model shows i_C on an active bond.

The other modulators in Fig. 13-6 are all primarily 1-port R elements with active-bond modulating variables. For the electric switch and the hydraulic valve the modulating variable is mechanical position; for the transistor the modulating variable is a flow (current); and for the mechanical clutch it is an effort (force). For several of the devices, a junction typically used with the 1-port R is also illustrated.

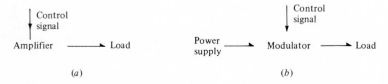

(a) (b)

Figure 13-5 General form of amplifier and typical decomposition.

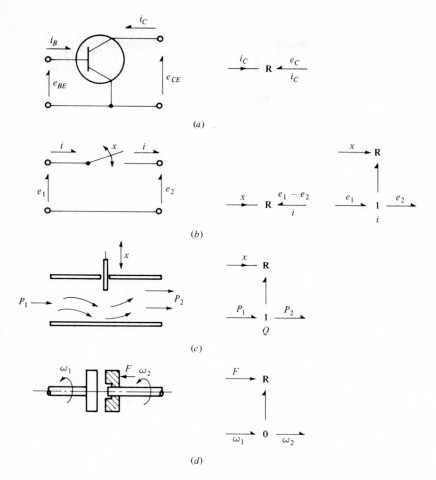

Figure 13-6 Some modulated resistance elements.

Since these modulated resistance elements all dissipate power, amplifiers constructed by adding sources to these modulators achieve power amplification at the expense of power loss. The result may be perfectly acceptable in a low-power electronic amplifier, but it may result in the "heat death" of the device, as it would if you tried to hold your car still for long on an upgrade by slipping the clutch.

However, most resistive modulators can be operated in a switched mode (like the electric switch), which has very low power loss. In one state the resistance is so low that even large flows are accompanied by only small efforts, while in the other mode the resistance is so high that even large efforts produce very small flows. Amplifiers based on switching between the high- and low-resistance states can be power-efficient. Often the jerky behavior this switching entails can be filtered with dynamic elements so that the resulting amplifier has a rather smooth output signal.

Many other devices can be used to make amplifiers, and some have more complicated constitutive laws than those shown in Fig. 13-6, but the general idea of a modulator and a source combining into an amplifier still applies. Let us consider an example of such an amplifier made out of hydraulic valves.

Example 13-2 Figure 13-7 shows a schematic diagram of an amplifier constructed of a four-way valve connected to a hydraulic pressure source. The unique feature of this valve, shown as a spool valve, is that four hydraulic resistances are modulated by the spool position z simultaneously. (Note that the normalized position $y = z/z_{max}$, which varies between $+1$ and -1, is sometimes more convenient in making plots than z itself.) The resistances are formed by the four edges of the spool lands, A, B, C, and D, and corresponding lands in the valve body. For $z > 0$, the flow Q_s passes through B, through the motor, and to the sump through D. For $z < 0$ the motor flow Q_m reverses.

The four resistors are connected in a Wheatstone-bridge circuit, and you can recognize the benzene-ring pattern in the bond graph of Fig. 13-7b. The R elements are put on the inside of the ring this time so that the common modulating variable z can be indicated easily. Assuming that the return pressure is zero, there are only three distinct pressures in the valve, P_s, P_1, and P_2, and each appears on a 0-junction. The motor pressure is $P_m = P_1 - P_2$. At the top of the ring Q_B is a function of $P_s - P_1$, and Q_A is a function of $P_s - P_2$. At the bottom of the graph Q_m is given by two expressions, $Q_B - Q_C$ and $Q_D - Q_A$. From our assumption that the return pressure is zero, Q_C is a function only of P_1, and Q_D is a function of P_2. With these considerations you can see that the junction structure corresponds to the schematic diagram.

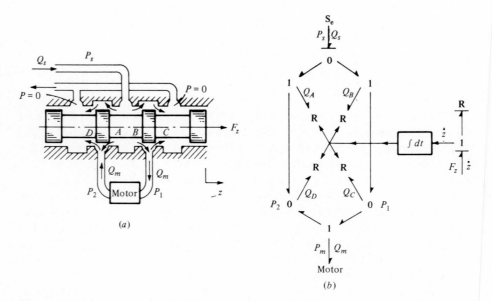

(a)

(b)

Figure 13-7 Four-way-valve amplifier (Example 13-2).

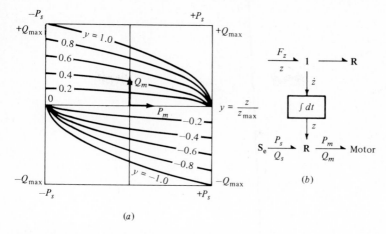

Figure 13-8 Reduced form of hydraulic amplifier.

In practice, Q_C and Q_A practically vanish for positive z while Q_D and Q_B vanish for negative z, simplifying the analysis. The individual resistors are nonlinear, the pressure drops being proportional to flow squared (or the flow proportional to the square root of pressure). For a closed-center valve, i.e., one which blocks the flows Q_A and Q_B when $z = 0$, one can work out the relationships between P_m and Q_m with z as a parameter; this is plotted in Fig. 13-8a. The algebraic reduction of the bridge circuit can be represented in a simplified way (Fig. 13-8b). The constitutive laws for the modulated 2-port R are just those shown in Fig. 13-8a. In both bond-graph representations we show not only the active modulation but also a piece of structure indicating that the force required to stroke the valve F_z is not really zero. In our simple model we just show that at least F_z must overcome seal friction to move the spool. More detailed models would include other force components.

Example 13-3 An even simpler model of our amplifier can be found by considering normal operation fairly near the point $P_m = 0$, $Q_m = 0$. Then you can see that y (or z) sets the flow if $P_m = 0$. For values of P_m not too large compared with P_s, the effect of P_m is to change Q_m somewhat along curves which are fairly straight until P_m and P_s are almost equal. This allows us to make a linearized approximation appropriate for preliminary system design.

Figure 13-9 shows the middle portion of the P_m, Q_m plane for an open-center valve. Such valves can have a steady small flow through both sides of the valve when $z = 0$ because the spool lands do not quite fill the grooves in the body. This has the disadvantage of a steady power loss, which the closed-center valve did not have, but it has the advantage of increasing the size of the approximately linear region. In the region we can write

$$Q_m \approx Q(y) - \frac{1}{R} P_m \approx K_y y - \frac{1}{R} P_m \approx K_z z - \frac{1}{R} P_m \qquad (13\text{-}16)$$

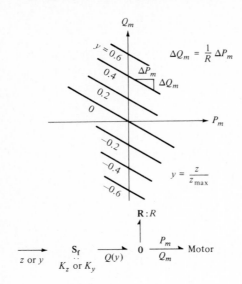

$$\Delta Q_m = \frac{1}{R} \Delta P_m$$

$$y = \frac{z}{z_{max}}$$

R : *R*

Figure 13-9 A linearized hydraulic-amplifier model for an open-centered valve.

where $Q(y)$, $K_y y$, or $K_z z$ represents the flow at $P_m = 0$ as a function of spool position and $1/R$ is the slope of the Q_m-versus-P_m curves near $P_m = 0$. (The slope $1/R$ is more nearly constant for open- than for closed-center valves.) The bond-graph representation of this approximation is quite simple. A position-controlled flow source is attached to a hydraulic resistance which enforces Eq. (13-16). Here we see the typical amplifier as a controlled-source model. This model would be good for many purposes, but for extreme fluctuations in pressure and flow one of the models of Fig. 13-7 or 13-8 would be more accurate.

13-4 INSTRUMENTS AND SIGNAL TRANSDUCERS

The basic models of instruments and signal transducers are similar to those already used for amplifiers, but by amplifier we normally mean a device with large controllable output power. In this section we restrict ourselves to devices which are intended to operate at low power levels.

Instruments are supposed to extract information about a system without disturbing it. Properly applied instruments therefore can be represented ideally by active bonds and block diagrams. Often the output of an instrument, even though amplified, may not have power significance. For example, a digital voltmeter should be used only on circuits which are little affected by the small input current of the voltmeter when it is attached. (A good voltmeter is said to have a high *input impedance,* which means that it will draw only a very small current.) Thus we might indicate the meter's input voltage as coming from a 0-junction on an active bond with negligible current. This display of the voltmeter may require some power to actuate the display devices, which means that there is amplification in the instrument, but the readings of the instrument may simply be recorded without causing any effect on the system. In this case, there is no need to consider finite power even in the output. It is simpler just to remain in the signal or block-diagram domain.

Many transducers can be considered to be instruments essentially staying in the signal domain and having zero power efficiency. For example, strain gages and accelerometers transduce mechanical variables into electrical ones, normally at power levels much smaller than those in the measured system. These devices can usefully be represented in a system model using active bonds and transfer functions or block diagrams.

Figure 13-10 indicates a general approach to including instruments and signal-level transducers in a system model. A column-mounted mass is to be shaken on a testing machine, which can supply $V_0(t)$ regardless of the load. An accelerometer is mounted on the mass, and a resistance strain gage is mounted on the column. Since neither device is supposed to affect the dynamics of the system, if the simple model of the system shown in Fig. 13-10b is adequate without the instruments, it also is adequate with them.

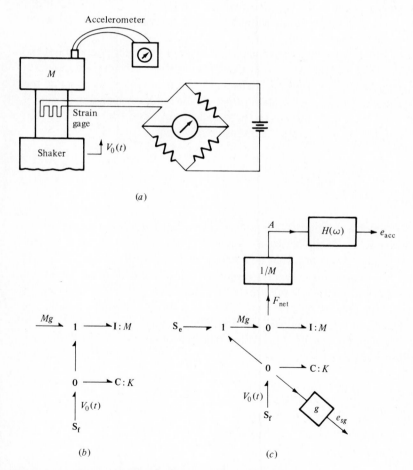

Figure 13-10 Signal transducers added to a system model.

Figure 13-10c shows models of the instruments. While we could model the strain-gage bridge in detail, it is probably not necessary. We just assume that there is a voltage across the display meter e_{sg} which is proportional to the measured strain. In our model we have assumed that the column is an elastic member with spring constant K. Thus we assume that the force in the column is proportional to its overall deflection. Since the force exists conveniently on a 0-junction, we extract the force on an active bond. Between the column force and the meter reading (or meter voltage) there are several factors: a strain-gage sensitivity, a proportionality factor between force and local strain, and a factor between resistance change and meter voltage—all combined into the indicated gain g. The units of g are volts per newton. In this case, we assume no dynamics of interest in the instrumentation system.

For the accelerometer, we add an extra 2-port 0-junction on the I bond so that the net force on the mass F_{net} is available on an active bond. Dividing by M, we have the acceleration. Now we assume that a frequency-response function $H(j\omega)$ is available which relates acceleration to a voltage signal e_{ac}. The units of H are v/m·s^2. If there are significant dynamic effects at frequencies of interest, we can use the methods of Sec. 13-2 to find an approximate rational transfer function and to include state equations for the instrument with those for the rest of the system.

Finally, instruments or signal transducers can be used in feedback control of the system dynamics. Then the output signals must ultimately determine the inputs to amplifiers connected back to the system. Between the instruments or *sensors* and the amplifiers or *effectors, actuators* or *manipulators* can be realized as analog or digital control systems. Such systems remain in the signal domain and are well described by signal-flow graphs, block diagrams, or transfer functions.

13-5 CONCLUSIONS

This chapter discusses a number of situations where signal descriptions are appropriate and convenient. These signals are a complement to the case in physical systems where bilateral signal flows are common and correlated with significant power exchanges between components. We have indicated how signal-level input-output models can be combined with bond graphs so that each part of the system is described in the most convenient and insightful way. Because complex systems are usually studied using computer analysis and simulation, we have shown how artificial state-variable descriptions can be derived for transfer-function and block-diagram models, so that our basic state-space description will apply to all our models.

PROBLEMS

13-1 Consider the transfer function

$$H(s) = \frac{V(s)}{V_0(s)} = \frac{bs + k}{ms^2 + bs + k}$$

which might have come from physical state equations. Convert this back into a set of state and output equations using the methods of Sec. 13-2. Make a block diagram first. (Notice the simple form; i.e., the numerator order is 1 and the denominator order is 2.)

13-2 Extend the notation of Eq. (13-7) to fifth order and write state and output equations directly by extending the pattern of Eqs. (13-10) and (13-11).

13-3 Part of a servomechanism has been tested experimentally and appears to act as a first-order system. The magnitude of the frequency-response function $H(\omega) = Y(\omega)/U(\omega)$ is plotted on a log-log plot in the Figure. At low frequencies $|H(\omega)| \to g_0$. At high frequencies $|H(\omega)| \to g_0/\tau\omega$, where the break frequency $\omega_b = 1/\tau$ and τ is a time constant. Construct a transfer function to fit the frequency response in terms of τ or ω_b and g_0. Then show a set of state and output equations which represent the transfer function.

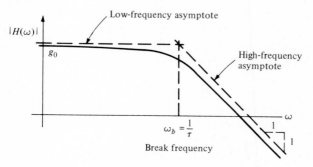

Figure **P13-3** Frequency-response plot on log-log scales.

13-4 The block diagram contains two nonlinear functions involving the absolute value of the input signal to the block. In one, the sign (or signum) function is represented as the ratio of the absolute value divided by the actual value of the input. In the other, a square law with sign correction is represented by the product of the absolute value times the value. (The first block might represent coulomb friction and the second turbulent pressure loss at an orifice.) Using the block diagram, write state and output equations.

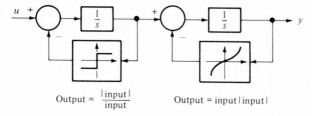

$$\text{Output} = \frac{|\text{input}|}{\text{input}} \qquad \text{Output} = \text{input}\,|\text{input}|$$

Figure P13-4

13-5 Show that the illustrated connection of two separately excited dc motors functions as an amplifier by making a bond-graph model. Motor 1 is driven at constant speed ω_0 by a large prime mover, and its

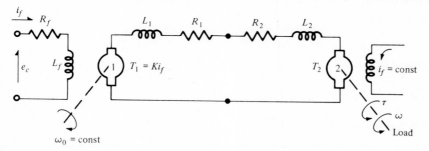

Figure P13-5

field is controlled by the voltage $e_c(t)$. The transduction coefficient is $T_1 = Ki_f$. Motor 2 has constant field current and transduction coefficient T_2.

13-6 The clutch is to be used to accelerate the large flywheel. Power comes from a driver with nearly constant angular speed, and the clutch is controlled by the force $F(t)$. The force can be varied without much power since the plates hardly move as F is varied. Make a bond graph for this system, and write state equations (note that this is an amplifier system).

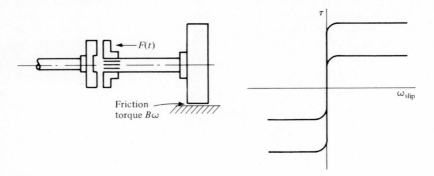

Figure P13-6

$$\tau \approx K(\text{sgn } \omega_{\text{slip}}) \frac{F(t)}{F_0} \approx \frac{K}{F_0} \frac{|\omega_{\text{slip}}|}{\omega_{\text{slip}}} F(t)$$

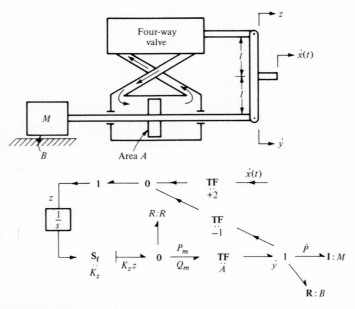

Figure P13-7 Note that ──── 1 ──── implies that $e_2 = 0$ since the activation means that e_1 should be neglected and the sum of the efforts on a 1-junction must add to zero: $e_1^0 + e_2 = 0$.

13-7 A four-way valve (Fig. P13-7) is characterized as in Fig. 13-9 or Eq. (13-16). The motor is a ram driving a large inertial and frictional load. A following servomechanism is made by connecting a mechanical feedback linkage between the output motion y, the valve stroke z, and the input motion $x(t)$. The ports have been arranged differently from those in Fig. 13-7, so that positive z causes positive Q_m and positive $\dot{y}$. You can understand the operation by imagining moving x an amount to the right. The immediate result is that z moves twice as far to the right. Then the ram moves to the right as Q_m flows, and finally the $\dot{y}$ movement through the linkage makes z return to zero, when the system stops moving. Augment the bond graph, noting that the entire transformer structure for the linkage has been activated since we have assumed that it takes no force to move the valve spool. Write the state equations and derive a transfer function relating y to $x(t)$ or $\dot{y}$ to $\dot{x}(t)$.

13-8 Suppose it takes a force $F = F_0$ sgn $\dot{z}$ to move the spool in the servo of Prob. 13-7. Modify the bond graph as in Fig. 13-7 to reflect this fact. Imagine now that the input motion $\dot{x}(t)$ is supplied by a velocity source. Write the state equations and compare with the results of Prob. 13-7.

13-9 An accelerometer is mounted on a mass-spring-damper oscillator excited by a shaker with velocity input $V_0(t)$. The accelerometer delivers a voltage $e(t)$ to a recorder, but there is a dynamic transfer function between the acceleration $\dot{V}(t)$ amd $e(t)$ due to the nature of the device

$$\frac{e(s)}{V(s)} = H(s) = \frac{g_0 \omega_n^2}{s^2 + 2\zeta\omega_n s + \omega_n^2}$$

where ω_n = natural frequency of accelerometer
ζ = damping ratio of accelerometer
g_0 = low frequency gain of accelerometer

In the bond graph we have found $\dot{V}(t)$ by extracting the net force on the mass by inserting an active bond and an extra 0-junction and then dividing by the mass. Write state equations for p and q and add state equations and an output equation for the transfer function, so that e could be found by simulating these equations on a computer.

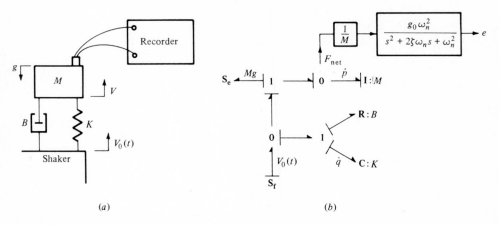

(a)   (b)

Figure P13-9

PART
FOUR

STATE EQUATIONS AND SYSTEM RESPONSE

In Part Two we discussed practical methods for formulating state equations and obtaining their time responses based on Laplace transforms. These methods were efficient for low-order linear systems with constant coefficients, probably the most important case for most practicing engineers. A low-order model also serves as a logical starting point for a more general study of physical-system dynamics. In Part Four we introduce a framework that includes most types of lumped-parameter (or finite-dimensional) systems that can arise. The formulation tools are matrix-based, and we include a discussion of nonlinear systems. The emphasis is on conceptual clarity and the use of digital computing as a practical method for obtaining specific results.

FOURTEEN

GENERAL EQUATION-FORMULATION TECHNIQUES

In this chapter we identify the major types of dependencies that occur between system equations and discuss the reduction of system equations to state-equation form.

14-1 ASSIGNING CAUSALITY TO BOND GRAPHS

We state here a general causality-assignment procedure for orienting bond graphs. If this procedure is followed, considerable insight into the structure of system equations can be obtained and causality will provide a valuable guide in formulating state equations. The procedure derives from recognizing a certain hierarchy or ordering among element constraints in causal terms and using that ordering systematically.

Among the nine basic bond-graph elements there are several whose causal properties are fixed by their definitions. In particular, recall the definitions of the sources S_e and S_f. These elements must be oriented causally

$$S_e \longmapsto \quad \text{and} \quad S_f \longmapsto$$

otherwise the mathematics and physics of the elements are inconsistent.

Other elements whose definitions imply necessary causal conditions are TF and GY; each has two acceptable causal forms and two forms that are inconsistent. And of course the two junction elements 0 and 1 have specific causal properties. Perhaps the easiest way to remember them is this: a 0-junction has *exactly one effort* as input, and a 1-junction has *exactly one flow* as input.

The remaining elements C, I, and R fall into a different category, causally speaking. Both C and I have a *preferred* causality, but the physics of a model does not require that the preferred causality be achieved. Preferred causality is called *integral causality* and is denoted

$$C \; \underline{\hspace{1.2cm}} | \quad \text{and} \quad I \; |\underline{\hspace{1.2cm}}$$

The equations for such C and I elements have the standard forms discussed earlier:

For C: $\qquad\qquad\qquad e = \phi_C(q) \qquad$ and $\qquad \dot{q} = f$

For I: $\qquad\qquad\qquad f = \phi_I(p) \qquad$ and $\qquad \dot{p} = e$

The nature of a particular system may force a C or I to have *derivative causality*, denoted by

$$C \; |\underline{\hspace{1.2cm}} \quad \text{and} \quad I \; \underline{\hspace{1.2cm}}|$$

In this case the describing equations must be inverted, to yield:

For C: $\qquad\qquad\qquad q = \psi_C(e) \qquad$ and $\qquad f = \dot{q}$

For I: $\qquad\qquad\qquad p = \psi_I(f) \qquad$ and $\qquad e = \dot{p}$

where the ψ functions are the inverse of the ϕ functions used in integral causality. For linear elements the inversion of relationships poses no difficulties, but for some nonlinear cases the situation may be rather difficult computationally.

Finally, although the R element generally is indifferent to the causality assigned to it, there are special cases of nonlinear R characteristics where the nature of the function interacts with an assigned causality for better or for worse. We shall see an example later.

The sequential causal-assignment procedure (SCAP) merely organizes the causal observations just discussed into an orderly pattern at the system level. It consists of up to three main steps, each carried out in smaller segments.

The Sequential Causal-Assignment Procedure

Level 1
1. Choose an input source (S_e or S_f) and causally direct it by applying a causal stroke to the appropriate end of the bond.
2. Immediately extend the causal implications throughout the graph, using 0, 1, TF, and GY elements.
3. Repeat steps 1 and 2 until all sources have been directed. If the graph is not causally complete, continue.

Level 2
4. Choose a storage element (I or C) and direct it with integral causality.
5. Immediately extend the causal implications, using 0, 1, TF, and GY elements.
6. Repeat steps 4 and 5 until all I's and C's have been directed. If the graph is not causally complete, continue.

Level 3
7. Choose an R element and direct it arbitrarily.
8. Immediately extend the causal implications, using 0, 1, TF, and GY elements.
9. Repeat steps 7 and 8 until all R's have been directed.

In the unlikely event that causality still has not been completed after level 3, this means that the graph contains an algebraic loop in the junction structure itself. That loop must be formulated and solved as part of the overall reduction process. It is rare in an engineering problem to encounter such a situation.

Now let us interpret some of the information available from using the SCAP. If the graph is causally completed after level 1, all system variables are determined algebraically by the inputs. Such a system is not really a dynamic system, since knowledge of the inputs alone at any instant is sufficient to determine the complete state of the system. In mechanics such problems are called *kinematics problems;* in hydraulics and heat-transfer circuits they often correspond to resistive networks. Even if dynamic elements (C and I elements) are present in the model, their state variables are simply algebraic functions of the input forcing variables.

We now study the organization of equations for systems which terminate in levels 2 and 3 in the SCAP.

14-2 SYSTEMS WITH INDEPENDENT STORAGE ELEMENTS

The storage elements are the I and C elements. So far in our examples and problems we have dealt with situations where the proper application of causality led each I and C element to have integral causality; i.e., efforts were inputs to I's and the corresponding flows were derived from the momenta, or flows were inputs to C's and the corresponding efforts were derived from the displacements.

When every C and I element has integral causality, each storage variable corresponds to an independent state variable. Therefore, the order of the dynamic system can be predicted from the causal graph, and the state and input vectors can be defined.

Example 14-1 Consider the mechanical system and its bond-graph model shown in Fig. 14-1. If SCAP is applied, the result after level 1 is as shown in Fig. 14-1c. In this case, only bonds 7 and 8 have been causally directed. When level 2 is completed, the graph has complete (and of course consistent) causality (Fig. 14-1d). The particular bonds of importance at level 2 are 1, 2, 4, and 5. The SCAP is finished.

At this point we can identify the state vector X and the input vector U as

$$X = \begin{bmatrix} p_1 \\ q_2 \\ p_4 \\ q_5 \end{bmatrix} \quad \text{and} \quad U = \begin{bmatrix} m_1 g \\ m_4 g \end{bmatrix}$$

where p_1 and p_4 are mass momenta and q_2 and q_5 are spring deflections. Since the number of storage ports with integral causality is four, there are four state variables. Thus we say that the system is of fourth order with two inputs.

Formulation of state equations can proceed rather directly when all I and C elements have integral causality. Bear in mind the eventual goal of equation formulation, which for a linear system is

$$\dot{X} = AX + BU \tag{14-1}$$

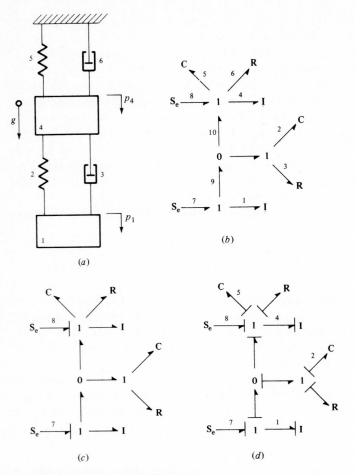

Figure 14-1 A linear mechanical example.

where X = state vector
U = input vector
A = square matrix relating state to its derivative
B = matrix relating inputs to state derivative

For the mechanical system of Fig. 14-1 a, if we assume constant spring-stiffness, mass, and damping parameters for each of the springs, masses, and dampers, a set of coupled state equations can be derived from the graph of Fig. 14-1 b simply by following the causal marks, as we did in Part Two. We obtain

$$\dot{p}_1 = -b_3 m_1^{-1} p_1 - k_2 q_2 \qquad + b_3 \ m_4^{-1} p_4 \qquad + m_1 g \quad (14\text{-}2a)$$

$$\dot{q}_2 = \qquad m_1^{-1} p_1 \qquad\qquad\qquad - \ m_4^{-1} p_4 \qquad\qquad\quad (14\text{-}2b)$$

$$\dot{p}_1 = -b_3 m_1^{-1} p_1 - k_2 q_2 \qquad + b_3\, m_4^{-1} p_4 \qquad + m_1 g \qquad (14\text{-}2\,a)$$

$$\dot{q}_5 = \qquad\qquad\qquad m_4^{-1} p_4 \qquad\qquad\qquad (14\text{-}2\,d)$$

Since all the constitutive laws have been assumed to be linear, we can arrange the equations in a matrixlike form.

As a useful check on the correctness of the state equations, we compute the quantity $\dot{p}_1 + \dot{p}_4$, the net change in system momentum

$$\dot{p}_1 + \dot{p}_4 = -b_6 m_4^{-1} p_4 - k_5 q_5 + m_1 g + m_4 g \qquad (14\text{-}3)$$

This shows that the system rate of change of momentum depends upon the sum of external forces, as it should.

It is reasonable to extrapolate from the example above to conclude that systems with integral causality present no major difficulty in obtaining the state equations, even if such systems are fairly large. Now let us consider the influence of nonlinearity on the formulation of system and state equations.

Example 14-2 Figure 14-2a depicts a hydraulic system with line friction and inertia, a capacitance near the end of the line, and a resistive restriction. The system is driven by a pressure source P_1, the fluid volume in the accumulator is V_5, and the pressure-momentum variable for the inertia is Γ_2.

A bond-graph model with integral causality is given in Fig. 14-2b. We assume the following constitutive equations for the hydraulic effects:

I_2:	$Q_2 = I_2^{-1}\Gamma_2$	$(14\text{-}4a)$
R_3:	$P_3 = \phi_3(Q_3)$	$(14\text{-}4b)$
C_5:	$P_5 = \phi_5(V_5)$	$(14\text{-}4c)$
R_6:	$P_6 = \phi_6(Q_6)$	$(14\text{-}4d)$

That is, elements R_3, C_5, and R_6 are nonlinear. The state and input vectors are, respectively,

$$X = \begin{bmatrix} \Gamma_2 \\ V_5 \end{bmatrix} \quad \text{and} \quad U = [P_1]$$

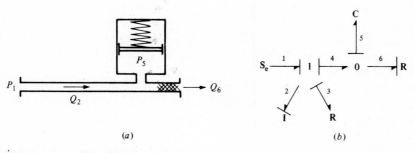

(a)   (b)

Figure 14-2 A nonlinear hydraulic example.

The state equations are

$$\dot{\Gamma}_2 = -\phi_3(I_2^{-1}\Gamma_2) - \phi_5(V_5) + P_1 \qquad (14\text{-}5a)$$

and

$$\dot{V}_5 = I_2^{-1}\Gamma_2 - \phi_6^{-1}[\phi_5(V_5)] \qquad (14\text{-}5b)$$

Notice that the order of the system (2) could have been predicted from the causality in Fig. 14-2b and that the describing equation for the R_6 element must be inverted for use in the state equations. We use $Q_6 = \phi_6^{-1}(P_6)$ to mean the inverse of $P_6 = \phi_6(Q_6)$.

It is reasonable to conclude that even if some element functions are nonlinear, formulation of state equations for a system having integral causality will proceed without any great difficulty. Here we summarize briefly the situation so far. For a given bond-graph model, put on causality using the SCAP. If the procedure terminates at level 2 and if all I and C elements have integral causality, the state vector corresponds to the energy- or storage-variable set and the number of such variables is the order of the system. Equation formulation typically is straightforward, and state equations can be derived in the form of Eq. (14-1) if the system is linear or in the following form if it is nonlinear:

$$\dot{X} = \phi(X, U) \qquad (14\text{-}6)$$

where ϕ is a vector of nonlinear functions of the state and input variables.

14-3 SYSTEMS WITH DEPENDENT STORAGE ELEMENTS

In the course of extending causal implications in level 2 after choosing some I or C to have integral causality suppose that one is forced to assign derivative causality to an I or C element; i.e., we must impose an effort onto a C or flow into an I. Such a case indicates that the energy variable on the element in derivative causality is not dynamically independent from the others.

Example 14-3 Consider the bond graph of Fig. 14-3a, which has no causality. After all source elements are assigned and extended, the graph has partial causality, as shown in Fig. 14-3b. Now proceed to level 2. As a result of imposing integral causality on the I_8 element, the graph has the form shown in Fig. 14-3c. What we observe is that a knowledge of inputs f_1 and e_9 together with variable p_8 is sufficient to determine f_4 and thus p_4 *in algebraic terms*. The system is of order 2 because only C_2 and I_8 are in integral causality. Physical interpretations of the bond-graph model in Fig. 14-3d and e demonstrate the occurrence of derivative causality is not necessarily obvious from the physics.

To see that p_4 is indeed algebraically related to the independent state variables and the input variables let us obtain the algebraic relation from the causal graph. Assume that all elements have constant coefficients and that m is the modulus of the transformer ($f_5 = mf_6$). Then

$$f_4 = f_5 = mf_6 = mf_8 \qquad \text{or} \qquad I_4^{-1}p_4 = mI_8^{-1}p_8$$

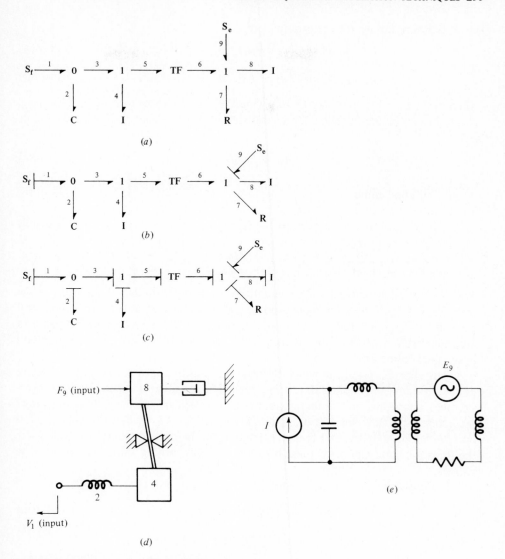

Figure 14-3 An example of derivative causality.

or, finally,

$$p_4 = I_4 m I_8^{-1} p_8 \qquad (14\text{-}7)$$

Thus, knowledge of p_8 at any time is sufficient to determine the value of p_4 directly. Hence p_4 is not an independent state variable, we do not need a state equation for it, and the system equations can be formulated in terms of the two state variables q_2 and p_8 and two inputs f_1 and e_9.

System equations for the example are

$$\dot{q}_2 = \quad\quad - m I_8^{-1} p_8 + f_1 \tag{14-8a}$$

$$\dot{p}_8 = m C_2^{-1} q_2 - b_7 I_8^{-1} p_8 + e_9 - m \dot{p}_4 \tag{14-8b}$$

where we have traced our way by following the causal strokes to e_4 or p_4. Using Eq. (14-7) or continuing to follow the causal strokes, we find that p_4 is related to e_8 or $\dot{p}_8$

$$\dot{p}_4 = I_4 m I_8^{-1} \dot{p}_8 \tag{14-8c}$$

Eliminating $\dot{p}_4$ from the equations in terms of $\dot{p}_8$ and then solving for $\dot{p}_8$ alone, we get

$$\dot{q}_2 = - m I_8^{-1} p_8 + f_1 \tag{14-9a}$$

$$\dot{p}_8 = \frac{m C_2^{-1}}{(1 + m^2 I_4 I_8^{-1})} q_2 - \frac{b_7 I_8^{-1}}{(1 + m^2 I_4 I_8^{-1})} p_8 + \frac{e_9}{(1 + m^2 I_4 I_8^{-1})} \tag{14-9b}$$

This is a set of second-order state equations, as causally indicated. As you can see from Eq. (14-7), if we know p_4, we know p_8; therefore if we had first applied integral causality to I_4, p_4 would be a state variable and p_8 would have been eliminated as a dependent state variable. Thus, there is some choice of which state variable to retain in the final formulation.

This class of system, in which derivative causality appears, can be much more difficult to reduce to explicit state-space form. If the system is linear, a matrix inversion will accomplish this reduction, but if it is nonlinear, algebraic difficulties may preclude simple state-space formulation. Some problems in nonlinear mechanics fall into the latter category, and their formulation can be quite difficult. We may know the form the state equations must take, but actually writing the equations may be tedious.

The generalization of Eq. (14-8) is

$$\dot{X}_i = \phi_i(X_i, \dot{X}_d, U) \tag{14-10a}$$

and
$$X_d = \phi_d(X_i, U) \tag{14-10b}$$

where X_i = set of independent energy variables corresponding to integral causality
X_d = set of dependent energy variables corresponding to derivative causality
U = input set

Note that it is the time derivative of X_d that appears in Eq. (14-10a), while it is X_d itself that is coupled to X_i in Eq. (14-10b).

Consider the reduction of Eq. (14-10) to explicit form, as defined by Eq. (14-1) or (14-6) for the linear or nonlinear case, respectively. We need to obtain an expression for $\dot{X}_d$ from Eq. (14-10b), namely,

$$\dot{X}_d = \phi_{dd}(X_i, \dot{X}_i, U, \dot{U}) \tag{14-11}$$

in general, where ϕ_{dd} denotes a new function derived from ϕ_d. When the results of Eq. (14-11) are substituted into Eq. (14-10a), thereby eliminating $\dot{X}_d$, we get

$$\dot{X}_i = \phi_{ii}(X_i, \dot{X}_i, U, \dot{U}) \tag{14-12}$$

where ϕ_{ii} denotes a new function. Notice that $\dot{X}_i$ now appears on both sides of the equation. This type of vector equation is termed *implicit*. It may or may not be possible to perform the algebra required to move $\dot{X}_i$ to the left side of Eq. (14-13), thereby obtaining the explicit form of Eq. (14-6).

One other observation is in order for this class of models. Notice the appearance of the U vector in Eqs. (14-11) and (14-12). Since $U(t)$ is known, its derivative certainly can be computed; hence, $\dot{U}(t)$ is also known. But what if $U(t)$ has a discontinuity, as occurs with the step function? Then $\dot{U}$ is infinite at points of discontinuity, with possibly disastrous effects on a physical system.

Example 14-4 As a very simple case, consider the circuit and its graph shown in Fig. 14-4. With causality assigned, we find that C_3 has derivative causality. Now consider the implications of a step change in input voltage $V_1(t)$. Such a change is imposed directly on the C_3 element, leading to a step change in energy state, as defined by the charge q_3. But that corresponds to a step change in energy, which can occur only if infinite power is available. In fact, it is not physically realistic to assume that a voltage source can apply an ideal step to a capacitor.

The appearance of a $\dot{U}$ term in a set of state equations is often a signal that a modeling error may have occurred. In any event, it bears careful consideration before proceeding.

To accommodate the case of derivative causality in linear systems, we generalize the standard form of state equations from Eq. (14-1) to

$$\dot{X} = AX + BU + E\dot{U} \tag{14-13}$$

Generally, linear systems can be converted from the form of Eq. (14-10) into the form of Eq. (14-13) without major difficulty.

Summary of the Formulation Procedure when Derivative Causality Occurs

1. Identify the X_i and X_d vectors. The energy variables from I and C elements in integral causality form X_i and from those in derivative causality form X_d. Identify the U vector from the sources.

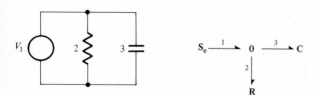

Figure 14-4 Example 4-4

2. Write the system equations in the initial form

$$\dot{X}_i = \phi_1(X_i, \dot{X}_d, U) \tag{1}$$

and

$$X_d = \phi_2(X_i, U) \tag{2}$$

3. Compute $\dot{X}_d$ from X_d in Eq. (2). Substitute the result into Eq. (1) to obtain

$$\dot{X}_i = \phi_3(X_i, \dot{X}_i, \dot{U}, U) \tag{3}$$

4. If possible, reduce the equation for $\dot{X}_i$ into explicit form

$$\dot{X}_i = \phi_4(X_i, U, \dot{U}) \tag{4}$$

Many times only a few of the elements of X_i and none of the elements of U appear in (2). Then only a subset of (3) actually needs to be manipulated to achieve the desired form (4). The ENPORT programs take care of derivative causality automatically for the linear case, producing the form (4) directly from the bond graph.

14-4 SYSTEMS WITH COUPLED DISSIPATION ELEMENTS

Sometimes the SCAP will reveal a system with coupled R elements, as indicated by the need to continue to level 3 in the assignment procedure. A simple example can illustrate both how the situation arises and how to handle the resulting equations. A circuit diagram and its bond graph are shown in Fig. 14-5. After level 2 the causal graph appears as in Fig. 14-5c. Completing the causality yields the graph in Fig. 14-5d. The causality was completed by assigning bond 2 arbitrarily and extending the information. Thus we see that the variables of elements R_2 and R_5 are tied together causally by the 0 and 1 elements.

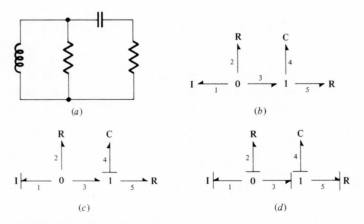

Figure 14-5 An example of level 3 causality.

The state vector is

$$X = \begin{bmatrix} p_1 \\ q_4 \end{bmatrix}$$

and there is no input. We shall also define an auxiliary vector H as

$$H = \begin{bmatrix} e_2 \\ f_5 \end{bmatrix}$$

Note that H consists of one variable from each R element. They are the variables which the R elements supply to the rest of the system, as the causal marks show. The system yields a set of equations

$$\dot{p}_1 = e_2 \tag{14-14a}$$

$$\dot{q}_4 = f_5 \tag{14-14b}$$

and
$$e_2 = R_2(-I_1^{-1}p_1 - f_5) \tag{14-15a}$$

$$f_5 = R_5^{-1}(e_2 - C_4^{-1}q_4) \tag{14-15b}$$

As causality foretold, the variables e_2 and f_5 are algebraically coupled and must be solved for together. Doing so, we find that

$$e_2 = -\frac{R_2 R_5}{R_2 + R_5} I_1^{-1} p_1 + \frac{R_2 C_4^{-1}}{R_2 + R_5} q_4 \tag{14-16a}$$

and

$$f_5 = -\frac{R_2}{R_2 + R_5} I_1^{-1} p_1 - \frac{C_4^{-1}}{R_2 + R_5} q_4 \tag{14-16b}$$

Then, using the results of Eq. (14-16) in Eq. (14-14), we get

$$\dot{p}_1 = -\frac{R_2 R_5}{R_2 + R_5} I_1^{-1} p_1 + \frac{R_2}{R_2 + R_5} C_4^{-1} q_4 \tag{14-17a}$$

and

$$\dot{q}_4 = -\frac{R_2}{R_2 + R_5} I_1^{-1} p_1 - \frac{1}{R_2 + R_5} C_4^{-1} q_4 \tag{14-17b}$$

Notice how the R elements act in a coupled fashion in the state equations. For comparison, you might wish to modify the graph of Fig. 14-5b by interchanging the 0 and 1 elements and repeating the procedure.

Summary of the Formulation Procedure for Coupled Dissipation Elements

1. Identify the X and U vectors. Identify an H vector whose elements are the outputs from the R ports included in the coupled dissipation fields.

2. Write system equations for X and H, namely,

$$\dot{X} = \phi_1(X, U; H) \tag{1}$$

and $$H = \phi_2(X, U; H) \tag{2}$$

3. Solve for H explicitly from Eq. (2), when possible, to get

$$H = \phi_3(X, U) \tag{3}$$

4. Substitute for H directly into Eq. (1) to get

$$\dot{X} = \phi_4(X, U) \tag{4}$$

It is possible for the same system to include both dependent storage elements and coupled dissipation elements simultaneously. When causality reveals this situation, it is best to formulate the equations by using H (auxiliary) variables for the R fields and separately anticipate the elimination of certain dependent storage variables from the state equations. An alternative approach based on element equivalences is discussed in the next section.

14-5 SOME USEFUL EQUIVALENCES

Certain types of element combinations show up in engineering work with some regularity, making it convenient on occasion to modify the model in a way different from the system procedure just described.

Some 1-Port Equivalences

Consider the simple mechanical model and its bond graph in Fig. 14-6a and b. Causality assignment shows that masses 2 and 3 are coupled and mass 2 has been chosen as independent. Instead of using the formal system-reduction procedure, we first redraw the bond graph slightly (Fig. 14-6c). Now we investigate the subgraph containing bonds 2, 3, and 10, separated in Fig. 14-5d. The input is e_{10}; there is one state variable p_2. The system equations are

$$\dot{p}_2 = e_{10} - \dot{p}_3 \tag{14-18a}$$

$$f_3 = f_2 \tag{14-18b}$$

hence $$m_3^{-1}p_3 = m_2^{-1}p_2 \tag{14-19}$$

Therefore $$p_3 = m_3 m_2^{-1}p_2 \tag{14-20}$$

and $$\dot{p}_3 = m_3 m_2^{-1}\dot{p}_2 \tag{14-21}$$

The state equation becomes

$$\dot{p}_2 = [(1 + m_3 m_2^{-1})^{-1}]e_{10} \tag{14-22}$$

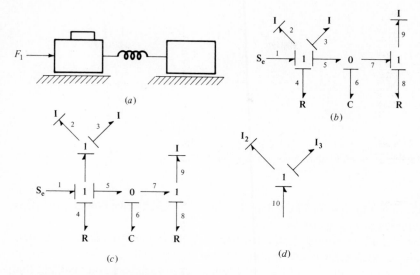

Figure 14-6 A two-mass equivalence example.

Had we chosen I_3 as independent, we would have obtained

$$\dot{p}_3 = [(1 + m_2 m_3^{-1})^{-1}]e_{10}$$

showing the symmetry of the roles of m_2 and m_3.

Another way to develop the equivalence is to imagine the graph of Fig. 14-6d with flow in on bond 10. Then we could write

$$p_2 = m_2 f_2 = m_2 f_{10} \qquad (14\text{-}23a)$$

and

$$p_3 = m_3 f_3 = m_3 f_{10} \qquad (14\text{-}23b)$$

Since

$$e_{10} = \dot{p}_2 + \dot{p}_3 \qquad (14\text{-}24)$$

we have

$$\int e_{10} \, dt = p_{10} = p_2 + p_3 \qquad (14\text{-}25)$$

and

$$p_{10} = (m_2 + m_3)f_{10} \qquad (14\text{-}26)$$

or

$$f_{10} = \frac{1}{m_2 + m_3} p_{10} \qquad (14\text{-}27)$$

In other words, the masses combine as a sum to act as one equivalent mass. *The generalization of this example is that I-element inertia parameters combine on a 1-junction as a sum.*

Now we can replace I_2 and I_3 by their combined effect in the original graph to

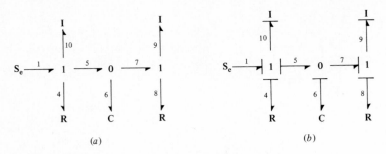

Figure 14-7 Effect of the mass equivalence on the system.

get the picture in Fig. 14-7a. The value of I_{10} is $I_2 + I_3$. Observe how the causality has become integral in Fig. 14-7b.

The method of identifying dependent storage elements and reducing the system by replacing such element combinations by their equivalents is fairly common in practice. Of course, one does not have the original p or q variables available, but that may not be an important factor.

A second common example is that of C elements on a 0-junction. We shall consider that case directly, as shown in Fig. 14-8a and b. Let us develop these equations for the equivalent C at port 3, considering the C elements to be nonlinear

$$f_3 = \dot{q}_1 + \dot{q}_2 \tag{14-28a}$$

$$q_3 = q_1 + q_2 \tag{14-28b}$$

Also, $\qquad e_1 = \phi_1(q_1) \qquad \text{and} \qquad e_2 = \phi_2(q_2) \tag{14-29}$

Hence $\qquad q_3 = \phi_1^{-1}(e_3) + \phi_2^{-1}(e_3) \tag{14-30}$

which characterizes port 3 as a nonlinear C element.

If both C elements are linear, we can write

$$q_3 = C_1 e_3 + C_2 e_3 \tag{14-31}$$

or $\qquad e_3 = \dfrac{1}{C_1 + C_2} q_3 \tag{14-32}$

Thus *capacitances add on a 0-junction.* Similar results can be derived for other 1-ports on 0 and 1-junctions.

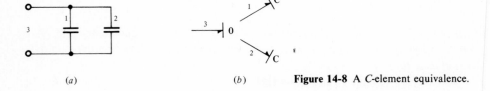

(a)   (b)   **Figure 14-8** A C-element equivalence.

The major results for combinations of C, I, and R elements are shown in Table 14-1. It is assumed that all elements are linear. Consistent use of these results will help transform many systems involving derivative causality to all-integral causality, thereby saving on formulation effort.

Some TF Equivalences

Another important type of equivalence is illustrated in Fig. 14-9a, where a spring is addressed through a lever arrangement. The state equations derived from Fig. 14-9b

Table 14-1 C, I, and R equivalences

Combination	Equivalent graph	Equivalent parameter
3 → 1 ⟨ I_1 , I_2	3 → I	$I_3 = I_1 + I_2$
3 → 0 ⟨ C_1 , C_2	3 → C	$C_3 = C_1 + C_2$
3 → 0 ⟨ I , I	3 → I	$\dfrac{1}{I_3} = \dfrac{1}{I_1} + \dfrac{1}{I_2}$
3 → 1 ⟨ C , C	3 → C	$\dfrac{1}{C_3} = \dfrac{1}{C_1} + \dfrac{1}{C_2}$
3 → 1 ⟨ R , R	3 → R	$R_3 = R_1 + R_2$
3 → 0 ⟨ R , R	3 → R	$\dfrac{1}{R_3} = \dfrac{1}{R_1} + \dfrac{1}{R_2}$

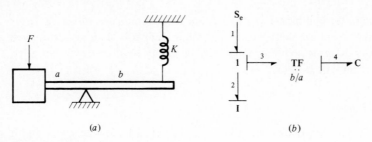

(a)   (b)

Figure 14-9 A TF-equivalence example.

are

$$\dot{p}_2 = -\frac{b}{a} k_4 q_4 + F_1 \tag{14-33a}$$

$$\dot{q}_4 = \frac{b}{a} m_2^{-1} p_2 \tag{14-33b}$$

The natural frequency is $(b/a) \sqrt{k_4/m_2}$.

Another way to treat the system is to first consider just the C element and its TF. That is, consider the equivalent effect by addressing port 3 into the TF. Then we write

$$F_3 = \frac{b}{a} F_4 \tag{14-34a}$$

$$F_4 = k_4 q_4 \tag{14-34b}$$

and

$$\dot{q}_4 = \frac{b}{a} k_3 = \frac{b}{a} \dot{q}_3 \tag{14-34c}$$

Thus,

$$F_3 = \left(\frac{b}{a}\right)^2 k_4 q_3 \tag{14-35}$$

and equivalent *stiffness* parameter is

$$k_3 = \left(\frac{b}{a}\right)^2 k_4$$

Now the mass can be thought of as being attached to an equivalent spring with stiffness k_3. Calculation of the natural frequency will yield the same result as before.

A summary of transformer equivalents is shown in Table 14-2. Care must be taken to ensure that the TF modulus is properly defined in each case.

Some GY Equivalences

Perhaps the most interesting simple equivalences are those relating to the gyrator. Consider the electric-network fragment of Fig. 14-10a, consisting of a capacitor

Table 14-2 Some TF equivalences†

Combination	Equivalent	Assumed form of transformer law
$\overset{2}{-}\vert$ TF $\overset{1}{-}\vert$ C (n_1)	$C_2 = n_1^2 C_1$	$f_2 = n_1 f_1$ $e_1 = n_1 e_2$
$\vert\overset{2}{-}$ TF $\vert\overset{1}{-}$ I (n_2)	$I_2 = n_2^2 I_1$	$e_2 = n_2 e_1$ $f_1 = n_2 f_2$
$\vert\overset{2}{-}$ TF $\vert\overset{1}{-}$ C (n_3)	$\dfrac{1}{C_2} = n_3^2 \dfrac{1}{C_1}$	$e_2 = n_3 e_1$ $f_1 = n_3 f_2$
$\overset{2}{-}\vert$ TF $\overset{1}{-}\vert$ I (n_4)	$\dfrac{1}{I_2} = n_4^2 \dfrac{1}{I_1}$	$f_2 = n_4 f_1$ $e_1 = n_4 e_2$
$\vert\overset{2}{-}$ TF $\vert\overset{1}{-}$ R (n_5)	$R_2 = n_5^2 R_1$	$e_2 = n_5 e_1$ $f_1 = n_5 f_2$
$\overset{2}{-}\vert$ TF $\overset{1}{-}\vert$ R (n_6)	$\dfrac{1}{R_2} = n_6^2 \dfrac{1}{R_2}$	$f_2 = n_6 f_1$ $e_1 = n_6 e_2$

† In each case n is defined in concert with the causality; for example, $e_2 = n_2 e_1$ and $f_1 = n_2 f_2$ for the second row.

addressed through a gyrator. The resistive modulus of the GY is r. We seek an expression for the element at port 2. Thus

$$f_2 = r^{-1} e_1 \tag{14-36a}$$

$$e_1 = C_1^{-1} q_1 \tag{14-36b}$$

and

$$\dot{q}_1 = r^{-1} e_2 \tag{14-36c}$$

If we integrate the last equation,

$$q_1 = r^{-1}\textstyle\int e_2 \, dt = r^{-1} p_2 \tag{14-37}$$

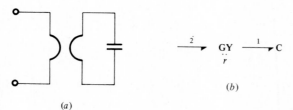

(a)

(b)

Figure 14-10 A GY-equivalence example.

and combine, we may obtain

$$f_2 = r^{-1}(C_1^{-1})r^{-1}p_2 \tag{14-38}$$

or

$$f_2 = \frac{1}{r^2 C_1} p_2 \tag{14-39}$$

This shows that the equivalent element at port 2 is an *inductor* with parameter value $r^2 C_1$.

The gyrator has the interesting property of making a C behave like an I and an I behave like a C. As H. M. Paynter observed, circa 1965, "The gyrator is the mirror by which I 'C' my 'I'." We have in fact just used a gyrator to "eye" our "see." This idea is used in practice in making microcircuits. Winding tiny inductors is impossible, but it is possible to make small gyrators and capacitors, and their combination creates the inductive effect.

Table 14-3 gives a set of gyrator equivalences. It dualizes the characteristics of the energy-storage elements, but once a dissipator, always a dissipator, so the combination $GY \rightarrow R$ is just another R.

14-6 CLASSIFICATION OF STATE EQUATIONS

In this section we consider some of the different types of state equations that arise in system dynamics. Three concepts of importance involve forcing, linearity, and explicitness.

Forced and Unforced Systems

A system subject to one or more independent inputs which are given functions of time is said to be a *forced system*. Otherwise the system is said to be *unforced*. Sometimes the terms nonautonomous (forced) and autonomous (unforced) are used. In engineering practice most systems have some disturbance or command forcings. Nonetheless, one can often gain much insight into the dynamic nature of a system

Table 14-3 Some GY equivalences

Combination	Equivalent element	Equivalent parameter
$\overset{1}{-} \overset{\overset{r}{\cdots}}{GY} \overset{2}{-} C$	$\overset{1}{-} I$	$I_1 = r^2 C_2$
$\overset{1}{-} \overset{\overset{r}{\cdots}}{GY} \overset{2}{-} I$	$\overset{1}{-} C$	$C_1 = r^2 I_2$
$\overset{1}{-} \overset{\overset{r}{\cdots}}{GY} \overset{2}{-} R$	$\overset{1}{-} R$	$R_1 = \dfrac{r^2}{R_2}$

by considering its unforced behavior first and then proceeding to the forced case. This is particularly true for linearized systems.

In the state equations, forced systems usually have a $U(t)$ vector and perhaps a related $\dot{U}$ vector. In fact, in formal terms, forced systems can be recognized when a t (independent) variable appears explicitly in the state equations. Thus we have

$$\phi(X, \dot{X}, t) = 0 \qquad (14\text{-}40a)$$

as forced, and

$$\phi(X, \dot{X}) = 0 \qquad (14\text{-}40b)$$

as unforced, where the ϕ's represent vectors of functions.

Linearity

When $\dot{X}$ can be expressed as a linear function of the states and the input variables, the state equations are linear. Otherwise, they are nonlinear. It is essential to know whether your model is linear or not because many powerful methods can be used only with linear systems. These methods involve Laplace transforms, frequency response, eigenvalue calculations, and modal analysis. Furthermore, it is true in general only for linear systems that the complete response can be assembled from the separate forced and unforced responses.

For linear state equations we distinguish two cases. If all elements in the A, B, and E matrices are constant, we have linear, time-invariant state equations. Such equations arise when the physical-system model has constant parameters for all I, C, and R elements and constant moduli for all TF and GY elements. On the other hand, if some of the elements of A, B, or E change with time, we have linear, time-varying state equations. An example occurs when physical parameters vary with temperature and the temperature is an independent time function.

Table 14-4 summarizes the common cases regarding forcing and linearity. The state equations are expressed in standard engineering notation. On occasion, the linear, time-varying case is said to be *parametrically forced* because one or more coefficients, derived from the physical parameters, vary independently.

Implicit Equations

As we have seen from our earlier discussions, it is not always possible to obtain state equations in explicit form from the system equations. The two practical cases of this

Table 14-4 Classification of state equations

Type	Unforced	Forced
Linear, time-invariant	$\dot{X} = AX$	$\dot{X} = AX + BU + \dot{E}U$
Linear, time-varying	$\dot{X} = A(t)X$	$\dot{X} = A(t)X + B(t)U + E(t)\dot{U}$
Nonlinear	$\dot{X} = \phi(X)$	$\dot{X} = \phi(X, U)$

sort that can cause problems are nonlinear dependent storage elements and coupled nonlinear dissipation elements. The system equations can be obtained in the form

$$\dot{X}_i = \phi_1(X_i, \dot{X}_d, U, H) \tag{14-41a}$$

$$X_d = \phi_2(X_i, U) \tag{14-41b}$$

$$H = \phi_3(X_i, U, H) \tag{14-41c}$$

where X_i, X_d, U, and H are the key vectors defined previously on the basis of causality. Furthermore, Eq. (14-41 b) can be converted into

$$\dot{X}_d = \phi_4(X_i, \dot{X}_i, U, \dot{U}) \tag{14-42}$$

by differentiating X_d with respect to time. Now the set of Eqs. (14-41 a), (14-42), and (14-41 c) is coupled in terms of X_i, $\dot{X}_i$, U, $\dot{U}$, H, and $\dot{X}_d$. If H and $\dot{X}_d$ can be eliminated, we obtain the implicit state equation

$$\dot{X}_i = \phi_5(X_i, \dot{X}_i, U, \dot{U}) \tag{14-43}$$

Finally, it may be possible to solve for X_i explicitly, yielding

$$\dot{X}_i = \phi_6(X_i, U, \dot{U}) \tag{14-44}$$

Since this is the easiest form to work with computationally, it is the natural objective of an equation-formulation process. Although this form is theoretically attainable, the reduction may be very difficult when nonlinear relations are involved.

To summarize, we observe that causality properly applied to a bond-graph model will help predict the structure of the state equations that may arise. Furthermore, the information thus gained can be used to guide the formulation and reduction process, as system equations are transformed into state equations.

PROBLEMS

14-1 Assign causality to each of the following bond graphs. In each case, interpret the graph as a mechanical system or an electric circuit and discuss your causal findings.

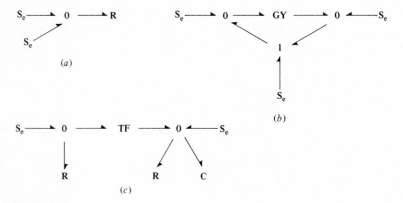

Figure P14-1

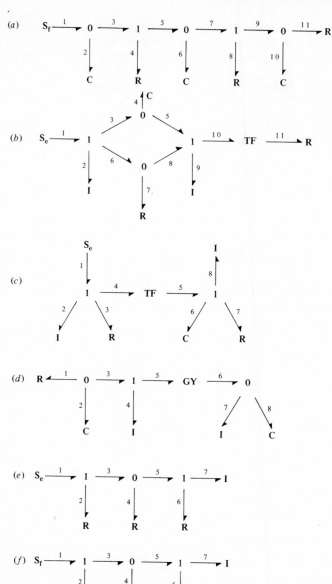

Figure P14-2

14-2 For each of the following bond graphs, assign causality and state whether the system has integral causality and whether R fields are present. Define X and U vectors. Define X_d and H vectors if necessary.

14-3 Write state equations for each of the bond graphs of Prob. 14-2. Assume that each element has a constant parameter. Put the state equations into explicit form.

14-4 For each of the bond graphs parts (a), (c), and (e) of Prob. 14-2 assume that all elements are nonlinear. That is, I, C, and R are described by $f = \phi_I(p)$, $e = \phi_C(q)$, and $e = \phi_R(f)$, respectively. Write a set of compact system equations. Where possible, write state equations.

14-5 A mechanical model of a muscle has been proposed as shown in the figure. A force F_1 is applied to deflect the muscle.

  (a) Make a bond-graph model.
  (b) Assign causality and define the state and input vectors.
  (c) Write state equations, assuming constant parameters for the elements.
  (d) Comment on the effects of a step change in $F_1(t)$.

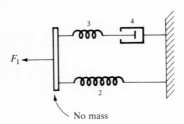

No mass   Figure P14-5

14-6 Show that the illustrations correspond to kinematic systems or resistive networks. Assume a rigid rod for A. Neglect the masses for all parts.

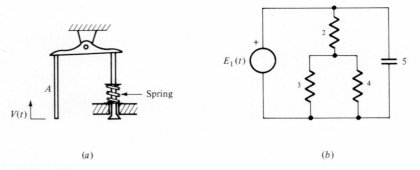

(a)   (b)

Figure P14-6

14-7 Given a linear system with constant coefficients in the form

$$\dot{X}_i = S_{11}X_i + S_{12}\dot{X}_d + S_{13}U \quad \text{and} \quad X_d = S_{21}X_i + S_{22}U$$

find A, B, and E, as defined by Eq. (14-13). What complications arise if S_{21} and S_{22} depend upon time t?

14-8 For the bond-graph model of Fig. P14-2c assume that all elements have constant parameters. Simplify the graph by using an equivalent element to remove the dependency between storage elements. Then write state equations and compare them with the solution for part (c) of Prob. 14-3.

14-9 A rack-and-pinion mechanism is shown. Assume that the pinion and rack have inertias J_p and M_R, respectively. The parameter for viscous friction is b. A torque input drives the pinion, and the rack works against a load force.

  (a) Make a bond-graph model of the system.
  (b) Assign causality and interpret the results.
  (c) How small should the pinion inertia be before it can be neglected compared with the mass of the rack?
  (d) How sensitive is the steady-state velocity to the inertia ratio for constant input torque and constant load force? That is, when can the pinion inertia be neglected safely?

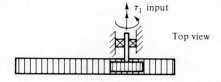

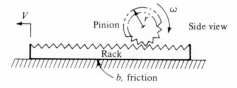

Figure P14-9

14-10 For part (*a*) of Fig. P14-10 make a bond-graph model, assign causality, and find an equivalent model with complete integral causality. *Hint:* The result requires a 2-port I element.

   (*b*) Repeat (*b*), assuming small angular motion only.

   (*c*) Repeat part (*b*), assuming that there is a torque input at the pin joint P and F_3 is zero. *Hint:* Define angular velocity for the bar as positive in the direction of positive torque input.

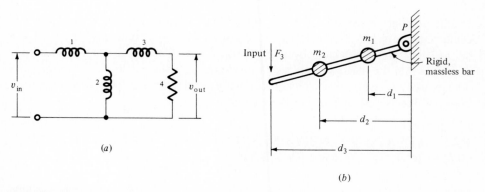

(*a*)

(*b*)

Figure P14-10

14-11 A hydraulic delivery system is shown. Assume that each numbered line has both inertia and resistance effects. Treat the tanks as simple capacitances with constant parameters.

   (*a*) Make a bond-graph model of the system.

   (*b*) Assign causality and interpret the results.

   (*c*) If line-inertia effects are negligible, repeat part (*b*). Derive state equations.

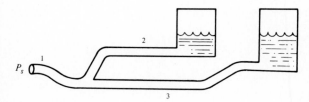

Figure P14-11

TIME AND FREQUENCY RESPONSE OF LINEAR SYSTEMS WITH CONSTANT COEFFICIENTS

In this chapter we consider the analysis of linear dynamic systems with constant coefficients. First we convert the state equations from the t domain (time) to the s domain (complex frequency) by Laplace transforms. Then we extract a maximum of insight into the linear-system response without actually calculating it in the time domain using several useful techniques. Finally, we develop frequency-response functions and investigate them using Bode diagrams.

15-1 LAPLACE-TRANSFORM SOLUTION OF CONSTANT-COEFFICIENT LINEAR SYSTEMS

Matrix methods are very useful in studying constant-coefficient linear systems (CCLS). Our investigation relies on matrix representation and a few matrix operations. In every case a completely worked example is given. Appendix B outlines the matrix operations necessary to use the methods in this chapter.

The Transformed State Equations

Consider the state equations for a CCLS in matrix form

$$\dot{X}(t) = AX(t) + BU(t) \tag{15-1}$$

where $X(t) = n \times 1$ column vector
$A = n \times n$ matrix
$U(t) = m \times 1$ column vector
$B = n \times m$ matrix

Matrices A and B have constant elements, $X(t)$ is the state vector, and $U(t)$ is the input vector. We say this is an nth-order system with m inputs.

As a specific example, consider the mechanical CCLS, bond graph, and causality in Fig. 15-1. From the causal graph we see that state and input vectors can be defined as

$$X(t) \equiv \begin{bmatrix} p_1 \\ p_2 \\ d_3 \end{bmatrix} \quad \text{and} \quad U(t) \equiv \begin{bmatrix} F_4(t) \\ F_9(t) \end{bmatrix}$$

where p_1, p_2 = momenta
d_3 = deflection of spring
F_4, F_9 = applied forces

With m, b, and k as symbols for mass, damping-constant, and spring-constant parameters, the state equations are

$$\dot{p}_1 = -b_5 m_1^{-1} p_1 \qquad\qquad - k_3 d_3 + 1 F_4 \tag{15-2a}$$

$$\dot{p}_2 = \qquad\qquad -b_6 m_2^{-1} p_2 + k_3 d_3 \qquad -1 F_9 \tag{15-2b}$$

$$\dot{d}_3 = \qquad m_1^{-1} p_1 - m_2^{-1} \; p_2 \tag{15-2c}$$

Observe that A is 3×3 and B is 3×2, so that n equals 3 and m equals 2. The matrices are

$$A = \begin{bmatrix} -b_5 m_1^{-1} & 0 & -k_3 \\ 0 & -b_6 m_2^{-1} & k_3 \\ m_1^{-1} & -m_2^{-1} & 0 \end{bmatrix} \quad \text{and} \quad B = \begin{bmatrix} 1 & 0 \\ 0 & -1 \\ 0 & 0 \end{bmatrix} \tag{15-3}$$

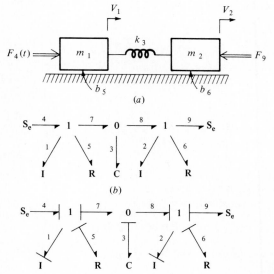

Figure 15-1 Mechanical example of a CCLS: (*a*) schematic, (*b*) bond graph, and (*c*) graph with causality.

To compute a response of the system we must know not only the inputs but also the initial conditions; i.e., we must know that

$$X(0) = \begin{bmatrix} p_1(0) \\ p_2(0) \\ d_3(0) \end{bmatrix} = \begin{bmatrix} p_{10} \\ p_{20} \\ d_{30} \end{bmatrix} \tag{15-4}$$

Observe that knowledge of p_{10}, p_{20}, and d_{30} fixes the system's energy at the initial time.

Now we proceed to Laplace-transform the state equations, first in general matrix form and then for the example. Define the corresponding vectors as

$$X(s) = \mathcal{L}\{X(t)\} \quad \text{and} \quad U(s) = \mathcal{L}\{U(t)\}$$

When no confusion between $X(t)$ and $X(s)$ will result, we shall drop the argument. When Eq. (15-1) is transformed, we get

$$sX(s) - X(0) = AX(s) + BU(s) \tag{15-5}$$

Collecting the unknowns on the left and the knowns on the right results in

$$(sI - A)X(s) = X(0) + BU(s) \tag{15-6}$$

where I is the $n \times n$ identity matrix. The known information is the initial state $X(0)$ and the transformed input vector $U(s)$. The unknowns are the elements of $X(s)$, which are the transformed state variables.

Applying the transformation to our example, Eqs. to (15-4), we obtain one vector-matrix equation in the s domain. Here we shall use A to construct $sI - A$ and write Eq. (15-6) formally. The reader is encouraged to obtain the same results by solving Eq. (15-2) in a longhand algebra fashion, having first Laplace-transformed each state equation. The savings in effort in using Eq. (15-6) will become evident. Return now to $sI - A$. We write

$$sI - A = \begin{bmatrix} s & 0 & 0 \\ 0 & s & 0 \\ 0 & 0 & s \end{bmatrix} - \begin{bmatrix} -b_5 m_1^{-1} & 0 & -k_3 \\ 0 & -b_6 m_2^{-1} & k_3 \\ m_1^{-1} & -m_2^{-1} & 0 \end{bmatrix} \tag{15-7}$$

$$= \begin{bmatrix} s + b_5 m_1^{-1} & 0 & k_3 \\ 0 & s + b_6 m_2^{-1} & -k_3 \\ -m_1^{-1} & m_2^{-1} & s \end{bmatrix}$$

Many useful results are obtained by studying this array.

When Eq. (15-6) is constructed, by substitution we obtain

$$(s + b_5 m_1^{-1})p_1(s) + k_3 d_3(s) = p_1(0) + 1F_4(s) \tag{15-8a}$$

$$(s + b_6 m_2^{-1})p_2(s) - k_3 d_3(s) = p_2(0) - 1F_9(s) \tag{15-8b}$$

$$-m_1^{-1} p_1(s) + m_2^{-1} p_2(s) + s d_3(s) = d_3(0) \tag{15-8c}$$

Transfer Functions

We shall return to the main theme shortly, but first we pause for an important digression. In engineering systems work it is frequently useful to ask how a particular variable responds to a given input. That is a single-input, single-output type of characterization. For example, in the mechanical system of Fig. 15-1 a we may wish to know how the spring deflects under a given input force $F_4(t)$ if the load force $F_9(t)$ is assumed zero. Furthermore, we may assume that all initial conditions are zero. For a CCLS this is not restrictive because we can use superposition to construct the response to any linear combination of inputs and initial conditions. In effect we consider influences one at a time.

We define *transfer function* in a CCLS as the ratio of the Laplace transform of a particular output to the Laplace transform of a particular input when all initial conditions and all other inputs are assumed to be zero. In the formal results of Eq. (15-6) the matrix equation becomes

$$(sI - A)X(s) = bu(s) \tag{15-9}$$

where b is the column vector of B corresponding to the chosen input element u (For example, u_2 and column 2 of B).

To solve Eq. (15-9) for a specific unknown, say $X_3(s)$, we can use Cramer's rule

$$X_3(s) = \frac{\det (sI - A)_3^*}{\det (sI - A)} \tag{15-10}$$

where $(sI - A)_3^*$ is the array $sI - A$ with column 3 replaced by the column vector $bu(s)$ and det denotes the determinant. In our example we have

$$d_3(s) = \frac{\det \begin{bmatrix} s + b_5m_1^{-1} & 0 & 1F_4(s) \\ 0 & s + b_6m_2^{-1} & 0 \\ -m_1^{-1} & m_2^{-1} & 0 \end{bmatrix}}{\det (sI - A)} \tag{15-11}$$

In expanded form

$$d_3(s) = \frac{(s + b_6m_2^{-1})m_1^{-1}F_4(s)}{\Delta_1}$$

or

$$\frac{d_3(s)}{F_4(s)} = \frac{(s + b_6m_2^{-1})m_1^{-1}}{\Delta_2} \tag{15-12}$$

where

$$\Delta_1 = (s + b_5m_1^{-1})(s + b_6m_2^{-1})s + m_1^{-1}(s + b_6m_2^{-1})k_3 + m_2^{-1}(s + b_5m_1^{-1})k_3$$

and

$$\Delta_2 = s^3 + s^2(b_5m_1^{-1} + b_6m_6^{-1}) + s(b_5m_1^{-1} + b_6m_2^{-1} \\ + k_3m_1^{-1} + k_3m_2^{-1}) + (b_5 + b_6)m_1^{-1}m_2^{-1}k_3$$

The polynomial ratio on the right-hand side of Eq. (15-12) is the desired transfer function.

It is of course possible to solve for $d_3(s)$ as a function of $F_4(s)$ by elimination from Eqs. (15-8), with initial conditions and F_9 set to zero. The important point is that the Laplace-transformed state equations can be used to generate all desired transfer functions. We shall return to the use of transfer functions when we discuss frequency response.

The Laplace Solution

The solution to Eq. (15-6) is given by

$$X(s) = (sI - A)^{-1}X(0) + (sI - A)^{-1}BU(s) \qquad (15\text{-}13)$$

or

$$X(s) = \phi(s)X(0) + \phi(s)BU(s) \qquad (15\text{-}14)$$

where

$$\phi(s) \equiv (sI - A)^{-1} = \text{resolvent matrix} \qquad (15\text{-}15)$$

Recall that the inverse of a matrix can be calculated as the adjoint divided by the determinant, i.e.,

$$(sI - A)^{-1} = \frac{\text{adj } (sI - A)}{\text{det } (sI - A)} \qquad (15\text{-}16)$$

The solution for $X(s)$ becomes

$$X(s) = \frac{\text{adj } (sI - A)}{\text{det } (sI - A)} [X(0) + BU(s)] \qquad (15\text{-}17)$$

where the numerator is equivalent to an $n \times 1$ column vector and the denominator is a scalar (in fact, it is an nth-order polynomial).

Here we focus our attention on the denominator. Observe that it does not depend upon either inputs or initial conditions but only upon A. The size of A determines the order of the polynomial in s. This polynomial, which shows up in the Laplace solution for every state (and system) variable, is called the *characteristic polynomial* $CP(s)$ (Chap. 5). We have

$$CP(s) = \text{det } (sI - A) \qquad (15\text{-}18)$$

For the example problem, the characteristic polynomial is

$$CP(s) = s^3 + s^2(b_5 m_1^{-1} + b_6 m_6^{-1}) + s(b_5 m_1^{-1} + b_6 m_2^{-1} \\ + k_3 m_1^{-1} + k_3 m_2^{-1}) + (b_5 + b_6)m_1^{-1}m_2^{-1}k_3 \qquad (15\text{-}19)$$

which can be obtained by evaluating the determinant of the matrix of Eq. (15-7). Why is this polynomial so important? The characteristic polynomial $CP(s)$ is important because it leads to a determination of the characteristic values (eigenvalues) of the CCLS by solving

$$CP(s) = 0 \qquad (15\text{-}20)$$

the *characteristic equation.*

The characteristic values of a CCLS are important because they show exactly which time functions will build up any particular unforced response in the given system. As we saw in Chap. 5, the roots of the characteristic equation are involved in the partial-fraction expansion of the Laplace transform of the solution for all state variables. Thus the eigenvalues determine the exponential or sinusoidal response components which add up to make the system response. Table 5-1 relates the eigenvalues to the time functions. Eigenvalues tell whether a system is stable or unstable and whether or not it can oscillate, for example. The eigenvalues typically provide insight into appropriate computing intervals when digital integration methods are used. They are also useful in assessing the potential effects of inputs. For example, if a system has a tendency to oscillate at a certain frequency, as shown by its eigenvalues, a sinusoidal input at or near that frequency will cause the system to resonate and may result in a large-amplitude response. Far better is it to know this early than to discover it later.

Consider our mechanical-system example. Let us set both damping parameters b_5 and b_6 to zero (a very slick surface indeed!) and assume that each inertia has a mass of 1 kg and the spring has a stiffness of 8 N/m. Referring to Eqs. (6-19) and (6-20), with these values we see that the characteristic equation is

$$s^3 + s^2(0) + s(16) + 0 = 0 \qquad (15\text{-}21a)$$

or
$$s(s^2 + 16) = 0 \qquad (15\text{-}21b)$$

The three solutions to Eq. (15-21 b) are

$$s_1 = 0 \qquad s_2 = +4j \qquad s_3 = -4j \qquad (15\text{-}22)$$

which are the characteristic values or eigenvalues. The complex-conjugate pair with zero real part, s_2 and s_3, corresponds to an oscillation. In a partial-fraction expansion, a denominator term of $s^2 + 16$ corresponds to a sinusoidal motion of circular frequency 4 rad/s. The zero value s_1 corresponds to a rigid-body motion of the entire system; i.e., the center of mass can translate freely when there is no damping and no force.

If damping is present, how does that alter the system response characteristics? Assume that both b_5 and b_6 equal 1 N·s/m. The characteristic equation is

$$s^3 + s^2(2) + s(18) + (8) = 0 \qquad (15\text{-}23)$$

One root can be found by trial and error as $s_1 = -0.463$. The factored form is then

$$(s + 0.463)(s^2 + 1.575 + 17.279) = 0$$

If you imagine a partial-fraction expansion with denominators $s + 0.463$ and $s^2 + 1.575s + 17.279$, you can see from Laplace transforms in Table 5-1 that the response will include a real decaying exponential as well as a damped sinusoidal term. The actual eigenvalues are

$$s_1 = -0.463 \qquad s_2 = -0.769 + 4.085j \qquad s_3 = -0.769 - 4.085j$$

As you can see, if you can find eigenvalues, you can predict the general form the components of the system response will take. Eigenvalues with negative real parts always have decaying exponential factors in their time functions. Eigenvalues with positive real parts have increasing exponential factors in their time functions and thus correspond to unstable systems. Pure imaginary eigenvalues correspond to pure sinusoidal forms, while complex eigenvalues correspond to decaying or growing sinusoidal forms, depending upon whether the real parts are negative or positive. Thus the general behavior of a linear system is exhibited by the type of eigenvalues it has.

15-2 THE TIME RESPONSE

Although matrix solution methods of analysis discussed here are not directly useful for hand calculation, they provide an excellent basis for understanding the structure of linear-system response. They also are valuable for designing automated solution procedures for implementation on digital computers.

The Unforced Case

Consider first the unforced case of CCLS. We have

$$X(t) = AX(t) \tag{15-24a}$$

$$X(0) = X_0 \tag{15-24b}$$

The Laplace solution is

$$X(s) = (sI - A)^{-1}X_0 \tag{15-25}$$

or

$$X(s) = \phi(s)X_0$$

If the $n \times n$ *resolvent matrix* $\phi(s)$ is inverted element by element, an $n \times n$ matrix results; i.e.,

$$\phi(t) = \mathcal{L}^{-1}\{\phi(s)\}$$

where $\phi(t)$ is called the *state transition matrix*. It plays a key role in constructing the response of CCLSs.

Example 15-1 Suppose we have the state equations

$$\dot{x}_1 = -5x_1 + 3x_2 \qquad x_1(0) = x_{10}$$

$$\dot{x}_2 = -2x_1 \qquad x_2(0) = x_{20}$$

Evaluate $\phi(t)$ and find $X(t)$. The resolvent matrix is

$$\phi(s) = (sI - A)^{-1} = \begin{bmatrix} s+5 & -3 \\ 2 & s \end{bmatrix}^{-1}$$

or

$$\phi(s) = \frac{1}{s^2 + 5s + 6} \begin{bmatrix} s & 3 \\ -2 & s+5 \end{bmatrix}$$

[You should check that $(sI - A)\phi(s) = I$, the identity matrix.] We note in passing that the denominator of every element in $\phi(s)$ is $CP(s)$, the characteristic polynomial.

To obtain $\phi(t) = \mathcal{L}^{-1}\{\phi(s)\}$ we must perform four Laplace inversions. For example, to find element $\phi_{11}(t)$ we write

$$\phi_{11}(t) = \mathcal{L}^{-1}\{\phi_{11}(s)\} = \mathcal{L}^{-1}\left\{\frac{s}{s^2 + 5s + 6}\right\}$$

Since

$$\phi_{11}(s) = \frac{s}{(s + 3)(s + 2)} = \frac{C_1}{s + 3} + \frac{C_2}{s + 2}$$

which is satisfied by $C_1 = 3$ and $C_2 = -2$, we find

$$\phi_{11}(s) = \frac{3}{s + 3} + \frac{-2}{s + 2}$$

Inverting term by term, we get

$$\phi_{11}(t) = 3e^{-3t} - 2e^{-2t}$$

The complete $\phi(t)$ matrix is

$$\phi(t) = \begin{bmatrix} 3e^{-3t} - 2e^{-2t} & -3e^{-3t} + 3e^{-2t} \\ 2e^{-3t} - 2e^{-2t} & -2e^{-3t} + 3e^{-2t} \end{bmatrix}$$

The time response is given by

$$X(t) = \phi(t)X_0$$

or

$$\begin{bmatrix} x_1(t) \\ x_2(t) \end{bmatrix} = \begin{bmatrix} \phi_{11}(t) & \phi_{12}(t) \\ \phi_{21}(t) & \phi_{22}(t) \end{bmatrix} \begin{bmatrix} x_{10} \\ x_{20} \end{bmatrix}$$

One quick check on the correctness of $\phi(t)$ is available by observing that $\phi(t = 0) = I$. (Do you see why?) In our case we have

$$\phi(t = 0) = \begin{bmatrix} 3 - 2 & -3 + 3 \\ 2 - 2 & -2 + 3 \end{bmatrix} = \begin{bmatrix} 1 & 0 \\ 0 & 1 \end{bmatrix}$$

so $\phi(t)$ checks.

The Forced Case

When a CCLS undergoes arbitrary forcing, we return to our Laplace solution, repeated for convenience,

$$X(s) = \phi(s)X(0) + \phi(s)BU(s) \tag{15-14}$$

Since the inputs enter into the solution as the product of Laplace-transform functions, their inversion leads to the convolution operation in the time domain. (It is well known that multiplication in the s domain implies convolution in the t domain.) The complete solution is

$$X(t) = \phi(t)X_0 + \int_0^t \phi(t - \tau)BU(\tau) \, d\tau \qquad (15\text{-}26)$$

in which the principle of superposition for linear systems allows us to add up the response due to initial conditions and the response due to the forcing terms. Notice that the state transition matrix $\phi(t)$ is involved in the convolution integral as well as the initial condition response.

Example 15-2 Consider the same state equations as in the previous example. Assume that the initial conditions are zero but there is a single input $u(t)$. Let

$$\dot{x}_1 = -5x_1 + 3x_2 + 1u(t) \qquad x_1(0) = 0$$

$$\dot{x}_2 = -2x_1 \qquad x_2(0) = 0$$

Since the initial conditions are zero, the formal solution becomes

$$X(t) = \int_0^t \phi(t - \tau) \begin{bmatrix} 1 \\ 0 \end{bmatrix} u(\tau) \, d\tau \qquad (15\text{-}27)$$

where the column vector is B from Eq. (15-26) and ϕ is a 2×2 matrix.

A very important special input is the unit impulse $\delta(t)$. This function is zero everywhere except when its argument is zero but is infinite when $t = 0$. It also has the property that any (time) integral that includes zero has the value 1. That is,

$$\int_{-\infty}^{\infty} \delta(t) \, dt = 1 \qquad (15\text{-}28)$$

where $\delta(t) = 0$ for $t \neq 0$.

This is certainly an interesting function, which may be thought of as the limit of a very tall, narrow pulse, and it is valuable in studying the time response of CCLSs.

If $\delta(t)$ is introduced into Eq. (15-27) in place of $u(t)$, we get

$$X(t) = \int_0^t \phi(t - \tau) \begin{bmatrix} 1 \\ 0 \end{bmatrix} \delta(\tau) \, d\tau \qquad (15\text{-}29)$$

or

$$X(t) = \phi(t) \begin{bmatrix} 1 \\ 0 \end{bmatrix} (1) \qquad (15\text{-}30)$$

where the integral is zero almost everywhere except in the vicinity of $t = 0$. Notice that the time response to an impulse is just a weighted (by B and the magnitude of u) version of the state transition matrix $\phi(t)$. For this reason $\phi(t)B$ is also called the *impulse-response function*.

In general, any input function can be calculated from the general solution, Eq. (15-26), but this is not usually practical. Several important types of input functions occur repeatedly in practice. A knowledge of how CCLSs respond to them provides an excellent basis for obtaining insight into response of arbitrary inputs and allows you to master computer-simulation results instead of being at their mercy (see Chap. 16).

The three major input types are impulse, step, and sinusoid. The impulse was illustrated in Example 15-2. Discussion of sinusoidal response will be taken up in the next section. That brings us to step response. A unit step, written $\mu(t)$, is a function that is zero for a negative argument and unity for an argument greater than or equal to zero. These three input types are defined in Table 15-1.

Step Response

If $u(t)$ is a single unit-step input $\mu(t)$, Eq. (15-26) becomes, for zero initial conditions,

$$X(t) = \int_0^t \phi(t - \tau)B(1) \, d\tau \tag{15-31}$$

where B is now a column vector. This can be expressed as

$$X(t) = \int_0^t \phi(t - \tau) \, d\tau \, B \tag{15-32}$$

Table 15-1 Some important input functions

Name	Notation	Definition	Sketch
Unit impulse	$\delta(t - \tau)$	$\delta = 0, \ t \neq \tau,$ $\int_{-\infty}^{\infty} \delta(\tau) \, d\tau = 1$	
Unit step	$\mu(t - \tau)$	$\mu = \begin{cases} 0 & t < \tau \\ 1 & t \geq \tau \end{cases}$	
Unit sinusoid	$\sin \omega t$	$\sin \omega t$	

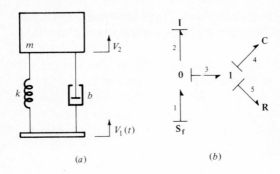

Figure 15-2 A mechanical oscillator: (*a*) mechanical model and (*b*) bond-graph model.

(*a*)   (*b*)

so that, once again, the key to the time response is the state transition matrix. This result can be applied to our example, assuming that B is $\begin{bmatrix} 1 \\ 0 \end{bmatrix}$ again, to yield

$$X(t) = \int_0^t \phi(t - \tau) \begin{bmatrix} 1 \\ 0 \end{bmatrix} d\tau \tag{15-33}$$

or

$$X(t) = \int_0^t \begin{bmatrix} \phi_{11}(t - \tau) \\ \phi_{21}(t - \tau) \end{bmatrix} d\tau \tag{15-34}$$

from which we obtain the result

$$X(t) = \begin{bmatrix} -e^{-3t} + e^{-2t} \\ -\tfrac{1}{3} - \tfrac{2}{3}e^{-3t} + e^{-2t} \end{bmatrix} \tag{15-35}$$

after substituting the previously obtained functions. We note that the steady-state response $X(t \to \infty)$ is given as $\begin{bmatrix} 0 \\ -\tfrac{1}{3} \end{bmatrix}$ from Eq. (15-35).

Steady-State Response

If a stable CCLS has only step inputs, it is reasonable to ask: What set of steady, i.e., constant, values does the state approach? (Unstable systems have solutions which diverge exponentially.) To answer this question we could use the direct-integration method just illustrated, but that requires both a knowledge of $\phi(t)$ and a fair amount of simple integration.

An alternative is to return to Eq. (15-14), the Laplace solution for the forced case, and apply the final-value theorem to it. If $u(t)$ is considered as a single step input $\mu(t)$ and B is taken to be the same as in our previous example, we get

$$X(s) = \phi(s)X(0) + \phi(s) \begin{bmatrix} 1 \\ 0 \end{bmatrix} \frac{1}{s} \tag{15-36}$$

Notice that the Laplace transform for the unit step has been introduced into Eq. (15-36). The final-value theorem applied to the equation gives, assuming zero initial conditions,

$$X(t \to \infty) = \lim_{s \to 0} \{sX(s)\} = \lim_{s \to 0} s \left\{ \phi(s) \begin{bmatrix} 1 \\ 0 \end{bmatrix} \frac{1}{s} \right\}$$

$$= \lim_{s \to 0} \left\{ \phi(s) \begin{bmatrix} 1 \\ 0 \end{bmatrix} \right\} = \phi(0) \begin{bmatrix} 1 \\ 0 \end{bmatrix} \qquad (15\text{-}37)$$

Recall that $\phi(s) = (sI - A)^{-1}$, so that $\phi(0) = (-A)^{-1}$. Then we have

$$X(t \to \infty) = \begin{bmatrix} 5 & -3 \\ 2 & 0 \end{bmatrix}^{-1} \begin{bmatrix} 1 \\ 0 \end{bmatrix} = \begin{bmatrix} 0 & \frac{1}{2} \\ -\frac{1}{3} & \frac{5}{6} \end{bmatrix} \begin{bmatrix} 1 \\ 0 \end{bmatrix} = \begin{bmatrix} 0 \\ -\frac{1}{3} \end{bmatrix} \qquad (15\text{-}38)$$

This result agrees with that obtained in the previous section by the direct time-integration method but is far less work. In fact, it does not require the calculations of $\phi(t)$ if only steady-state information is required. If the system is stable, the initial-condition response will die away to zero, and so this procecure is quite general. The generalization of the result is

$$X(t \to \infty) = \lim_{s \to 0} \left\{ s\phi(s) BU_c \frac{1}{s} \right\} \qquad (15\text{-}39)$$

where U_c is the vector of step magnitudes. This yields

$$X(t \to \infty) = \phi(0) BU_c \qquad (15\text{-}40)$$

for systems with multiple step inputs. Since $\phi(0) = -A^{-1}$, this gives

$$X(t \to \infty) = -A^{-1} BU_c \qquad (15\text{-}41)$$

Of course, if A is singular, i.e., det A is zero, Eq. (15-41) is not applicable.

The third major input type, sinusoidal, is important enough to deserve a section to itself.

15-3 FREQUENCY RESPONSE

Motivation

Many important phenomena in the study of CCLSs can be understood best by considering the response to sinusoidal inputs. We do not have space here to do more than mention a few situations that arise frequently. In acoustics it is useful to characterize problems in terms of types and amounts of sinusoidal signals in the environment. In noise and vibrational disturbance problems in general the same is true. For example, the vibrations induced in a machine due to an out-of-balance rotating mass are most easily analyzed and corrected by frequency-response methods. Many types of electric circuits and electronic networks are designed with respect to frequency-response criteria. This is true of filters, signal-processing equipment, and many electromechanical instruments. The ride-quality analysis of automobiles used in designing suspensions to provide good passenger comfort often proceeds according to frequency-response analysis and design methods. In all these areas and many others the basic question to be answered is: Given a CCLS with a single sinusoidal input, what is the nature of the steady-state response?

We have already developed a tool that can be used to answer the question in a formal way. Laplace transforms can be applied to the system equations, a transfer function can be derived relating the desired response variable to the particular input, and the time response can be found. Since we are interested in the steady-state sinusoidal response after transients due to initial conditions have died out, some special techniques can be used.

Example 15-3 Consider the mechanical oscillator with its bond graph shown in Fig. 15-2. The state equations are

$$\dot{p}_2 = -bm^{-1}p_2 + kx_4 + bV_1(t) \qquad (15\text{-}42a)$$

$$\dot{x}_4 = -m^{-1}p_2 + \qquad\qquad 1\,V_1(t) \qquad (15\text{-}42b)$$

Suppose we are interested in the response of the mass velocity v_2 as $V_1(t)$ varies sinusoidally. If p_2 is replaced by its v_2 equivalent, we get

$$\dot{v}_2 = -bm^{-1}v_2 + km^{-1}x_4 + bm^{-1}V_1(t) \qquad (15\text{-}43\,a)$$

$$\dot{x}_4 = -1 \quad v_2 \qquad\qquad\quad +1 \quad V_1(t) \qquad (15\text{-}43\,b)$$

Now solve for $V_2(s)$ in terms of $V_1(s)$, with initial conditions zero, to get

$$V_2(s) = \frac{s+1}{s^2+s+1}\,V_1(s) \qquad (15\text{-}44)$$

[Of course we could have used Eq. (15.42a) to solve for $P_2(s)$ and then found $V_2(s)$ from an output equation.] If

$$V_1(t) = C_1 \sin \omega_1 t$$

then

$$V_1(s) = \frac{C_1\omega_1}{s^2+\omega_1^2}$$

so that

$$V_2(s) = \frac{(s+1)C_1\omega_1}{(s^2+s+1)(s^2+\omega_1^2)} \qquad (15\text{-}45)$$

Let us find $v_2(t)$ after any decaying terms have died out for $\omega_1 = 2$ rad/s:

$$V_2(s) = \frac{(s+1)2C_1}{(s^2+s+1)(s^2+4)} \qquad (15\text{-}46)$$

or
$$V_2(s) = \frac{\%_{13}C_1 s + \%_{13}C_1}{(s + \frac{1}{2})^2 + (\sqrt{3}/2)^2} + \frac{-\%_{13}C_1 s + \%_{13}C_1}{s^2 + 2^2}$$

after much laborious calculation to find the numerator constants. The first term above corresponds to a time function which approaches zero as time approaches infinity, because $e^{-t/2}$ is a multiplying factor. We see this from a Laplace-transform table. The second term, the steady-state sinusoidal response, can be inverted by table lookup to give

$$v_2(t) = -\%_{13}C_1 \cos 2t + \%_{13}C_1 \sin 2t$$

or
$$v_2(t) = \frac{\sqrt{65}}{13} C_1 \sin (2t - 1.56) \qquad (15\text{-}47)$$

Note that the ratio of the steady-state response amplitude to the input amplitude is $\sqrt{65}/13$ and the response is shifted relative to the input by -1.56 rad or $89.4°$. This negative phase angle between the input sinusoid and the steady-state response sinusoid is called a *phase lag*.

Imagine that we would like to know the response not just to a frequency of 2 rad/s but over a wide range, say 0.1 to 100 rad/s. Since it would be tedious to try to map the response by Laplace calculations, we shall develop special techniques for these calculations.

The Frequency-Response Function

The frequency-response function for a given CCLS is the transfer function with s replaced by $j\omega$. The frequency-response function gives the steady-state response due to a sinusoidal input for a stable CCLS. Space does not permit us to justify this statement, but we can at least show here that it does work, by application to the previous example.

First, observe from Eq. (15-47) that the key information about the steady-state response of $v_2(t)$ was the magnitude ratio of output to input $\sqrt{65}/13$ and the phase shift -1.56 rad $= -89.4°$. Now consider the Laplace solution, Eq. (15-44). The transfer function is

$$T(s) = \frac{V_2(s)}{V_1(s)} = \frac{s + 1}{s^2 + s + 1} \qquad (15\text{-}48)$$

Again let $V_1(t) = C_1 \sin \omega_1 t$. Now find the frequency-response function $T(s = j\omega)$

$$T(j\omega) = \frac{j\omega + 1}{(j\omega)^2 + j\omega + 1} = \frac{1 + j\omega}{(1 - \omega^2) + j\omega} \qquad (15\text{-}49)$$

If we let $\omega = 2$ rad/s, as in our example,

$$T(j2) = \frac{1 + 2j}{-3 + 2j} \qquad (15\text{-}50)$$

The amplitude ratio of V_2 to V_1 is given by the magnitude of $T(j\omega)$; thus

$$|T(j2)| = \frac{|1 + 2j|}{-3 + 2j|} = \frac{(1 + 4)^{1/2}}{(9 + 4)^{1/2}}$$

or

$$|T(j2)| = \frac{5^{1/2}}{13^{1/2}} = \frac{\sqrt{65}}{13}$$

This is the same value we found earlier with a great deal more effort.

The phase shift is given by the argument of $T(j\omega)$; thus

$$\arg T(j2) = \tan^{-1}\frac{2}{1} - \tan^{-1}\frac{2}{-3} = -1.56 \text{ rad} = -89.4°$$

Here we have used the idea that the angle of T is the angle of the numerator complex number minus the angle of the denominator number. Again, this matches the result found earlier.

Suppose we wish to evaluate the steady-state response at 1 rad/s. Simply return to Eq. (15-49) and let $\omega = 1$. Repeat the calculations shown above to find that

$$T(j1) = \frac{1 + j1}{j1} = 1 - j$$

Then

$$|T(j1)| = |1 - j| = \sqrt{2}$$

and

$$\arg T(j1) = \tan^{-1}\frac{-1}{1} = \frac{-\pi}{4} \text{ rad} = -45°$$

The response, v_2, is $\sqrt{2}\, C_1 \sin (1t - \pi/4)$, where C_1 is the amplitude of the input $V_1(t)$.

Bode Diagrams

In considering CCLS steady-state response to sinusoidal inputs, we commonly are interested in behavior over a wide range of frequency. We might like to predict response for $0 < \omega < \infty$, in fact, just as in the time domain we imagine solving for $v_2(t)$ for $0 < t < \infty$.

For example, consider the frequency-response function

$$T(j\omega) = \frac{1 + j\omega}{(1 - \omega^2) + j\omega} \tag{15-49}$$

The magnitude of $T(j\omega)$ is

$$|T(j\omega)| = \left|\frac{1 + j\omega}{(1 - \omega^2) + j\omega}\right| = \frac{|1 + j\omega|}{|(1 - \omega^2) + j\omega|}$$

or

$$|T(j\omega)| = \frac{\sqrt{1 + \omega^2}}{\sqrt{(1 - \omega^2)^2 + \omega^2}} = \frac{\sqrt{1 + \omega^2}}{\sqrt{1 - \omega^2 + \omega^4}} \tag{15-51}$$

This function can be sketched as ω varies from zero to infinity to show the magnitude ratio.

The phase angle is the argument of the complex number $T(j\omega)$. It is the angle of the number in the complex plane

$$\arg\ T(j\omega)\ =\ \arg \frac{1\ +\ j\omega}{(1\ -\ \omega^2)\ +\ j\omega}$$

or

$$\arg\ T(j\omega)\ =\ \arg\ (1\ +\ j\omega)\ -\ \arg\ [(1\ -\ \omega^2)\ +\ j\omega] \qquad (15\text{-}52)$$

The phase-angle function could be sketched by evaluating Eq. (15-52) at various values of ω.

Because of the polynomial nature of the expressions for $T(j\omega)$, a useful form for presenting the data is a *Bode diagram*. We make two related sketches. One is of the magnitude ratio versus frequency, and one is of the phase shift versus frequency. Furthermore, we use the logarithm of frequency (to the base 10) as the abscissa, and for the magnitude ordinate we use the logarithm of $|T(j\omega)|$. For the phase ordinate we use arg $T(j\omega)$ directly in radians or degrees. This way of plotting simplifies sketching the frequency-response functions for wide ranges of frequency. The Bode diagram of the example of Fig. 15-2 in Fig. 15-3 shows the log of $|T(j\omega)|$ and arg $T(j\omega)$. Notice how they are related by the frequency axis.

Logarithmic scales have several advantages: a wide range of frequencies can be covered, and we do not actually have to consider the full range 0 to ∞ for ω because

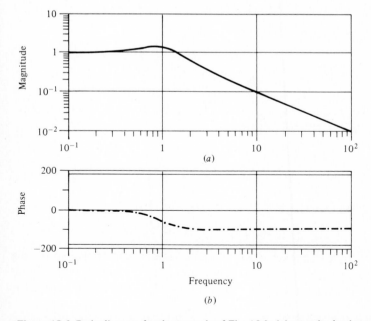

(a)

(b)

Figure 15-3 Bode diagram for the example of Fig. 15-2: (a) magnitude plot and (b) phase plot.

at both low and high frequencies the log-log curve approaches straight-line asymptotes. Let us examine this last statement more closely. Suppose we have a transfer function $T(s)$ given. It can be written in polynomial-ratio form as

$$T(s) = \frac{K(s^m + a_1 s^{m-1} + \cdots + a_{m-1} s + a_m)}{s^n + b_1 s^{n-1} + \cdots + b_{n-1} s + b_n} \tag{15-53}$$

where m = order of numerator
$\quad\quad\;\; n$ = order of denominator
$\quad\quad\;\; K$ = constant

The frequency-response function is

$$T(j\omega) = \frac{K[(j\omega)^m + a_1(j\omega)^{m-1} + \cdots + a_{m-1}(j\omega) + a_m]}{(j\omega)^n + b_1(j\omega)^{n-1} + \cdots + b_{n-1}(j\omega) + b_n} \tag{15-54}$$

For very low frequencies $|T(j\omega)|$ approaches the form

$$|T(j\omega)|_{\text{lo}} = \left| K \frac{a_m}{b_n} \right| \tag{15-55}$$

which is a finite constant provided b_n is not zero. For very high frequencies $|T(j\omega)|$ approaches the form

$$T|(j\omega)|_{\text{hi}} = K \frac{|(j\omega)^m|}{|(j\omega)^n|} = K \frac{\omega^m}{\omega^n}$$

or

$$|T(j\omega)|_{\text{hi}} = K\omega^{m-n} \tag{15-56}$$

Now consider the logarithm of Eq. (15-55)

$$\log |(j\omega)|_{\text{lo}} = \log \left| K \frac{a_m}{b_n} \right| = \log |K| + \log \left| \frac{a_m}{b_n} \right|$$

which is a straight line of zero slope; i.e., it does not depend upon log ω. On the other hand, the logarithm of Eq. (15-54) becomes

$$\log |T(j\omega)|_{\text{hi}} = \log |K\omega^{(m-n)}|$$

or

$$\log |T(j\omega)|_{\text{hi}} = \log |K| + \log | \omega^{m-n}$$
$$= \log |K| + (m-n) \log |\omega|$$

Thus we see that at high frequencies the log magnitude is proportional to the log frequency, with an integer slope of value $m-n$. Since typical engineering systems have m less than n, the magnitude of the response falls off at high frequencies in the usual case.

Example 5-4 The transfer function for a band-pass filter is

$$T(s) = \frac{100s}{s^2 + 51s + 50}$$

The frequency-response function is

$$T(j\omega) = \frac{100j\omega}{(j\omega)^2 + 51j\omega + 50}$$

For low frequency $|T(j\omega)| \to 0$ because the numerator goes to zero as ω goes to zero. The limiting form is $T(j\omega) \to 100j\omega/50$ for small values of ω. For high frequency $T(j\omega) \to 0$, because the $(j\omega)^2$ term in the denominator becomes much greater than the $100j\omega$ term in the numerator for large values of ω. The limiting form is $T(j\omega) \to 100/j\omega$ for large values of ω. A Bode diagram of the frequency-response function is shown in Fig. 15-4. Can you see the band-pass nature of the filter? Outside the frequency range $1 < \omega < 50$ there is a dramatic suppression of the input signal. Can you see now why the log magnitude versus log frequency has a $+1$ slope at low frequencies and a -1 slope at high frequencies? Think of the asymptotic forms and how they would look on a log-log plot.

Sketching the Bode Diagram

We now present some simple rules for sketching the Bode diagram over a wide frequency range. The ability to sketch the log-magnitude response helps not only in

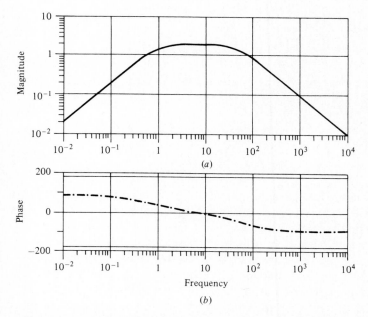

Figure 15-4 Bode diagram for filter example: (*a*) magnitude plot and (*b*) phase plot.

designing systems but also in interpreting Bode plots generated either analytically or experimentally. The sketching method is based on a straight-line approximation to the log-magnitude–versus–log-frequency expression.

Consider the transfer function of Eq. (15-53), rewritten in factored form

$$T(s) = K \frac{N(s)}{D(s)} = K \frac{(s + Z_1)(s + Z_2) \cdots (s + Z_m)}{(s + P_1)(s + P_2) \cdots (s + P_n)} \quad (15\text{-}57)$$

where the Z_i are negatives of the zero locations and the P_i are the negatives of the pole locations of $T(s)$. For now, we assume that P's and Z's are real numbers. K is a constant. The associated frequency-response function is

$$T(j\omega) = K \frac{(j\omega + Z_1)(j\omega + Z_2) \cdots (j\omega + Z_m)}{(j\omega + P_1)(j\omega + P_2) \cdots (j\omega + P_n)}$$

or, in slightly different form,

$$T(j\omega) = K \frac{\displaystyle\prod^m Z_i \left(1 + j\frac{\omega}{Z_1}\right)\left(1 + j\frac{\omega}{Z_2}\right) \cdots \left(1 + j\frac{\omega}{Z_m}\right)}{\displaystyle\prod^n P_i \left(1 + j\frac{\omega}{P_1}\right)\left(1 + j\frac{\omega}{P_2}\right) \cdots \left(1 + j\frac{\omega}{P_n}\right)} \quad (15\text{-}58a)$$

where Π stands for a product. Now the magnitude of $T(j\omega)$ is given by

$$|T(j\omega)| = \left| K \frac{\displaystyle\prod^m Z_i}{\displaystyle\prod^n P_i} \right| \frac{\left|1 + j\dfrac{\omega}{Z_1}\right| \cdots \left|1 + j\dfrac{\omega}{Z_m}\right|}{\left|1 + j\dfrac{\omega}{P_1}\right| \cdots \left|1 + j\dfrac{\omega}{P_n}\right|}$$

or, after denoting the constant term by K' and taking the log of $|T(j\omega)|$,

$$\log |T(j\omega)| = \log |K'| + \log \left|1 + j\frac{\omega}{Z_1}\right| \cdots + \log \left|1 + j\frac{\omega}{Z_m}\right|$$

$$- \log \left|1 + j\frac{\omega}{P_1}\right| \cdots - \log \left|1 + j\frac{\omega}{P_n}\right| \quad (15\text{-}58b)$$

Equation (15-58b) shows that the log magnitude of the frequency-response function is given by a sum of factors. If each factor can be sketched easily, the result is obtained just by adding the factors together, using the proper signs, of course.

Clearly, the constant term K' contributes a straight line with zero slope to the T-versus-ω plot since it does not depend upon ω. The sketch of a typical first-order factor can be obtained by approximating its behavior for both low- and high-frequency ranges and joining the approximations. Consider the factor $|1 + j\omega/Z_1|$, for example. If $\omega \ll Z$, the factor approaches unity; if $\omega \gg Z_1$, the factor goes as $|j\omega/Z_1|$. We also note that the two straight-line approximations intersect at $\omega = Z_1$. Figure 15-5a shows the sketch. Notice that the slope for high frequencies is $+$

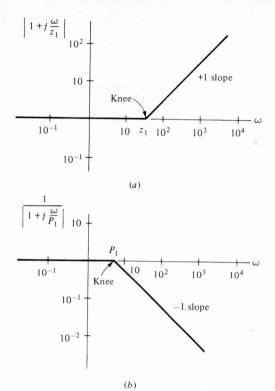

Figure 15-5 Straight-line approximation of first-order factors: (*a*) numerator factor and (*b*) denominator factor.

1, since the factor enters the log $T(j\omega)$ expression with a plus sign. A first-order denominator factor can be sketched in a similar fashion. Consider $-\log |1 + j\omega/P_1|$. We have

$$-\log \left| 1 + j\frac{\omega}{P_1} \right| \approx \begin{cases} -\log |1| & \omega \ll P_1 \\ -\log |\omega| + \log |P_1| & \omega \gg P_1 \end{cases} \quad (15\text{-}59a)$$

$$(15\text{-}59b)$$

At $\omega = P_1$ the expressions are equal. Figure 15-5*b* shows a sketch of a denominator factor. Of course, the two straight lines are nearly exact only at very low and very high frequencies. Near the knee of the diagram the actual curve for the factor deviates from the asymptotic expression.

All first-order factors of the form $|1 + j\omega/\omega_1|$ have low-frequency approximations of $|1|$, that is, a line of slope 0 at magnitude 1. This continues to the *break frequency* ω_1, whereupon the approximation becomes a straight line of slope $+1$ or -1 for a numerator or denominator factor, respectively.

The maximum error in the straight-line approximation is at the knee of the curve, i.e., at $\omega = \omega_1$. Let us calculate the true value. For a numerator factor we have

$$\left|1 + j\frac{\omega}{Z_1}\right|\Bigg|_{\omega=Z_1} = \left|1 + j\frac{Z_1}{Z_1}\right| = |1 + j| = \sqrt{2}$$

compared with an approximate factor of 1. For a denominator factor we have

$$\frac{1}{|1 + j\omega/P_1|}\Bigg|_{\omega=P_1} = \left|\frac{1}{1 + jP_1/P_1}\right| = \frac{1}{|1 + j|} = \frac{1}{\sqrt{2}}$$

showing that the true curve is below 1. Figure 15-6 gives the classic first-order denominator factor.

Example 15-5 Let

$$T(s) = \frac{10,000(s + 1)}{(s + 10)(s + 100)}$$

or

$$T(j\omega) = \frac{10(1 + j\omega/1)}{(1 + j\omega/10)(1 + j\omega/100)}$$

The log-magnitude expression is

$$\log |T(j\omega)| = \log |10| + \log \left|1 + j\frac{\omega}{1}\right| - \log \left|1 + j\frac{\omega}{10}\right| - \log \left|1 + j\frac{\omega}{100}\right|$$

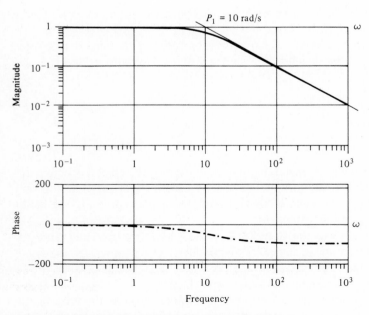

Figure 15-6 Bode diagram for first-order denominator factor.

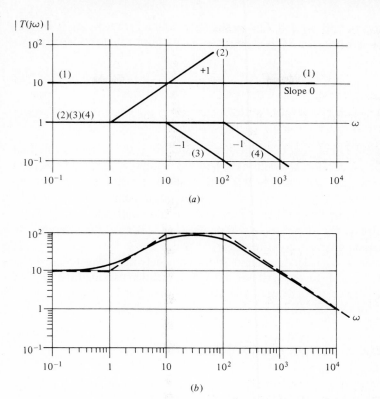

$|T(j\omega)|$

(a)

(b)

Figure 15-7 An example: (*a*) factor sketches and (*b*) composite sketch and exact solution.

The several factors are sketched separately and combined in Fig. 15-7. The straight-line approximation is superimposed on an actual plot. Notice that the greatest errors occur at the knees of the sketch. You can verify for yourself that the low- and high-frequency asymptotes are correct, using the reasoning of the previous section.

It is always a good idea to check the sketch at a few key points. Let us make an accurate calculation for $\omega = 10$. Then

$$|T(j\omega)| = \left|\frac{10{,}000(j\omega + 1)}{(j\omega + 10)(j\omega + 100)}\right| = \left|\frac{10^4(1 + j10)}{(10 + j10)(100 + j10)}\right|$$

or

$$|T(j\omega)| = \frac{10^4|1 + j10|}{|10 + j10|\,|100 + j10|} = \frac{10^4\,\sqrt{101}}{(10\,\sqrt{2})(10\,\sqrt{101})}$$

or

$$|T(j\omega)| = \frac{100}{\sqrt{2}}$$

And the result agrees with both the sketch and the computed solution of Fig. 15-7*b*.

Two other types of factors should be considered in order to round out the straight-line approximation method. One is the pure s factor, that is, $Z_i = 0$ or $P_i = 0$. Consider the expansion of $T(s)$ when it contains a pure s factor, for example,

$$T(s) = K \frac{N(s)}{D(s)} = K \frac{\Pi(s + Z_i)}{\Pi(s + P_i)} \frac{1}{s} \tag{15-60}$$

Then

$$\log |T(j\omega)| = \log \left| K \frac{\Pi(j\omega + Z_i)}{\Pi(j\omega + P_i)} \right| - \log |j\omega| \tag{15-61}$$

The new factor has a minus sign here because it came from the denominator. If it had come from the numerator, it would have had a plus sign. A sketch of the term $-\log |j\omega| = -\log |\omega|$ is a straight line with a slope of -1 for all ω, passing through the point where $|T(j\omega)| = 1$ at $\omega = 1$. (The origin point $T = 0$, $\omega = 0$ is, of course, at $-\infty$, $-\infty$ on log-log plots.) Pure s factors have both a low- and a high-frequency role to play; the slope persists over all ω. In this case the straight-line approximation method gives an exact result.

The final factor to consider is a quadratic factor, corresponding to a case of complex-conjugate Z_i or P_i factors. All our previous discussion assumed that Z_i and P_i were real. Now we consider the case when they are not. For brevity we examine only the denominator case. Consider

$$T(s) = \frac{1}{s^2 + 2\zeta\omega_n s + \omega_n^2} \tag{15-62}$$

where $0 \le \zeta \le 1$ and ω_n real.

The parameter ζ is the damping ratio. For $\zeta < 1$ and positive, the factor has complex roots and represents damped oscillatory behavior. Then

$$T(j\omega) = \frac{1/\omega_n^2}{1 + 2\zeta j\omega/\omega_n + (j\omega/\omega_n)^2}$$

or

$$T(j\omega) = \frac{1/\omega_n^2}{[1 - (\omega/\omega_n)^2] + 2\zeta j\omega/\omega_n} \tag{15-63}$$

The undamped natural frequency ω_n appears to be a break frequency of some kind, but the damping ratio ζ also influences the behavior of the response. For example, let us evaluate $T(j\omega)$ for two values of ζ

$$|T(j\omega)| = \begin{cases} \dfrac{1/\omega_n^2}{1 - (\omega/\omega_n)^2} & \text{for } \zeta = 0 \tag{15-64a} \\[4mm] \dfrac{1/\omega_n^2}{|(1 + j\omega/\omega_n)^2|} & \text{for } \zeta = 1 \tag{15-64b} \end{cases}$$

To see how different these values can be, let us check them at $\omega = \omega_n$:

$$|T(j\omega_n)| = \begin{cases} \dfrac{1/\omega_n^2}{1 - 1} \to \infty & \text{for } \zeta = 0 \tag{15-65a} \\[4mm] \dfrac{1/\omega_n^2}{2} & \text{for } \zeta = 1 \tag{15-65b} \end{cases}$$

A complete family of curves that shows the exact magnitude ratio as the damping ratio ζ is varied is given in Fig. 15-8, along with the phase shift. Notice that the low-frequency asymptote for all curves is the same; the high-frequency asymptote falls off with a slope of -2. The break frequency is ω_n.

The usual way to sketch a quadratic factor is to use a straight-line slope of zero for $0 < \omega \leq \omega_n$ and a slope of -2 for $\omega_n \leq \omega$ (for a denominator factor). They

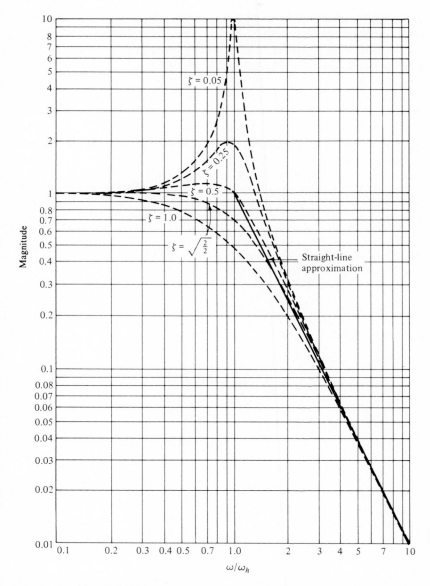

Figure 15-8 Magnitude plot of quadratic denominator factor $\omega_n^2/(S^2 + 2\omega_n S + \omega_n^2)$. *(From J. L. Melsa and D. G. Schultz, Linear Control Systems, McGraw-Hill Book Company, New York, 1969.)*

meet at $\omega = \omega_n$. Then use the ζ parameter to find the correction value at $\omega = \omega_n$, which can be very different from the knee value, especially if ζ is small.

We have now introduced all the types of factors to be expected in a rational transfer function. They include the constant, a real first-order factor, a quadratic factor, and a pure s. The quadratic terms need to be specially considered near $\omega = \omega_n$, depending on the value of ζ. Each of these factors has its own simple straight-line approximations. The set of approximations can be combined by addition on log-log axes to obtain a composite sketch, which can be refined at the break frequencies or knees to improve the accuracy, as required.

Frequency-response plots are widely used to describe systems since they are fairly easy to generate experimentally on prototype systems. They recur in Chap. 18 and 19, where we study vibrations and automatic control.

PROBLEMS

15-1 Since the state transition matrix $\phi(t)$ obeys $\phi(t = 0) = I$, show how the initial-value theorem can be used to check the correctness of $\phi(s)$, the resolvent matrix.

15-2 For the system with a characteristic polynomial given by Eq. (15-19) and for parameters $m_1 = m_6 = 1.0$ kg and $K_3 = 8.0$ N/m find values for b_5 and b_6 such that no oscillations are possible. Assume that b_5 and b_6 are equal. *Hint:* Find a value for b such that repeated roots occur.

15-3 An electric circuit used as a 2-port filter is shown in the figure. An effort source provides the voltage v_{in}, and no current flows at the output.
   (a) Find the transfer function between v_{out} and v_{in}.
   (b) Find the characteristic polynomial and the associated characteristic values.
   (c) Find the impulse-response function for the system.
   (d) Find $v_{out}(t)$ if $v_{in}(t)$ is a unit impulse at time zero and the capacitors are uncharged at $t = 0$.

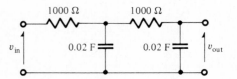

Figure P15-3

15-4 A hydraulic system has the state equations

$$\dot{P}_1 = -4P_1 \qquad\qquad + 10Q_4(t)$$

$$\dot{Q}_2 = 2P_1 - 3Q_2 - 4P_2$$

$$\dot{P}_3 = \qquad 1Q_2$$

where $Q_4(t)$ is an input flow.
   (a) Find the resolvent matrix.
   (b) Find the state transition matrix.
   (c) If the initial conditions are zero and the input $Q_4(t)$ is a unit step at time zero, find the steady-state response for P_1.
   (d) If the input is zero, find a set of initial conditions that makes $P_1(t = 1) = 1$. *Hint:* The answer is not unique.

15-5 A classical problem in the study of CCLS is the double oscillator, of which a mechanical version is shown.

(a) Make a bond-graph model. Assume no friction.

(b) Write state equations.

(c) Assume that all parameters are unity. Find the characteristic polynomial and the characteristic values. Interpret the system response.

(d) Write the general form of the unforced response.

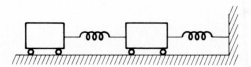

Figure P15-5

15-6 For the filter example discussed in Sec. 15-3 the Bode diagram is given in Fig. 15-4. If the input is 4 sin 10t, what is the response magnitude? The response phase shift? Verify your result by checking $T(j\omega)$ for $\omega = 10$.

15-7 For the Bode diagram in Fig. 15-3 what input frequency would probably show the greatest response? Show that if the input frequency is 100 rad/s, the phase shift is approximately $-90°$ or $\pi/2$ rad. Use $T(s)$ with $s = j\omega$.

15-8 Evaluate

$$T(s) = \frac{s + 1}{s^2 + 1s + 1}$$

for its magnitude at low and high frequencies. More specifically, show that $\log |T(j\omega)|$ versus $\log \omega$ agrees with the plot of Fig. 15-3a for these frequency ranges.

15-9 Suppose a CCLS is governed by the equation

$$\dot{x}(t) = -2x + 3u_1 + 1u_2 \qquad x(0) = -1$$

Let $u_1(t) = 6\mu(t)$ and $u_2(t) = 2\delta(t)$. Find the time response by superposition.

15-10 A CCLS is governed by

$$\dot{x}_1 = -2x_1$$

$$\dot{x}_2 = \qquad -3x_2$$

$$\dot{x}_3 = \qquad\qquad -10x_3$$

(a) Show that the resolvent matrix obeys the proper conditions. Find the eigenvalues.

(b) Show that the state transition matrix is decoupled. Hence, each solution is independent of the others, and the original system can be studied in separate pieces.

SIXTEEN

NUMERICAL METHODS

Numerical methods for studying the behavior of dynamic engineering systems have come to the fore in recent years for several reasons. Probably the most important is that both the power efficiency and cost efficiency of digital computation continue to increase, so that larger and more complex problems can be studied at a reasonable cost. A second major reason is that numerical methods are well suited to the solution of nonlinear problems, which arise frequently in engineering practice. Also, linear problems continue to grow in size, requiring a major computational effort to obtain specific results, such as eigenvalues and state trajectories. For example, the entire field of finite-element analysis depends upon the existence of suitable computational power for its implementation. Finally, once a digital computer has been used for the numerical-solution part of the problem, it is convenient to provide a related set of input and output commands for problem specification and display, respectively, leading to simulation languages, as described in Chap. 7.

Here we discuss the numerical integration of state equations, which for us is the most important application of numerical methods. Then we consider the numerical solution of nonlinear algebraic equations. Finally, we apply the methods to linear systems with constant coefficients, again considering integration of the state equations and other problems. For a more thorough treatment of numerical analysis the reader is referred to books on numerical methods.[1]

16-1 NUMERICAL INTEGRATION OF STATE EQUATIONS

The problem we are interested in can be stated as

$$\frac{dX(t)}{dt} = F(X(t), t) \tag{16-1}$$

given $$X(t_0) = X_0 \qquad (16\text{-}2)$$

where $X(t)$ = state vector of dimension n
t = continuous time variable
F = vector function of dimension n
X_0 = initial state
t_0 = initial time

In Eq. (16-1) we have indicated the presence of forcing functions $U(t)$ by the explicit inclusion of t in F. This form also includes time-varying systems, e.g., linear systems in which the A matrix changes with time, but not implicit problems.

Numerical Representation

For convenience we first consider the scalar case ($n = 1$) of Eqs. (16-1) and (16-2). Then

$$\frac{dx(t)}{dt} = f(x(t), t) \qquad (16\text{-}3)$$

with $$x(t_0) = x_0 \qquad (16\text{-}4)$$

Since in this chapter the generalization of a method to an n-dimensional state space will often be quite clear, for brevity we often use the scalar case for insight.

In a numerical approach to the solution of Eq. (16-3) subject to the initial condition (16-4) we agree to represent the continuous time variable t by a sequence of discrete points t_k, $k = 0, 1, \ldots, N$. These points are usually spaced at equal intervals h for convenience, but they need not be. When the interval is constant, we can write

$$t_k = kh \qquad (16\text{-}5)$$

and denote $x(t - kh) = x(t_k)$ by x_k. Another convenience we adopt is to make the initial time zero. Hence $k = 0$ is the first index.

The second element in the numerical representation is to characterize the true solution $x(t)$ by a set of points (t_k, x_k), where $x_k = x(t_k)$, as shown in Fig. 16-1. This representation has two features: (1) we seek to obtain x_k as a good approximation to the value the continuous solution would have at the time t_k, and (2) even if the x_k are found exactly, intermediate values of x are not known. That is, a value such as $x(t_k + h/2)$ can only be estimated from the solution values x_k. The main focus of our discussion will be on how to obtain reasonably accurate values for x_k. Often, if we choose h to make x_k accurate, we shall have enough points to estimate $x(t)$ between the computed points quite easily.

One-Step Methods

Euler There are many possible approaches to the numerical solution of Eq. (16-1) or (16-3). Let us focus on Eq. (16-3) and make a simple approximation. Using equal

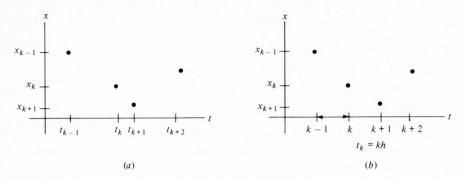

Figure 16-1 Numerical representation of $x(t)$: (a) arbitrary spacing of points t_k; (b) equal spacing of points t_k.

time intervals h between points and assuming that the slope dx/dt is approximately constant over such an interval, we write

$$x(t + h) = x(t) + h \left. \frac{dx}{dt} \right|_t$$

or
$$x_{k+1} = x_k + hf_k \qquad k = 0, 1, \ldots \qquad (16\text{-}6)$$

where f_k denotes $f(x_k, t_k)$. This is the Euler integration method. The procedure is quite straightforward. Starting with $k = 0$ ($t = 0$), evaluate the right-hand side of Eq. (16-6). This yields x_1 (x at h). Now repeat the process for $k = 1$, etc. Notice that each step requires only one evaluation of f, that is, f_k.

Although this method is computationally inexpensive, it has a major drawback in many practical cases. To obtain reasonable accuracy over a given time interval, the computing increment h must be made very small. This is bad for two reasons: (1) it increases the importance of roundoff error in the solution (this is the error that arises due to the finite word length and finite accuracy of computer representation of numbers); (2) it may cost too much in computation to advance over a reasonable time interval because so many steps are required.

Taylor series In an effort to remedy these deficiencies we can appeal to a Taylor-series expansion. If a solution is available at x_k, we can extrapolate to x_{k+1} by the relation

$$x_{k+1} = x_k + h \frac{dx}{dt} + \frac{h^2}{2} \frac{d^2x}{dt^2} + \cdots + \frac{h^p}{p!} \frac{d^px}{dt^p} + \cdots \qquad (16\text{-}7)$$

where the derivatives are all to be evaluated at the time t_k. When Eq. (16-3) is used in Eq. (16-7), we get

$$x_{k+1} = x_k + hf + \frac{h^2}{2} \frac{df}{dt} + \cdots + \frac{h^p}{p!} \frac{d^{p-1}f}{dt^{p-1}} + \cdots \qquad (16\text{-}8)$$

where again the derivatives are to be evaluated at time t_k. Observe that if p is unity, we recover the Euler integration rule. For any finite value of p we shall be ignoring a series whose first term has a factor h^{p+1} in it. This is called the *truncation error,* and the method is said to have a truncation error of order $p + 1$.

Unfortunately, there is a serious practical disadvantage to the use of Eq. (16-8) for $p > 1$ in that a set of derivatives of f must be calculated and stored. Consider just the first derivative, df/dt

$$\frac{df}{dt} = \frac{\partial f}{\partial x}\frac{dx}{dt} + \frac{\partial f}{\partial t}$$

or
$$\frac{df}{dt} = \frac{\partial f}{\partial x}f + \frac{\partial f}{\partial t} \tag{16-9}$$

From Eq. (16-9) it is apparent that both f and its two partial derivatives must be evaluated at each stage (k) if a direct Taylor-series approach with $p = 2$ is used. For a given h we may get a better value of x_{k+1} from x_k by using a truncated Taylor series, but we shall need more computations per step than in Euler's method.

Runge-Kutta Another one-step method, based on approximating the slope in the interval of interest, is the Runge Kutta method. It retains the important advantage that only one prior value of x is required to advance the solution by the amount h. Here we give a fourth-order approximation as an example. First, a set of slope estimates in the interval $[t_k, t_{k+1}]$ is calculated

$$s_1 = f(x_k, t_k) \tag{16-10a}$$

$$s_2 = f\left(x_k + \frac{h}{2}s_1, t_k + \frac{h}{2}\right) \tag{16-10b}$$

$$s_3 = f\left(x_k + \frac{h}{2}s_2, t_k + \frac{h}{2}\right) \tag{16-10c}$$

$$s_4 = f(x_k + hs_3, t_k + h) \tag{16-10d}$$

Except for s_1 each new slope estimate s_i uses information derived in the previous step. The new value of x is derived using a weighted average of the s_i from the relation

$$x_{k+1} = x(t_k + h) = x_k + \frac{h}{6}(s_1 + 2s_2 + 2s_3 + s_4) \tag{16-11}$$

This particular method has a truncation error of order h^4 and requires four evaluations of f per time increment h. It is frequently used in practice since it represents a reasonable trade-off between accuracy and computing efficiency for many problems and it is self-starting; i.e., we can find x_{k+1} knowing only x_k and doing some computations. Other methods which are not self-starting require values of x at more than one time before the solution can be projected.

Example 16-1 Consider the differential equation

$$\dot{x} = x \qquad x(0) = 1 \tag{16-12}$$

The analytic solution is $x(t) = e^t$. An Euler approach yields the equation

$$x_{k+1} = x_k + hx_k \qquad k = 0, 1, 2, \ldots, n \tag{16-13}$$

where h is the step size and $f_k = x_k$ in this case. It is possible to solve Eq. (16-13) with $x_0 = 1$ to get

$$x_k = (1 + h)^k \tag{16-14}$$

in this special case. This solution can be compared with the exact solution to gain some insight into the Euler approximation. Terms of the Euler result are given by Eq. (16-14), which compares with the exact solution at $t = kh$

$$e^{kh} = 1 + kh + \frac{(kh)^2}{2!} + \cdots + \frac{(kh)^m}{m!} + \cdots \tag{16-15}$$

For any value of k greater than zero the result of Eq. (16-14) matches that of Eq. (16-15) for only the first two terms. Truncation error begins with the h^2 terms, as expected. As you can see, if h is small, the analytical solution almost matches the computational solution for modest values of k. But as the time $t_k = kh$ gets large, the two solutions inevitably diverge.

From this example we can observe another point. Even evaluating the true solution requires truncation of an infinite series such as that given in Eq. (16-15). Thus it is sometimes more helpful to have an accurate numerical solution in plotted or tabular form available for inspection than to have a complex analytic expression for a given result.

Example 16-2 Consider the system of state equations

$$\dot{x} = -y^3 \qquad x(0) = 1 \tag{16-16a}$$

$$\dot{y} = x \qquad y(0) = 0 \tag{16-16b}$$

which correspond to a mass-spring oscillator with a cubic spring law. The Euler method for this problem leads to

$$x_{k+1} = x_k + h(-y_k^3) \tag{16-17a}$$

$k = 0, 1, 2, \ldots, m$

$$y_{k+1} = y_k + h(x_k) \tag{16-17b}$$

Some results are given in the top portion of Table 16-1 for $h = 0.1$. For nonlinear equations such as these we usually do not know an analytical solution and must infer the accuracy of the solution by studying several numerical solutions.

The Runge-Kutta method of fourth order can be applied to this example by first noting the general form of the algorithm, which is given by

$$s_1 = F(X_k, t_k) \tag{16-18a}$$

Table 16-1 Integration for Example 16-2†

k	t_k	x_k	y_k
		Euler, $h = 0.1$	
0	0.0	1.000000	0.000000
1	0.1	1.000000	0.100000
2	0.2	0.999900	0.200000
3	0.3	0.999100	0.299999
4	0.4	0.996400	0.399900
5	0.5	0.990005	0.499540
		Runge-Kutta, $h = 0.1$	
0	0.0	1.000000	0.000000
1	0.1	0.999975	0.100000
2	0.2	0.999600	0.199984
3	0.3	0.997976	0.299879
4	0.4	0.993612	0.399489
5	0.5	0.984480	0.498443
		Euler, $h = 0.025$	
0	0.0	1.000000	0.000000
4	0.1	0.999986	0.100000
8	0.2	0.999694	0.199992
12	0.3	0.998299	0.299922
16	0.4	0.994382	0.399632
20	0.5	0.985942	0.498798

† Data significant to six decimals.

$$s_2 = F\left(X_k + \frac{h}{2} s_1, t_k + \frac{h}{2} \right) \tag{16-18b}$$

$$s_3 = F\left(X_k + \frac{h}{2} s_2, t_k + \frac{h}{2} \right) \tag{16-18c}$$

$$s_4 = F(X_k + hs_3, t_k + h) \tag{16-18d}$$

where $\dot{X} = F(X, t)$ is given by Eq. (16-16). The quantities s_i are vectors of appropriate dimension, and the slope estimate is

$$S_k = \tfrac{1}{6}(s_1 + 2s_2 + 2s_3 + s_4) \tag{16-19}$$

from which the new state can be computed from the relation

$$X_{k+1} = X_k + hS_k \tag{16-20}$$

(A second-order state version of the Runge-Kutta method is given in Ref. 2.)

The center part of Table 16-1 presents Runge-Kutta results, which can be compared with those in the top part of the table over the closed interval 0 to 0.5. Although the Runge-Kutta results are more accurate, they also require about 4 times more computing effort. Let us reduce the time increment h to 0.025 and run the Euler method again. The comparable results are given in the bottom part of the table. Now the accuracy is improved considerably, but the computing cost has increased. It is a matter of experience and judgment to select a suitable method paired with a suitable interval to obtain accurate results efficiently.

Multistep Methods

In an effort to reduce the number of evaluations of the vector function F that defines the system, we turn our attention to multistep methods, which use a certain number of prior values of the solution in order to obtain accuracy in extending the solution by an increment h. When the state trajectory is reasonably smooth, this is an efficient way to proceed.

Multistep methods have been developed along the following lines. Consider a system governed by Eqs. (16-3) and (16-4). Convert Eq. (16-3) into integral form

$$x(t) = x(0) + \int_0^t f(x(t), t) \, dt \tag{16-21}$$

From an interval from t_k to t_{k+1} separated by h we get

$$x_{k+1} = x_k + \int_{t_k}^{t_{k+1}} f(x(t), t) \, dt \tag{16-22}$$

If the integrand $f(x, t)$ is approximated over the required interval by a known approximating function, say a polynomial, the integration can be carried out numerically. If f is reasonably smooth and a number of prior values are available for fitting, Eq. (16-22) can yield an accurate result for x_{k+1}.

One area of choice in multistep methods is the selection of approximating functions for use in Eq. (16-22). A second area is in the selection of points to be included. Here we distinguish between *explicit methods,* in which only available points, i.e., data through t_k, are used, and *implicit methods,* in which at least data for t_{k+1} are involved.

Milne predictor-correction method Choosing from the many multistep methods available, we present the Milne predictor-corrector method. The equations for a fourth order approximation are

Predictor: $$X_{k+1} = X_{k-3} + \frac{4h}{3} (2F_{k-2} - F_{k-1} + 2F_k) \tag{16-23}$$

Corrector:
$$X_{k+1} = X_{k-1} + \frac{h}{3}(F_{k-1} + 4F_k + F_{k+1}) \qquad (16\text{-}24)$$

From Eq. (16-23) we observe that the method of prediction is explicit and involves four prior points. Equation (16-24) is used to "correct" the result X_{k+1} by using the predicted values in the evaluation of the term F_{k+1}. The difference between the predicted and corrected values gives an insight into the accuracy of the results.

The function F is evaluated only twice per step, once as F_k in Eq. (16-23) and once as F_{k+1} in Eq. (16-24). The first value of k that can be used is 3; this requires the previous computation of X_1, X_2, and X_3 together with the initial condition X_0. Thus a one-step method must be used to start the solution. It is important that the starting values be calculated accurately: the predictor-corrector method will not be able to compensate for an inaccurate start.

The Art of Numerical Integration

New methods of numerical integration are constantly being devised. Several journals regularly publish new methods as well as comparisons and compilations of existing methods. Even long and careful study of available methods would not necessarily lead to a precise answer to the reasonable practical question: Here is my problem; which is the best method to use? What, then, should you do?

Typically you would try one or two of the methods available to you with perhaps several time-step values and compare the results. If they seem to converge and appear reliable, that is good evidence. Any preliminary analysis and physical insight should be brought to bear. If the problem is large or must be run repeatedly with variations, computing efficiency joins accuracy as an important factor.

Finally, we have not yet addressed another aspect of numerical integration, stability. It is possible for some methods to generate unstable results even when the actual system is stable. Some integration methods may become unstable, e.g., some multistep implicit methods, while some may not. Probably the best practical control for this phenomenon is to vary the time increment h and observe the influence on the results $X(t)$. When h is small enough, usually the solution is almost unchanged when an even smaller h is used.

When you are considering using existing programs or developing your own implementation it is always wise to consult with an experienced numerical analyst if possible. In numerical integration, as in life, experience is the best teacher.

16-2 NONLINEAR ALGEBRAIC EQUATIONS

Nonlinear algebraic equations arise in the study of dynamic systems in several ways. One is in the solution of nonlinear R fields, such as occur in hydraulic piping networks and electric circuits. We can anticipate nonlinear algebraic equations from bond-graph models when input and state-variable causality leaves a set of R variables to be determined implicitly and when the R elements have nonlinear constitutive laws.

A second important circumstance is in the solution of models with nonlinear dependent storage elements, leading to nonlinear algebraic coupling between the independent and dependent energy variables. This is common in mechanical systems undergoing large angular motions, where the nonlinearities introduced by geometry often interact with the causally dependent part of the inertial storage field.

Our general problem can be stated as

$$F(X, t) = 0 \tag{16-25}$$

where X = vector of dimension n
   t = time
   F = vector of dimension n

Note that X represents any set of system variables, not necessarily state or energy variables. The appearance of t in Eq. (16-25) is not a source of difficulty since in fact we seek solutions for specific values of t, for example, t_k, $k = 0, 1, 2, \ldots$.

Techniques used for solving Eq. (16-25) numerically are variations of a trial-and-error approach, in which the error in the current estimate of X is used to guide the next value for X. Eventually, X should converge to a set of values which satisfy Eq. (16-25). The more we know about F the easier it is to find a satisfactory solution. For example, can the solution be shown to lie within a certain region? Is it known how many solutions there are? Is F monotonic and "smooth"? In nonlinear problems it is not unusual to have multiple solutions, in marked contrast to the linear case. While we cannot do more here than explore a few approaches, any standard numerical reference[3] will consider this problem in detail.

To give an idea of some of the methods available for solving Eq. (16-25), consider a vector of dimension 1. Then

$$f(x, t) = 0 \tag{16-26}$$

and we seek a value for x that satisfies the equation for a given value of t. If t is fixed at t' and f can be analytically inverted, we can write

$$x = g(t') \tag{16-27}$$

and we are done. Unfortunately, it is rarely possible in practice to invert f analytically, and in some cases f may not be specified analytically in the first place.

One way to approach Eq. (16-26) is to guess at a value for x (say x_0), substitute into Eq. (16-26), and evaluate the left-hand side. The difference from zero is the error. For example, let us search for x to satisfy

$$x^3 + 7.8 = 0 \tag{16-28}$$

If we try values of x from -3 to $+3$ in increments of 1, evaluate the left side of Eq. (16-28), and tabulate the error, we find the results below. Since we want the error to be as close to zero as possible, the best guess in the range appears to be -2. Now we can refine our increment, narrow the range, and repeat the process until a satisfactory result (a sufficiently precise x_k or a small enough e_k) is found.

Trial	k	1	2	3	4	5	6	7
Guess	x_k	-3	-2	-1	0	1	2	3
Error	e_k	-19.2	-0.2	6.8	7.8	8.8	15.8	34.8

Newton's method In order to avoid searching over a large region for many variables in the vector case, we need to make better use of local information. Newton's method makes use of the local derivatives of the algebraic functions. In formal terms we write

$$f(x_k + h, t') = f(x_k, t') + f'(x_k, t')h_k \qquad (16\text{-}29)$$

where x_k = current guess
 t' = constant
 h_k = change in x
 f' = derivative of f with respect to x

Since we seek a solution that has $f(x_k + h, t') = 0$, we can solve for h_k derived from Eq. (16-29) to get

$$h_k = \frac{-f(x_k, t')}{f'(x_k, t')} \qquad (16\text{-}30)$$

as long as $f' \neq 0$. Then we change x_k to $x_{k1} = x_k + h_k$.

In our example, $f = x^3 + 7.8$ and $f' = 3x^2$. Hence, Eq. (16-30) yields

$$h = \frac{-x^3 + 7.8}{3x^2} \qquad (16\text{-}31)$$

If we apply this corrector at the trial value $x_1 = -3$, our correction is

$$h_1 = -\frac{(-3)^3 + 7.8}{3(-3)^2} = 0.71$$

and the new estimate of x is $x_2 = -2.3$, which moves us closer to a satisfactory answer.

The generalization of Newton's method to a system of equations, called the *Newton-Raphson method,* requires the evaluation of the gradient f' at each trial step. Often this must be accomplished by numerical means, which can be expensive computationally. Sometimes a few iterations will yield a satisfactory solution, but sometimes the method is unstable, i.e., successive steps generate increasing errors.

Successive substitution Often a constraint can be partially "solved" for the unknown into the form

$$X = G(X, t') \qquad (16\text{-}32)$$

In this case it is useful to think of making a guess for X (say X_k), evaluating the right-hand side of Eq. (16-32), and deriving a left-hand value X'_k. The error e_k is defined by $X'_k - X_k$. Then this new value of X can be used as the next guess or can be combined with the first value to generate a new guess. Since the method involves only a single direct evaluation of G per step, it can be efficient when it works.

Consider the problem

$$2x^3 + 2x + 6 = 0 \tag{16-33}$$

which can be put into the form

$$x = -x^3 - 3 \tag{16-34a}$$

or the form

$$x = -(x + 3)^{1/3} \tag{16-34b}$$

Table 16-2 summarizes the results of successive substitution applied to Eq. (16-34b) for five steps, using the rules

$$x'_k = -1(x_k + 3)^{1/3} \tag{16-35a}$$

$$e_k = x'_k - x_k \tag{16-35b}$$

and $$x_{k+1} = 0.5(x_k + x'_k) \tag{16-35c}$$

Here we have averaged x_k and x'_k to produce x_{k+1}. It is not uncommon for this method to diverge if the initial guess is far off.

Optimization methods It is possible to restate the conditions of Eq. (16-25) in such a way that an optimization problem results. For example, given

$$f(x, y) = 0 \tag{16-36a}$$

$$g(x, y) = 0 \tag{16-36b}$$

as a coupled pair of equations, form the function

$$V(x, y) = f^2(x, y) + g^2(x, y) \tag{16-37}$$

Of course V is always positive except for values of x and y which satisfy Eq. (16-36). These values make V achieve its minimum value of zero.

Table 16-2 Successive-substitution method†

k	x_k	x'_k	e_k
1	0.000	−1.442	−1.442
2	−0.721	−1.316	−0.595
3	−1.018	−1.256	−0.238
4	−1.137	−1.230	−0.093
5	−1.184	−1.220	−0.036

† Best estimate is −1.202.

There are many numerical methods for minimizing (or maximizing) a given function. A standard numerical reference[3] will provide a detailed discussion.

Approximation by numerical integration Since we are discussing the solution of dynamics problems and have already considered numerical integration of initial-value problems, we point out here that it is possible to convert the algebraic problem of Eq. (16-25) into a related dynamics problem by introducing a set of derivatives in the form

$$\dot{X} = F(X, t') \qquad (16\text{-}38)$$

If we make an initial guess for X and solve Eq. (16-38) until the derivative vector $(\dot{X})$ is small enough, the corresponding value for X yields an approximate solution to Eq. (16-25).

Unfortunately, it is easier to construct an unstable dynamic system than a stable one, so Eq. (16-38) may well not settle down to a steady value of X. If it does not, you could try $\dot{X} = -F(X, t')$, but this may also be unstable. The message is that only with luck or insight can one solve algebraic equations by changing them into dynamic equations and looking for a steady state.

16-3 LINEAR SYSTEMS WITH CONSTANT COEFFICIENTS

Since linear systems with constant coefficients arise so frequently in practical engineering work (especially in control and vibrations studies), we shall discuss two special aspects of such systems, numerical integration and eigenvalue computation. There is an extensive literature on numerical methods for linear systems, principally based upon matrix techniques.

Integration of State Equations

The problem we are interested in can be stated as

$$\dot{X} = AX + BU(t) \qquad (16\text{-}39)$$

which is a special case of Eq. (16-1). We assume that the initial conditions are given, as in Eq. (16-2). A and B are matrices with constant elements. Of course, any numerical integration that applies to nonlinear systems will be applicable here. However, we are interested in improving the efficiency of our methods by taking advantage of the form of Eq. (16-39).

Exponential-matrix method An interesting method of numerical solution is based on the exponential matrix for A expressed in series form. Consider the unforced problem first, where

$$\dot{X} = AX \qquad X(0) = X_0 \qquad (16\text{-}40)$$

and X_0 represents the initial state. The solution to Eq. (16-40) is $X(t) = e^{At}X_0$, where

$$M = e^{At} = I + At + \frac{(At)^2}{2!} + \cdots + \frac{(At)^k}{k!} + \cdots \qquad (16\text{-}41)$$

and M denotes the exponential matrix.

To obtain values of the state at fixed intervals, we choose a time increment T and evaluate $M(T)$ by adding up the first K terms in the series

$$M(T) = e^{AT} \approx \sum_{k=0}^{K} \frac{(AT)^k}{k!} \qquad (16\text{-}42)$$

Now the matrix $M(T)$ will eventually converge for a given T, since the denominator continues to grow as higher-order terms are included. Convergence is faster (fewer terms are needed for a given degree of approximation) as T is taken smaller. Careful study has been made of the truncation error and convergence problems in evaluating M (see Ref. 4 for an entry into this literature).

Once $M(T)$ has been evaluated, the state solution follows readily as

$$X_{k+1} = MX_k \qquad k = 0, 1, 2, \ldots \qquad (16\text{-}43)$$

where X_k denotes $X(t_k)$ and $t_k = kT$. The solution is stepped along in time steps of length T by repetitive multiplication by M.

Suppose a forcing function $U(t)$ is present. It is possible to extend the exponential-matrix approach to this case. Here we make a particular approximation to $U(t)$ to derive a result. The exact solution to Eq. (16-39) is

$$X(t) = e^{AT}X_0 + \int_0^t e^{A(t-\tau)}BU(\tau)\,d\tau \qquad (16\text{-}44)$$

where the integral term is the convolution integral.

Suppose we consider a short interval T, small enough with respect to $U(t)$ for U to be approximately constant over the interval. Consider the interval from $0 \leq t \leq T$. Then we can write Eq. (16-44) in the form

$$X(T) = e^{AT}X_0 + e^{AT}\int_0^T e^{-A\tau}\,d\tau\,BU(0) \qquad (16\text{-}45)$$

Now the integral depends only upon $e^{-A\tau}$, and by a series approach we can find that

$$\int_0^T e^{-A\tau}\,dt = A^{-1}(I - e^{-AT}) \qquad (16\text{-}46)$$

The result can be written as

$$X(T) = e^{AT}X_0 + (e^{AT} - I)A^{-1}BU(0) \qquad (16\text{-}47)$$

or, more succinctly,

$$X(T) = MX_0 + NU_0 \qquad (16\text{-}48)$$

where $$N = (e^{AT} - I)A^{-1}B \qquad (16\text{-}49)$$

Computationally we can evaluate M and N from the pair of series expressions

$$M \approx \sum_{k=0}^{K} \frac{(AT)^k}{k!} \qquad (16\text{-}50)$$

and $$N \approx \sum_{k=0}^{K} \frac{(AT)^k}{(k+1)!} TB \qquad (16\text{-}51)$$

where K is introduced as the finite upper limit. In practice, it is an easy matter to compute each term for N as the corresponding term for M is being evaluated. In a constant-coefficient system the state results can be computed from the recursive relation

$$X_{k+1} = MX_k + NU_k \qquad k = 0, 1, 2, \ldots \qquad (16\text{-}52)$$

where the subscript k denotes the time ($t_k = kT$) in question.

An advantage of the exponential-matrix method is that no inversion of A is required; nor does one need to compute the eigenvalues. However, there is a distinct set of trade-offs involving the size of T, the nature of A, the particular $U(t)$, and the number of terms needed for accuracy in the series expansions for M and N. In particular, if A is stiff, meaning that A has widely separated eigenvalues, T must be quite small or K must be large for accuracy. Either way, a lot of computation is required to carry out a solution.

Example 6-3 We use the exponential-matrix method to evaluate the solution for the system

$$\dot{x} = -0.2x + 4 \qquad x(0) = 1 \qquad (16\text{-}53)$$

Since the input is constant ($U = 4$), we need not concern ourselves with its influence on a choice for T. Choosing $T = 2$ (somewhat blindly), we have

$$M = e^{AT} = 0.67032 \qquad \text{and} \qquad N = (e^{AT} - I)A^{-1}B = 1.6484$$

The solution can be generated by a relation of the form of Eq. (16-52)

$$x_{k+1} = Mx_k + NU_k = 0.67032x_k + 1.6484(4) \qquad (16\text{-}54)$$

Some results are shown in Table 16-3.

What if we use an interval $T = 10$? Then we get $M = 0.135335$ and $N = 4.32332$. The corresponding results in the third column of Table 16-3 are compared with the exact solution,

$$x(t) = e^{-0.2t} + 20(1 - e^{-0.2t}) = 20 - 19e^{-0.2t}$$

at intervals of 2. See the fourth column of the table. In this case we have accuracy to six figures with a relatively wide range of choice for T.

Table 16-3 Results for Example 16-3†

t_k	$x_k(T = 2)$	$x_k(T = 10)$	$x(t)$
0	1.0	1.0	1.0
2	7.26392		7.26392
4	11.4627		11.4627
6	14.2773		14.2773
8	16.1640		16.1640
10	17.4286	17.4286	17.4286

† Six significant figures in the data.

The exponential-matrix procedure for constant coefficient linear systems can be summarized as follows:

1. Select T (the time step) and K (the series truncation limit) based on A, $U(t)$, and accuracy considerations. One way to choose K is to keep adding terms until the terms in M and N do not change appreciably.
2. Use Eqs. (16-50) and (16-51) to evaluate M and N, respectively.
3. Use Eq. (16-52) together with $U(t)$ and the initial condition X_0 to compute the state X_k at the points $t_k = kT$.

This completes our brief discussion of numerical integration of linear constant-coefficient systems. We now turn our attention to some other problems associated with the dynamics of such systems.

Other Numerical Problems

Linear constant-coefficient systems have problems other than integration associated with their numerical analysis and solution. As the typical engineer or system analyst you would not need to write your own programs to solve these problems, but it is good to be aware of their existence.

Matrix inversion When resistance fields or dependent causality of certain types exist in a linear system, linear equations of the form

$$AX = Y \tag{16-55}$$

where X = vector of unknowns
A = square matrix
Y = vector of knowns

are implied. The formal solution to the problem is simply

$$X = A^{-1}Y \tag{16-56}$$

which is easy to write but may be difficult to compute. Consider the question: Does A^{-1} exist? That may seem easy—just compute the determinant of A and see whether it is zero. What if A is large, say 100×100, and the resulting determinant is a very small value, say 10^{-10}? Does this mean that the determinant is really zero and we are dealing with roundoff error? Or is it a valid result, showing that the inverse does indeed exist although the equations are badly conditioned?

Fortunately many reliable programs are now available to perform matrix inversion numerically, such as those contained in the International Mathematics and Statistics Library. You should have no difficulty in obtaining some suitable version for your use. Just remember that since all numbers are represented to finite precision in numerical problems, there will be problems which cannot be solved satisfactorily by any program even when you think there should be some solution. You can try double-precision calculation in such a case, or you may decide that your problem is not really sensible in the first place.

The eigenvalue problem Associated with a linear system governed by an equation of the form of Eq. (16-39) is an eigenvalue problem

$$(A - \lambda I)X = 0 \qquad (16\text{-}57)$$

Here we seek values of λ (the eigenvalues) together with a corresponding set of X vectors (the eigenvectors) that satisfy Eq. (16-57).

There are many methods for finding the desired information. One approach is to derive the characteristic polynomial $CP(\lambda)$ from the determinant of $A - \lambda I$. Solving $CP(\lambda) = 0$ yields the set of eigenvalues. Another approach is to operate directly on the algebraic equations (16-44). One method yields a single eigenvalue and eigenvector, which can then be eliminated from the problem. In other words, the results are generated one at a time. Another method yields all the eigenvalues simultaneously; then n algebra problems can be solved to generate the eigenvectors.

There are many programs available for the numerical solution of eigenvalue problems. You must know something about the type of A matrix in question to make a good selection. Here are some factors to consider: Is A symmetric? Is A real? Is A large and sparse, perhaps banded? What about accuracy versus computational cost? Because eigenvalue problems are so common, virtually every scientific computation center has a library of eigenvalue programs.

We conclude this chapter on numerical methods with a reminder that numerical analysis of dynamic systems has much art in it. The more you know about available options with regard to methods and computer programs and the more experience you have the better the results are likely to be. So compute away, but remember to think about your options.

PROBLEMS

16-1 Interpolating polynomials are used in numerical-integration formulas as well as in finding solution points that have not been computed directly. Suppose that the exact solution to a problem is $x(t) = 2e^t$ and that the solution is tabulated for the following points:

k	0	1	2	3	4	5
t	0	0.1	0.2	0.3	0.4	0.5
x	2	2.21	2.44	2.70	2.98	3.30

We wish to estimate x at $t = 0.25$.

(a) Use a polynomial of the form $y_a = C_1 t + C_0$. Fit C_1 and C_0 from the data and evaluate y_a ($t = 0.25$).

(b) Use a polynomial of the form $y_b = C_2 t^2 + C_1 t + C_0$. Repeat the procedure from part (a). *Hint:* There are two choices here. Use them both and compare.

(c) Use a truncated-series approach for e to approximate the solution. Compare the two- and three-term expansions to parts (a) and (b), respectively.

16-2 The solution to a system has been developed in the form:

t_k	0.7	0.8	0.9
x_k	0.4316	0.3377	0.2137

Estimate x at $t = 1.0$ by extrapolation.

16-3 The classical oscillator equations are $\dot{x}$ and y and $\dot{y} = -x$. Assume that $x(0) = 1$ and $y(0) = 0$.

(a) Use the Euler method to evaluate the response for $0 \le t \le 1$.

(b) Use the Milne predictor-corrector method to compute the response for $0 \le t \le 1$.

(c) Compare both sets of results to the exact solution.

16-4 Apply the successive-substitution method to

$$2x^3 - 2x + 6 = 0$$

with an initial guess of $x = 0$. Use the form $x = x^3 + 3$ first. What happens?

16-5 Evaluate the transition matrix M for $\dot{x} = -2x$ for $T = 0.5, 1,$ and 2. Compare the values for x at $t = 2$ computed from each M.

16-6 The pressure-flow relations for a hydraulic piping network are quadratic; i.e., for R, $P = bQ|Q|$. Find the pressure and flow distribution for the following model. Assume $P_1 = 1$, $P_2 = 0$, $P_3 = 0$ and $b_4 = b_5 = b_6 = 0.1$. Find P_7, Q_4, Q_5, and Q_6.

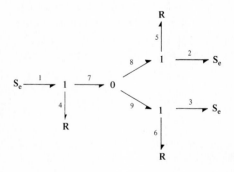

Figure P16-6

16-7 A mechanical system is driven by a force input. Spring 2 is linear, spring 3 is cubic, and damper 4 is linear

$$F_2 = k_2 x_2 \qquad F_3 = k_3 x_3^3 \qquad F_4 = b_4 \dot{x}_4 \qquad \text{where } k_2, k_3, b_4 = 1$$

If the system is initially relaxed $[x_2(0) = x_3(0) = 0]$ and $F(t) = 10, t \geq 0$, find $x_2(t)$.

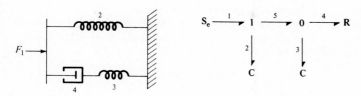

Figure P16-7

16-8 Consider the numerical integration of the following state equations:

$$\dot{x}_1 = x_3$$
$$\dot{x}_2 = x_4$$
$$\dot{x}_3 = -x_1$$
$$\dot{x}_4 = -10^4 x_2$$

(*a*) What are the eigenvalues?

(*b*) What integration methods seem appropriate?

(*c*) How would you choose a time step h for this problem? What difficulties might your choice present?

(*d*) What is the best way to solve this particular problem? *Hint:* It has a peculiar structure.

16-9 Hardening springs are used to limit the travel of masses, e.g., in automobile suspensions. Such an oscillator system is shown in the figure.

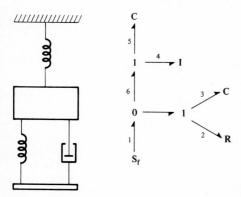

Figure P16-9

(*a*) Derive the state equations for the system.

(*b*) Assume that $m_4 = 1$, $k_3 = 1$, and $b_2 = 0.1$ and that

$$F_5 = \begin{cases} 1x_5 & x_5 \le 0.1 \\ 10x_5 - 0.9 & x_5 > 0.1 \end{cases}$$

where x_5 represents compression in spring 5. If the input is $V_1 = 1$, $t \ge 0$, find $x_5(t)$.

(*c*) Describe qualitatively the effects of moving the break point of spring 5 from 0.1 to 0.5.

16-10 Mechanical coulomb friction is characterized as shown in Fig. P16-10*a*, where V_{rel} is the relative velocity between surfaces and B is the breakaway force. An example system is shown in Fig. P16-10*b* with $p = F(t) - F_c$. Let the input force be $F(t)$ and let $B = 0.5$, with $m = 1$. Find the motion $V(t)$ of the mass and evaluate its position $x(t)$.

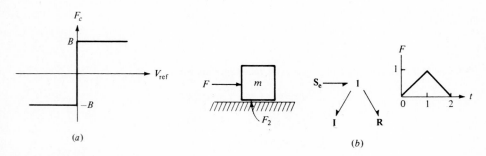

(*a*)   (*b*)

Figure P16-10

REFERENCES

1. P. Henrici: *Elements of Numerical Analysis*, Wiley, New York, 1964.
2. S. A. Hovanessian and L. A. Pipes: *Digital Computer Methods in Engineering*, McGraw-Hill, New York, 1969.
3. L. Kelley: *Handbook of Numerical Methods and Applications*, Addison-Wesley, Reading, Mass., 1967.
4. M. L. Liou: Evaluation of the Transition Matrix, *Proc. IEEE*, 228–229, February 1967.

FIVE

ENGINEERING APPLICATIONS

In this final part we first show how our modeling techniques can be extended to include fairly complex component types arising in engineering practice. Then we introduce two important topics, mechanical vibration and automatic control, which principally are based on linear dynamic-system models. You may well study these topics in formal courses or on your own some day, and we hope you will appreciate the role that system dynamics plays in these subjects. In fact, we expect you to find system-dynamics concepts and techniques useful in many areas throughout your career.

SEVENTEEN

EXTENDED MODELING METHODS FOR ENGINEERING SYSTEMS

To this point we have concentrated on building system models by assembling components into a working whole. In this chapter we shift our point of view and consider the entire system first. We examine the structure of multiport systems, both linear and nonlinear. Then we extend the definitions of our basic elements, C, I, and R. The result is that we get a very general and powerful extension to the methods we have studied thus far.

17-1 THE STRUCTURE OF MULTIPORT SYSTEMS

The set of nine elements we have been using to model engineering systems can be sorted into four groups or fields according to their power or energy properties: the dissipation field R, in which power is lost from the system; the storage fields C and I, which are energy-conservative; the general junction-structure group 0, 1, TF, and GY, which conserves power; and the source fields S_e and S_f. Now imagine that we are considering a rather large system model in bond-graph form. We can partition the graph into the four major groups just mentioned. This idea is represented in Fig. 17-1, where the dots represent the set of all bonds that join the junction structure to the given field. Bonds that connect the fields to the junction structure are called *external bonds,* and bonds that join one element of the junction structure to another are called *internal.*

Assume that causality has been assigned to the graph of Fig. 17-1, resulting in integral causality throughout. The result is shown symbolically in Fig. 17-2.

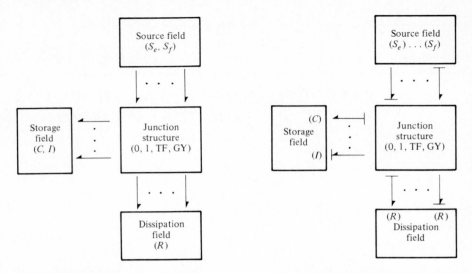

Figure 17-1 Partitioning of a bond graph.   **Figure 17-2** Complete integral causality.

The key vectors associated with the graph are defined as follows:

X = state vector (p on I, q on C)
$\dot{X}$ = time derivative of X
Z = complementary state vector (f on I, e on C)
U = source input vector (e on S_e, f on S_f)
D_{in} = input vector to R field
D_{out} = output vector from the R field

The causal bond graph of Fig. 17-2 converted into a symbolic vector diagram is shown in Fig. 17-3. We arrive at a structure with five key vectors shown and a sixth (V) implied. The elements of X, $\dot{X}$, and Z are arranged in the same order as the set of storage field ports in the arrays, and elements of D_{in} and D_{out} are arranged in the same order as the dissipation field ports. This convention will allow us to use vector-matrix notation effectively. At this point we want to develop the constraints between the vectors implied by the graph. We shall consider two different cases.

Constant-Coefficient Linear Systems

For CCLSs the storage- and dissipation-field constraints can be expressed in matrix form. For the dissipation field we write

$$D_{out} = LD_{in} \qquad (17-1)$$

where L is a square array the dimension of which is equal to the number of R-field ports. If all R elements are 1-ports, as we have studied so far, L is a diagonal matrix with R or R^{-1} parameters on the diagonal. For the storage field we write

$$Z = FX \qquad (17-2)$$

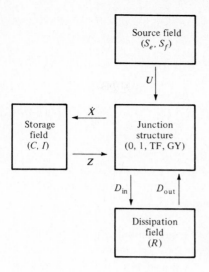

Figure 17-3 Key vectors in multiport systems with integral causality.

where F is a square array the dimension of which is equal to the number of C and I ports. When every C and I is a 1-port, the F array is diagonal.

Now we must turn our attention to the junction-structure part of the system. The set of inputs to the junction structure is Z, D_{out}, and U, as Fig. 17-3 shows. The set of outputs is X and D_{in}. Denoting the junction-structure matrix as S, we write

$$\begin{bmatrix} \dot{X} \\ D_{in} \end{bmatrix} = [S] \begin{bmatrix} Z \\ D_{out} \\ U \end{bmatrix}$$

or

$$\begin{bmatrix} \dot{X} \\ D_{in} \end{bmatrix} = \begin{bmatrix} S_{11} & S_{12} & S_{13} \\ S_{21} & S_{22} & S_{23} \end{bmatrix} \begin{bmatrix} Z \\ D_{out} \\ U \end{bmatrix} \qquad (17\text{-}3)$$

Here the S matrix is segmented into submatrices. S_{11} and S_{22} are square, but S_{12} (dimension of Z by dimension of D_{out}) and S_{21} (dimension of D_{in} by dimension of X) are not necessarily square. The expanded version of Eq. (17-3) is

$$\dot{X} = S_{11}Z + S_{12}D_{out} + S_{13}U \qquad (17\text{-}4a)$$

$$D_{in} = S_{21}Z + S_{22}D_{out} + S_{23}U \qquad (17\text{-}4b)$$

The elements of S are 0 and ± 1, which derive from 0- and 1-junctions and numbers derived from TF and GY moduli if such elements are present.

The set of four equations (17-1), (17-2), (17-4a) and (17-4b) can be reduced to state-space form. This requires that the vectors D_{in}, D_{out}, and Z be eliminated, to give

$$\dot{X} = \{[S_{11} + S_{12}(I - LS_{22})^{-1}LS_{21}]F\}X$$
$$+ [S_{13} + S_{12}(I - LS_{22})^{-1}LS_{23}]U \qquad (17\text{-}5)$$

where I represents unit matrices of appropriate size. From Eq. (17-5) we observe that any CCLS with integral causality has state-equation matrices given by

$$A = [S_{11} + S_{12}(I - LS_{22})^{-1}LS_{21}]F \qquad (17\text{-}6a)$$

and

$$B = S_{13} + S_{12}(I - LS_{22})^{-1}LS_{23} \qquad (17\text{-}6b)$$

provided the necessary inverse matrix exists.

The information in Eq. (17-5) is interesting because the state vector $\dot{X}$ is determined from a combination of the influence of X directly through part of the junction structure (S_1) and indirectly through the R field (L, S_{22}) and junction structure (S_{12}, S_{21}) plus the same types of influence from the input U.

Example 17-1 A familiar mechanical system and its augmented graph with integral causality are shown in Fig. 17-4a and b. The parameters are identified as m, k, and b for I, C, and R, respectively. In Fig. 17-4c the graph has been rearranged to emphasize the field structure of the system. Notice the correspondence to Fig. 17-2. We identify the key vectors as

$$X = \begin{bmatrix} p_1 \\ q_2 \end{bmatrix} \qquad Z = \begin{bmatrix} V_1 \\ F_2 \end{bmatrix} \qquad D_{\text{out}} = [F_3] \qquad D_{\text{in}} = [V_3] \qquad u = \begin{bmatrix} F_4 \\ V_5 \end{bmatrix}$$

where F and V denote force and velocity, respectively. In terms of the key vectors, we write the constraints

$$\begin{bmatrix} V_1 \\ F_2 \end{bmatrix} = \begin{bmatrix} m^{-1} & 0 \\ 0 & k \end{bmatrix} \begin{bmatrix} p_1 \\ q_2 \end{bmatrix} = [F] \begin{bmatrix} p_1 \\ q_2 \end{bmatrix} \qquad (17\text{-}7)$$

and

$$[F_3] = [b][V_3] = [L][V_3] \qquad (17\text{-}8)$$

for the storage and dissipation fields, respectively.

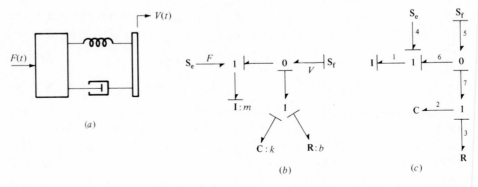

(a)

(b)

(c)

Figure 17-4 Example of the structure of a CCLS: (a) a mechanical diagram, (b) an augmented bond graph, and (c) field arrangement of the graph.

From the junction structure we obtain the S array

$$
\begin{bmatrix} \dot{p}_1 \\ \dot{q}_2 \\ \text{---} \\ V_3 \end{bmatrix} = \begin{bmatrix} 0 & 1 & \vdots & 1 & \vdots & 1 & 0 \\ -1 & 0 & \vdots & 0 & \vdots & 0 & 1 \\ \text{---} & \text{---} & \vdots & \text{---} & \vdots & \text{---} & \text{---} \\ -1 & 0 & \vdots & 0 & \vdots & 0 & 1 \end{bmatrix} \begin{bmatrix} V_1 \\ F_2 \\ \text{---} \\ F_3 \\ \text{---} \\ F_4 \\ V_5 \end{bmatrix}
\tag{17-9}
$$

The partition lines correspond to the structure shown in Eq. (17-3).

The formal result for A and B from Eqs. (17-6) is

$$
A = \left(\begin{bmatrix} 0 & 1 \\ -1 & 0 \end{bmatrix} + \begin{bmatrix} 1 \\ 0 \end{bmatrix} [1 - b0]^{-1}[b][-1 \quad 0] \right) \begin{bmatrix} m^{-1} & 0 \\ 0 & k \end{bmatrix}
$$

and

$$
B = \begin{bmatrix} 1 & 0 \\ 0 & 1 \end{bmatrix} + \begin{bmatrix} 1 \\ 0 \end{bmatrix} [1 - b0]^{-1}[b][0 \quad 1]
$$

or

$$
A = \begin{bmatrix} -bm^{-1} & k \\ -m^{-1} & 0 \end{bmatrix} \quad \text{and} \quad B = \begin{bmatrix} 1 & b \\ 0 & 1 \end{bmatrix}
\tag{17-10}
$$

The state equations are

$$
\dot{p}_1 = -bm^{-1}p_1 + kq_2 + 1F_4 + bV_5 \tag{17-11a}
$$
$$
\dot{q}_2 = -m^{-1}p_1 \qquad\qquad\quad + 1V_5 \tag{17-11b}
$$

At this point in the development you may be wondering why the matrix approach is useful, since our direct formulation methods of earlier chapters could be used to produce the same results more readily. There are two advantages to the matrix approach: (1) insight can be gained into the structural nature of CCLSs in a general way and results need not be derived more than once and (2) the results indicate how a computer can be used to great advantage in the automated formulation of state equations. This pays dividends for large systems, say with 50 bonds and perhaps 10 state variables. These matrix methods have been used in several versions of the ENPORT simulation programs.

Nonlinear Multiport Systems

We now turn our attention to nonlinear multiport systems to see what can be said about them. Examination of Figs. 17-1 to 17-3 shows that since no assumptions about linearity were made, they are still applicable; i.e., we are still considering five key vectors for systems with integral causality.

The nonlinear storage-field equation is a general algebraic relation

$$
Z = \phi_F(X) \tag{17-12}
$$

where each Z_i is a function of X_i alone if all C and I elements are 1-ports. The nonlinear dissipation-field equation is

$$D_{out} = \phi_L(D_{in}) \qquad (17\text{-}13)$$

where each $D_{out,i}$ is a function of $D_{in,i}$ alone, if all R elements are 1-ports. The junction-structure equation remains unchanged in form if we restrict the TF and GY moduli to be constant or time-varying. Then Eq. (17-3) still applies.

What about the reduction of Eqs. (17-12), (17-13), and (17-3) to state-equation form? If the vectors Z and D_{out} are eliminated, we get

$$\dot{X} = S_{11}\phi_F(X) + S_{12}\phi_L(D_{in}) + S_{13}U \qquad (17\text{-}14a)$$

$$D_{in} = S_{21}\phi_F(X) + S_{22}\phi_L(D_{in}) + S_{23}U \qquad (17\text{-}14b)$$

Only if D_{in} can be found from Eq. (17-14b) in terms of X and U can an explicit state form be found. If such is the case, we write

$$D_{in} = \phi_D(X, U) \qquad (17\text{-}15)$$

as the solution to Eq. (17-14b). Then Eq. (17-14a) can be made to yield the result

$$\dot{X} = S_{11}\phi_F(X) + S_{12}\phi_L[\phi_D(X, U)] + S_{13}U \qquad (17\text{-}16)$$

which is a form of explicit nonlinear state equation; i.e., $\dot{X}$ depends only upon X and $U(t)$.

If no explicit solution such as Eq. (17-15) can be found, we must appeal to the computer for help in terms of Eqs. (17-14). We may try to find D_{in} numerically each time the state and input variables change after an integration step. This is a fairly common case in practice and suggests why adequate computation is virtually a requirement for studying the response of nonlinear systems in general.

Example 17-2 A hydraulic system and its augmented bond graph are shown in Fig. 17-5a and b. We model the tanks and the valves as nonlinear elements. The system equations, derived from the field arrangement of the graph in Fig. 18-5c, are

$$\begin{bmatrix} P_2 \\ P_6 \end{bmatrix} = \begin{bmatrix} \phi_2(V_2) \\ \phi_6(V_6) \end{bmatrix} \qquad (17\text{-}17)$$

$$\begin{bmatrix} Q_4 \\ Q_7 \end{bmatrix} = \begin{bmatrix} \phi_4(P_4) \\ \phi_7(P_7) \end{bmatrix} \qquad (17\text{-}18)$$

$$\begin{bmatrix} \dot{V}_2 \\ \dot{V}_6 \end{bmatrix} = \begin{bmatrix} 0 & 0 & \vdots & -1 & 0 & \vdots & 1 \\ 0 & 0 & \vdots & 1 & -1 & \vdots & 0 \end{bmatrix} \begin{bmatrix} P_2 \\ P_6 \\ --- \\ Q_4 \\ Q_7 \\ --- \\ Q_1 \end{bmatrix} \qquad (17\text{-}19)$$

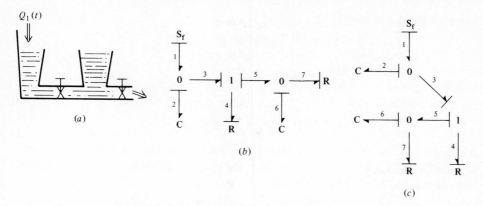

Figure 17-5 A nonlinear hydraulic example: (*a*) hydraulic schematic, (*b*) bond graph, and (*c*) field arrangement of the graph.

and

$$
\begin{bmatrix} P_4 \\ P_7 \end{bmatrix} = \begin{bmatrix} 1 & -1 & 0 & 0 & 0 \\ 0 & 0 & 0 & 0 & 0 \end{bmatrix} \begin{bmatrix} P_2 \\ P_6 \\ --- \\ Q_4 \\ Q_7 \\ --- \\ Q_1 \end{bmatrix} \tag{17-20}
$$

where V = volume
Q = volume flow rate
P = pressure

Equation (17-17) comes from the storage field (C_2 and C_6); Eq. (17-18) comes from the dissipation field (R_4 and R_7); and Eqs. (17-19) and (17-20) come from the junction structure. Note that Eq. (17-18) is written in inverse resistance form as required by the causality.

As we reduce to state-equation form, one step in the reduction involves Eqs. (17-18) and (17-19)

$$
\begin{bmatrix} \dot{V}_2 \\ \dot{V}_6 \end{bmatrix} = \begin{bmatrix} -1 & 0 \\ 1 & -1 \end{bmatrix} \begin{bmatrix} \phi_4(P_4) \\ \phi_7(P_7) \end{bmatrix} + \begin{bmatrix} 1 \\ 0 \end{bmatrix} [Q_1] \tag{17-21a}
$$

and, using Eqs. (17-17) and (17-20), we get

$$
\begin{bmatrix} P_4 \\ P_7 \end{bmatrix} = \begin{bmatrix} 1 & -1 \\ 0 & 1 \end{bmatrix} \begin{bmatrix} \phi_2(V_2) \\ \phi_6(V_6) \end{bmatrix} \tag{17-21b}
$$

from which we can derive

$$\begin{bmatrix} \dot{V}_2 \\ \dot{V}_6 \end{bmatrix} = \begin{bmatrix} -1 & 0 \\ 1 & -1 \end{bmatrix} \begin{bmatrix} \phi_4[\phi_2(V_2) - \phi_6(V_6)] \\ \phi_7[\phi_6(V_6)] \end{bmatrix} + \begin{bmatrix} 1 \\ 0 \end{bmatrix} [Q_1] \qquad (17\text{-}22)$$

and finally

$$\begin{bmatrix} \dot{V}_2 \\ \dot{V}_6 \end{bmatrix} = \begin{bmatrix} -\phi_4[\phi_2(V_2) - \phi_6(V_6)] \\ \phi_4[\phi_2(V_2) - \phi_6(V_6)] - \phi_7[\phi_6(V_6)] \end{bmatrix} + \begin{bmatrix} Q_1 \\ 0 \end{bmatrix} \qquad (17\text{-}23)$$

Although this last equation is in explicit nonlinear state form, it is hardly a simply equation. In fact, Eqs. (17.21) are perhaps the most useful forms, and computational assistance could be applied effectively to them.

Example 17-3 A circuit with nonlinear resistances is shown in Fig. 17-6a. The inductor and capacitor have constant parameters. A bond graph with partial causality is shown in Fig. 17-6b and the causality is completed in Fig. 17-6c. Here, v is voltage, i is current, λ is flux linkage, and q is change.

There are four key vectors because there are no inputs. The system equations in the pattern discussed above are

$$\begin{bmatrix} i_1 \\ v_3 \end{bmatrix} = \begin{bmatrix} L^{-1} & 0 \\ 0 & C^{-1} \end{bmatrix} \begin{bmatrix} \lambda_1 \\ q_3 \end{bmatrix} \qquad (17\text{-}24)$$

$$\begin{bmatrix} v_2 \\ i_4 \end{bmatrix} = \begin{bmatrix} \phi_2(i_2) \\ \phi_4(v_4) \end{bmatrix} \qquad (17\text{-}25)$$

$$\begin{bmatrix} \dot{\lambda}_1 \\ \dot{q}_3 \end{bmatrix} = \begin{bmatrix} 0 & 0 & | & 1 & 0 \\ 0 & 0 & | & 0 & 1 \end{bmatrix} \begin{bmatrix} i_1 \\ v_3 \\ v_2 \\ i_4 \end{bmatrix} \qquad (17\text{-}26a)$$

$$\begin{bmatrix} i_2 \\ v_4 \end{bmatrix} = \begin{bmatrix} -1 & 0 & | & p & -1 \\ 0 & -1 & | & 1 & 0 \end{bmatrix} \begin{bmatrix} i_1 \\ v_3 \\ v_2 \\ i_4 \end{bmatrix} \qquad (17\text{-}26b)$$

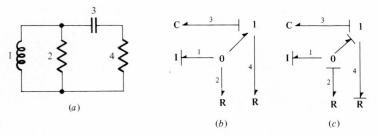

(a)

(b)

(c)

Figure 17-6 Electrical example with nonlinear elements: (a) circuit diagram, (b) bond graph with partial causality, and (c) bond graph with complete causality.

The vector containing the i_1 and v_3 elements can be eliminated by using Eq. (17-24) in Eqs. (17-26), but when we turn our attention to the R-field vectors, we see a problem. Equations (17-25) and (17-26 b) can be combined to yield

$$\begin{bmatrix} i_2 \\ v_4 \end{bmatrix} = \begin{bmatrix} -1 & 0 \\ 0 & -1 \end{bmatrix} \begin{bmatrix} i_1 \\ v_3 \end{bmatrix} + \begin{bmatrix} 0 & -1 \\ 1 & 0 \end{bmatrix} \begin{bmatrix} \phi_2(i_2) \\ \phi_4(v_4) \end{bmatrix} \qquad (17\text{-}27)$$

the solution to which must be obtained by numerical methods for given values of i_1 and v_3 unless both ϕ_2 and ϕ_4 are very tractable functions. Even so, the equations are coupled and must be solved simultaneously. That is, we get

$$\begin{bmatrix} i_2 \\ v_4 \end{bmatrix} = \begin{bmatrix} -i_1 \\ -v_3 \end{bmatrix} + \begin{bmatrix} -\phi_4(v_4) \\ \phi_2(i_2) \end{bmatrix}$$

to solve for i_2 and v_4 in terms of i_1 and v_3. This example shows that even simple nonlinear systems may result in difficult algebraic problems which preclude writing explicit state equations. On the other hand, some nonlinear systems which may seem complex yield simple explicit equations which are readily integrated though nonlinear. This is one reason why you should be prepared to experiment with alternative models of your physical system.

Now that we have taken a more general view of multiport systems, we are in a better position to extend the basic element-set definitions, permitting us to model more complex and realistic systems effectively.

17-2 THE STORAGE FIELD

The storage field in a multiport system is composed of all the C and I elements. Their ports are the ports through which the rest of the system exchanges energy with the field.

C Fields

The simple cantilever beam shown in Fig. 17-7a is subjected to both a vertical load F and a moment M at its end. From the related geometry in Fig. 17-7b the vertical deflection is y and the angular deflection is θ. When y is zero and θ is zero, the beam is straight in a horizontal position at the free end. We consider that both y and θ are small.

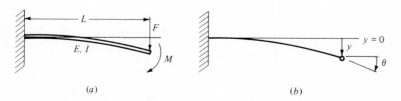

(a) (b)

Figure 17-7 A cantilevered beam: (a) loads and (b) geometry.

Suppose the beam is initially unloaded and we put a load force F on the beam. What will happen at the free end? We shall observe *both* a deflection and an angle at the end; i.e., both y and θ will be different from zero. Or suppose we want to deflect the end by 2 cm but cause no end angle. Then *both* a load force and a moment must be applied. The load-deflection characteristics of a cantilevered beam are predicted using an idealized theory as follows:

$$\begin{bmatrix} F \\ M \end{bmatrix} = EI \begin{bmatrix} \dfrac{12}{L^2} & \dfrac{-6}{L^2} \\ \dfrac{-6}{L^2} & \dfrac{4}{L} \end{bmatrix} \begin{bmatrix} y \\ \theta \end{bmatrix} \tag{17-28}$$

where E = Young's modulus

      I = beam-area moment of inertia

      L = beam length

This result is to be found in most elementary strength of materials texts.

The beam stores potential energy in a conservative manner. Since the input power is $F\dot{y} + M\dot{\theta} = Fv + M\omega$, we show it as a 2-port version of a C element

$$\underset{\dot{y} = v}{\overset{F}{\longrightarrow}} C \underset{\dot{\theta} = \omega}{\overset{M}{\longleftarrow}}$$

The energy variables are y and θ, and the other port variables are as shown. The constitutive equation describing the C is given by Eq. (17-28), where y and θ are the time integrals of the flows v and ω.

The generalization of a C element to n ports ($n \geq 1$) is shown in the first row of Table 17-1. The element is drawn as a C with n ports. On each port, power is oriented positive inward. The nonlinear integral-causality form for the equations is

Table 17-1 Multiport C, I, and R fields

Field type	Notation	Integral form	
		Nonlinear	Linear
C	$\overset{1}{-} C \overset{n}{\underset{2}{\cdot\cdot}}$	$e_i = \phi_i(\mathbf{q}),$ $i = 1, \ldots, n$	$e = K\mathbf{q}$
I	$\overset{1}{-} I \overset{n}{\underset{2}{\cdot\cdot}}$	$f_i = \phi_i(\mathbf{p}),$ $i = 1, \ldots, n$	$f = M^{-1}\mathbf{p}$
R	$\overset{1}{-} R \overset{n}{\underset{2}{\cdot\cdot}}$	$e_i = \phi_i(\mathbf{f}),$ $i = 1, \ldots, n$	$e = R\mathbf{f}$

shown symbolically in the third column. Each of the n port efforts e_i depends upon *all* n energy variables $\mathbf{q}$ in general. If the C element has constant parameters (or, more generally, is linear), the K matrix can be used to describe it. In mechanics the K matrix is called the *stiffness matrix;* see Eq. (17-28) as an example. Note that since the port variables are mixed translation and rotation types, the K-matrix elements have various units.

The linear spring of stiffness k in Fig. 17-8a is loaded by forces F_x and F_y in the x-y plane. The spring rotates freely about P and has a free length L_0. We wish to model the spring in the plane as a 2-port device in xy coordinates. In particular, we wish to find a characterization

$$F_x = \phi_x(x, y) \qquad F_y = \phi_y(x, y) \tag{17-29}$$

which has the bond graph shown in Fig. 17-8b.

Let us find Eq. (17-29) by an energy argument. The energy stored in the spring for any deflection (x, y) is

$$E(x, y) = \tfrac{1}{2}k[(x^2 + y^2)^{1/2} - L_0]^2 \tag{17-30}$$

To change the stored energy requires some power exchange; i.e.,

$$\frac{dE}{dt} = \frac{\partial E}{\partial x}\frac{dx}{dt} + \frac{\partial E}{\partial y}\frac{dy}{dt}$$

Observe that the expressions $\partial E/\partial x$ and $\partial E/\partial y$ must be the F_x and F_y forces, respectively, since the power dE/dt must be broken into force-times-velocity terms. Thus

$$F_x = \frac{\partial E}{\partial x} = kx\left[1 - \frac{L_0}{(x^2 + y^2)^{1/2}}\right] \tag{17-31a}$$

and

$$F_y = \frac{\partial E}{\partial y} = ky\left[1 - \frac{L_0}{(x^2 + y^2)^{1/2}}\right] \tag{17-31b}$$

Clearly, both forces are zero when the spring is at free length. And if, say, $y = 0$, then $F_x = k(x - L_0)$ and $F_y = 0$. Equation (17-31) has the form of Eq. (17-29),

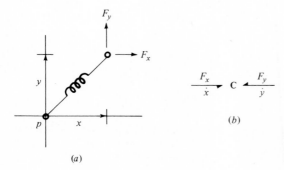

(a)

(b)

Figure 17-8 Example of a nonlinear C-field: (a) mechanical schematic and (b) bond graph.

our initial objective. The use of stored-energy expressions to derive constitutive laws is of general usefulness for C and I fields.

Suppose the 2-port C element representing the spring effect in the plane is coupled into a system in such a way that derivative causality is required on, say, the y port. Then we would want to express the constitutive equations as

$$F_x = \phi_x'(x, F_y) \qquad \text{and} \qquad y = \phi_y'(x, F_y) \qquad (17\text{-}32)$$

where the primes denote functions different from those in Eq. (17-29). Transforming Eq. (17-31) into the desired form may be impossible analytically, or at least impractical. So once again in a nonlinear situation we must rely upon the assistance of a computer. Such algebraic problems are much simpler if the constitutive laws are linear.

I Fields

The discussion of multiport I fields is similar to that for C fields. The second row of Table 17-1 shows the bond-graph notation, the general nonlinear for constitutive equations, and the linear case. The symbol M^{-1} stands for the inverse of the mass matrix in mechanics and for the inverse of the inductance matrix in electrical systems.

If the device of three coils wound on a common iron ring core (Fig. 17-9a) is operated in such a way that the core does not saturate, we can use a linear approximation to the device. The three energy variables are the flux linkages λ_1, λ_2, and λ_3, which are related to the three port currents i_1, i_2, and i_3. We write

$$\begin{bmatrix} \lambda_1 \\ \lambda_2 \\ \lambda_3 \end{bmatrix} = \begin{bmatrix} L_1 & M_{12} & M_{13} \\ M_{21} & L_2 & M_{23} \\ M_{31} & M_{32} & L_3 \end{bmatrix} \begin{bmatrix} i_1 \\ i_2 \\ i_3 \end{bmatrix} \qquad (17\text{-}33)$$

where L_i denotes self-inductance and M_{ij} denotes the mutual inductance between ports i and j. The signs for the ij terms depend upon relative coil orientations and the definition of positive current directions. The input power is $v_1 i_1 + v_2 i_2 + v_3 i_3$, where the voltages v_i are just the time derivatives of the flux-linkage variables λ_i. The device conserves energy and is a 3-port generalization of the 1-port I.

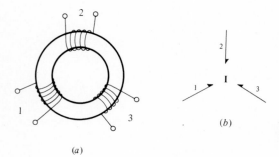

(a)

(b)

Figure 17-9 Mutual-inductance example: (*a*) electric schematic and (*b*) bond-graph element.

If a 2-port electric transformer is operated in a saturating region, the constitutive equations can be written

$$\lambda_1 = \phi_1(i_1, i_2) \qquad \text{and} \qquad \lambda_2 = \phi_2(i_1, i_2) \qquad (17\text{-}34)$$

which describe a nonlinear I field

$$\overset{1\qquad 2}{\longrightarrow I \longleftarrow}$$

in derivative form. The inverse form is

$$i_1 = \phi_3(\lambda_1, \lambda_2) \qquad \text{and} \qquad i_2 = \phi_4(\lambda_1, \lambda_2) \qquad (17\text{-}35)$$

where ϕ_3 and ϕ_{04} are related to ϕ_1 and ϕ_2 above.

I-field models are very useful in studying the behavior of masses in mechanics problems. For example, the rigid body in Fig. 17-10a is free to translate in the y direction and to undergo small rotations. Point P is the center of mass of the bar with mass m, and its moment of inertia about P is J_p. The two ports at which the bar interacts with its environment are shown as ports 1 and 2. Each has a load force acting upon it, and the y velocities are denoted v_1 and v_2, respectively. Consider only small angular motion.

The characterization we seek is that shown in Fig. 17-10b, a 2-port I field. The energy variables are the momenta p_1 and p_2. The equations we seek are of the form

$$\begin{bmatrix} v_1 \\ v_2 \end{bmatrix} = [M^{-1}] \begin{bmatrix} p_1 \\ p_2 \end{bmatrix} \qquad (17\text{-}36)$$

where $[M^{-1}]$ is the inverse mass matrix. The first thing we observe is that v_1 and v_2 are related to center-of-mass velocity v_p and angular velocity ω by

$$v_1 = v_p - \frac{L}{2}\omega \qquad \text{and} \qquad v_2 = v_p + \frac{L}{2}\omega \qquad (17\text{-}37)$$

for a uniform bar of length O, assuming small angular motion of the bar. Using the properties of point P, we next write

$$v_p = m^{-1}P_p \qquad \text{and} \qquad \omega = J_p^{-1}h_p \qquad (17\text{-}38)$$

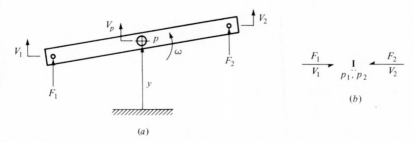

(a)

(b)

Figure 17-10 A bar in plane motion.

where P_p is the translational momentum and h_p is the angular momentum about P. Now $\dot{p}_p$ and $\dot{h}_p$ are related to the applied forces F_1 and F_2 by

$$\dot{p}_p = F_1 + F_2 \qquad \dot{h}_p = -\frac{L}{2} F_1 + \frac{L}{2} F_2 \qquad (17\text{-}39)$$

To obtain the desired result, we must integrate Eq. (17-39) to get

$$p_p = p_1 + p_2 \qquad \text{and} \qquad h_p = -\frac{L}{2} p_1 + \frac{L}{2} p_2 \qquad (17\text{-}40)$$

where the initial conditions have been chosen to make the integration constants vanish. Basically this means that if $p_p = h_p = 0$, $p_1 = p_2 = 0$ also. Now we collect Eqs. (17-37), (17-38), and (17-40) to get

$$\begin{bmatrix} v_1 \\ v_2 \end{bmatrix} = \begin{bmatrix} 1 & -\dfrac{L}{2} \\ 1 & \dfrac{L}{2} \end{bmatrix} \begin{bmatrix} m^{-1} & 0 \\ 0 & J_p^{-1} \end{bmatrix} \begin{bmatrix} 1 & 1 \\ -\dfrac{L}{2} & \dfrac{L}{2} \end{bmatrix} \begin{bmatrix} p_1 \\ p_2 \end{bmatrix}$$

or

$$\begin{bmatrix} v_1 \\ v_2 \end{bmatrix} = \begin{bmatrix} m^{-1} + \dfrac{L}{4} J_p^{-1} & m^{-1} - \dfrac{L}{4}^2 J_p^{-1} \\[2mm] m^{-1} - \dfrac{L}{4} J_p^{-1} & m^{-1} + \dfrac{L}{4}^2 J_p^{-1} \end{bmatrix} \begin{bmatrix} p_1 \\ p_2 \end{bmatrix} \qquad (17\text{-}41)$$

the desired result. To see this, compare Eqs. (17-41) and (17-36).

Mixed Fields

On occasion it is useful to think in terms of a mixed storage field; i.e., one port behaves like an I type and another port behaves like a C type. For example, in the solenoid shown in Fig. 17-11a the core is free to slide in the sleeve and the electrical part functions like an inductance. The device is a 2-port; one port involves electric current i and voltage v and the other port mechanical force F and velocity v. The

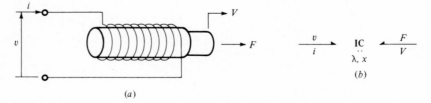

(a)

(b)

Figure 17-11 A mixed storage-field example: (a) solenoid schematic and (b) bond-graph model.

energy is known when λ, the flux linkage, and x, the core position relative to the sleeve, are specified.

Suppose the energy is $E(\lambda,x)$. Then a change in energy is caused by a net power flux

$$\frac{dE}{dt} = \frac{\partial E(\lambda,\ x)}{\partial \lambda} \frac{d\lambda}{dt} + \frac{\partial E(\lambda,\ x)}{dx} \frac{dx}{dt}$$

or

$$\frac{dE}{dt} = iv + FV$$

where we have used the facts that voltage and velocity are given by

$$v = \frac{d\lambda}{dt} \quad \text{and} \quad V = \frac{dx}{dt}$$

and we see that current and force must be given by

$$i = \frac{\partial E(\lambda,\ x)}{\partial \lambda} = i(\lambda,\ x) \quad \text{and} \quad F = \frac{\partial E(\lambda,\ x)}{\partial x} = F(\lambda,\ x) \quad (17\text{-}42)$$

To evaluate the constitutive equations (17-42) we start with the observation that the solenoid behaves in an electrically linear fashion under ordinary conditions of use; i.e., for a fixed position x we can write

$$i(\lambda,\ x) = \frac{1}{L(x)} \lambda \qquad (17\text{-}43)$$

where $L(x)$ is the inductance at a given core position x. The stored energy is the integral of i times $d\lambda$ if x is held constant. Since, for $\lambda = 0$, the force also equals zero, we have

$$E(\lambda,\ x) = \int_0^\lambda \frac{\lambda}{L(x)} d\lambda = \frac{\lambda^2}{2L(x)} \qquad (17\text{-}44)$$

To find the force we write

$$F(\lambda,\ x) = \frac{\partial E}{\partial x} = -\frac{dL(x)}{dx} \frac{\lambda^2}{2L^2(x)} \qquad (17\text{-}45)$$

Together Eqs. (17-43) and (17-45) are the desired constitutive equations in the form of Eq. (17-42).

When a mixed-storage-field element is inserted into a system graph, the normal causal conditions apply at each port. For the IC element model of the solenoid (Fig. 17-11 b), we should assign I-type integral causality to the electric port and C-type integral causality to the mechanical port if possible.

The use of stored energy to derive constitutive laws in this case is very useful. Any assumption we make for $E(\lambda,x)$ will result in a conservative IC field characterization in the form of Eqs. (18-42), but if we had simply made some assumptions about how i and F were related to λ and x, we might have created a model in which energy was not conserved. This could mean that our solenoid model could produce or absorb indefinite amounts of energy, which would certainly be wrong. In general, if we know that a multiport should be conservative, it is better to derive constitutive laws by differentiating a stored energy function than to construct constitutive laws directly. This is particularly true for mixed-energy-domain devices like the solenoid.

17-3 THE DISSIPATION FIELD

Just as C and I elements can be generalized from one port to n ports, so can the R element. The third row of Table 17-1 shows the notation for an n-port R field, gives the nonlinear equations in resistive form, and denotes the R matrix for linear multiport R fields.

One of the common ways for R fields to arise in practice is as a subgraph of a larger system graph. The idea is that embedded within a complete system model a part of the system may exhibit only dissipative behavior. For example, consider the electric circuit and bond-graph model with partial causality shown in Fig. 17-12a and b. At this point we observe that there are two resistors contained between ports 3 and 4. Since the entire effect of the 2-port must be dissipative, we seek a simplified representation along the lines of Fig. 17-12c. If the resistors have constant parameters, we seek an equivalent R matrix for the 2-port R element. Our problem is to find $[R]$ in

$$\begin{bmatrix} v_5 \\ v_8 \end{bmatrix} = [R] \begin{bmatrix} i_5 \\ i_8 \end{bmatrix} \tag{17-46}$$

We shall return to this question a moment. Note that the power direction on bond 6 in Fig. 17-12d is into the equivalent R element. This is for convenience in deriving the R matrix. To fit R_{eq} back into the system model we insert a 2-port 0-junction,

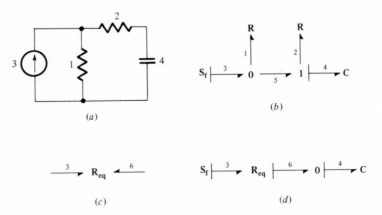

Figure 17-12 A dissipative-field example: (a) circuit diagram, (b) bond graph, (c) equivalent R field, and (d) system graph.

the purpose of which is to permit a power-direction change. In this case it is done by changing the current direction between bonds 4 and 6; that is, $i_6 = -i_4$.

To find R we first extract the R-field segment of the graph, as shown in Fig. 17-13a. The causality is in mixed form on the ports. In Fig. 17-13b causality has been completed. There are several ways to do this; a convenient one has been chosen. It will take us three steps to formulate the problem; then reduction can be done in two more steps to obtain the R matrix.

First, we express the inputs to the R elements in terms of their outputs and the port inputs (although we do it in matrix form, it could be written in direct algebraic fashion)

$$\begin{bmatrix} i_1 \\ v_2 \end{bmatrix} = \begin{bmatrix} 1 & 0 \\ 0 & 1 \end{bmatrix} \begin{bmatrix} i_3 \\ v_6 \end{bmatrix} + \begin{bmatrix} 0 & -1 \\ 1 & 0 \end{bmatrix} \begin{bmatrix} v_1 \\ i_2 \end{bmatrix} \tag{17-47}$$

Next, we characterize the R elements directly

$$\begin{bmatrix} v_1 \\ i_2 \end{bmatrix} = \begin{bmatrix} R_1 & 0 \\ 0 & R_2^{-1} \end{bmatrix} \begin{bmatrix} i_1 \\ v_2 \end{bmatrix} \tag{17-48}$$

Finally, we find the port outputs to complete the formulation

$$\begin{bmatrix} v_3 \\ i_6 \end{bmatrix} = \begin{bmatrix} 1 & 0 \\ 0 & 1 \end{bmatrix} \begin{bmatrix} v_1 \\ i_2 \end{bmatrix} \tag{17-49}$$

Now we use Eqs. (17-47) to (17-49) to eliminate the internal variables, leaving a relation between port variables. Solve first for the vector containing i_1 and v_2

$$\begin{bmatrix} i_1 \\ v_2 \end{bmatrix} = \frac{1}{1 + R_1 R_2^{-1}} \begin{bmatrix} 1 & -R_2^{-1} \\ R_1 & 1 \end{bmatrix} \begin{bmatrix} i_3 \\ v_6 \end{bmatrix} \tag{17-50}$$

Then find the vector containing v_1 and i_2

$$\begin{bmatrix} v_1 \\ i_2 \end{bmatrix} = \frac{1}{1 + R_1 R_2^{-1}} \begin{bmatrix} R_1 & -R_1 R_2^{-1} \\ R_1 R_2^{-1} & R_2^{-1} \end{bmatrix} \begin{bmatrix} i_3 \\ v_6 \end{bmatrix} \tag{17-51}$$

and finally obtain

$$\begin{bmatrix} v_3 \\ i_6 \end{bmatrix} = \frac{1}{1 + R_1 R_2^{-1}} \begin{bmatrix} R_1 & -R_1 R_2^{-1} \\ R_1 R_2^{-1} & R_2^{-1} \end{bmatrix} \begin{bmatrix} i_3 \\ v_6 \end{bmatrix} \tag{17-52}$$

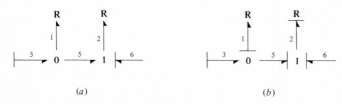

(a)   (b)

Figure 17-13 The R field isolated.

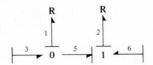

Figure 17-14 Causal alternative for the R field.

Let us convert the R-field description of Eq. (17-52) into resistance form. This requires switching the input–output roles of i_6 and v_6, that is, changing the causality on port 6. When Eq. (17-52) is solved in this way, we get the R-matrix result

$$\begin{bmatrix} v_3 \\ v_6 \end{bmatrix} = \begin{bmatrix} R_1 & -R_1 \\ -R_1 & R_1 + R_1 \end{bmatrix} \begin{bmatrix} i_3 \\ i_6 \end{bmatrix} \qquad (17\text{-}53)$$

Another strategy that can be used to obtain the desired result is to identify the equivalent R-field ports, as in Fig. 17-12b, isolate the R field, and then choose a port causality that is as useful as possible. Had that been done, we should have arrived at the causal graph of Fig. 17-14. In this case, the matrix result of Eq. (17-53) can be obtained by direct processing of the graph. Then the R-matrix form can be converted into the mixed causal form required for further system study.

17-4 MODULATED TRANSFORMERS AND GYRATORS

The definitions of the TF and GY elements, which are power-conserving with constant moduli, can be extended to include the modulated ideal 2-ports, written MTF and MGY, respectively. Figure 17-15a shows the notation for MTF, including a signal coupling to show that the modulus varies with some particular signal x. Similarly, Fig. 17-15b shows the MGY element with a varying parameter $r(x)$. Since it is still true for each element that $e_1f_1 = e_2f_2$, power is conserved at the ports.

An example of an MTF element is given in Fig. 17-16a, in which a link of length L acts as a converter from rotational to translational power. If we assume ideal conversion, we can write

$$T\omega = FV \qquad (17\text{-}54)$$

where the variables are indicated in the figure. Since the link is massless and rigid and there is no friction effect, we can write further

$$T = FL \cos \theta \qquad (17\text{-}55)$$

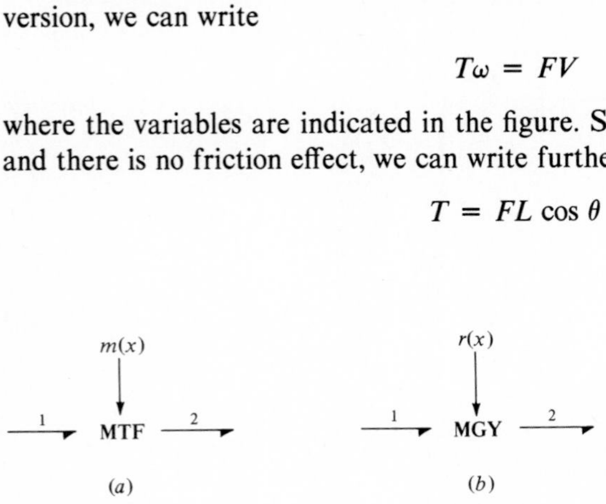

$$m(x)$$
$$\xrightarrow{1} \text{MTF} \xrightarrow{2}$$
$$(a)$$

$$r(x)$$
$$\xrightarrow{1} \text{MGY} \xrightarrow{2}$$
$$(b)$$

Figure 17-15 The ideal modulated two-ports: (a) MTF, $e_1 = m(x)e_2$ and $m(x)f_1 = f_2$; (b) MGY, $e_1 = r(x)f_2$ and $e_2 = r(x)f_1$.

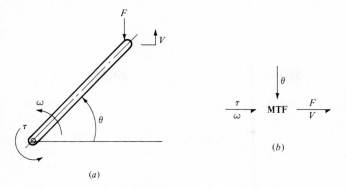

(a)

(b)

Figure 17-16 A modulated-transformer example: (a) mechanical schematic and (b) MTF model.

Since torque is proportional to force, an MTF is indicated, as shown in Fig. 17-16b. If Eq. (17-55) is substituted into Eq. (17-54), the result is

$$V = \omega L \cos \theta \qquad (17\text{-}56)$$

The modulus $m(x)$ is $L \cos \theta$, where the position variable θ plays the role of the signal x in Fig. 17-15.

Modulated transformers whose moduli arguments are displacement variables occur frequently in nonlinear mechanics. They are called *displacement-modulated transformers* and incorporate the effects of geometry associated with large angular motions.

A common and important electromechanical conversion effect is shown in Fig. 17-17a. A current i_f generates a magnetic field $B(i_f)$ across a gap in which a wire of length L is placed. If the gap faces are parallel and the wire is normal to the field lines, the basic vector coupling laws can be expressed in scalar form as

$$F = IB(i_f)L \qquad \text{and} \qquad v = VB(i_f)L \qquad (17\text{-}57)$$

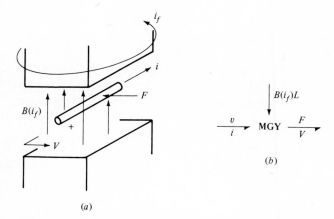

(a)

(b)

Figure 17-17 A modulated-gyrator example: (a) electromechanical schematic and (b) MGY model.

where F = force
V = velocity
v = voltage
i = current

A bond-graph model is given in Fig. 17-16b, where the modulus of the MGY is $B(i_f)L$. The observant reader will have noticed that $FV = vi$ provided the mechanical and electrical units of power are identical, as SI units are.

The electromechanical coupling just discussed is at the heart of many practical devices, such as electric motors and generators, galvanometer movements, and even certain types of loudspeakers and microphones. Permanent-magnetic devices have a constant field for B, and a simple GY will suffice to model them.

We shall find that MTF and MGY elements combined with 0 and 1 elements can make powerful building blocks for representing power-conserving couplings in dynamic systems.

17-5 JUNCTION STRUCTURES

The junction structure is that part of a bond graph consisting of the 0, 1, TF, and GY elements. Since all these elements are power-conserving, the junction structure is also power-conserving. The junction structure has been described as the "energy switchyard" of multiport systems since it shuttles power and energy around in a system without losing any.

We distinguish three types of junction structures:

Simple, consisting of only 0 and 1 elements
Weighted, consisting of 0, 1, and TF elements
General, including 0, 1, TF, and GY elements

Simple Junction Structures

A junction structure (JS) of 0 and 1 elements can be characterized in matrix form when suitable causality has been assigned. That is, a set of output variables at the ports can be expressed in terms of a set of input port variables. Consider the simple JS of Fig. 17-18a. It has four ports and one internal bond (5). A suitable causality has been assigned to the ports in Fig. 17-18b. We identify the input-vector elements as f_1, e_2, e_3, and f_4 and the output-vector elements as e_1, f_2, f_3, and e_4. The input and

(a) (b)

Figure 17-18 A simple junction structure: (a) the JS and (b) JS with causal orientation.

output elements are always to be chosen in the same port order. When the input–output equations are developed and expressed in matrix form, we get

$$
\begin{bmatrix} e_1 \\ f_2 \\ f_3 \\ e_4 \end{bmatrix} = \left[\begin{array}{cc|cc} 0 & 1 & 0 & 0 \\ -1 & 0 & 0 & 1 \\ \hline 0 & 0 & 0 & 1 \\ 0 & -1 & -1 & 0 \end{array} \right] \begin{bmatrix} f_1 \\ e_2 \\ e_3 \\ f_4 \end{bmatrix}
\tag{17-58}
$$

Suppose we reorder the input and output vectors, so that output efforts appear first. Rewriting Eq. (17-58) gives

$$
\begin{bmatrix} e_1 \\ e_4 \\ f_2 \\ f_3 \end{bmatrix} = \left[\begin{array}{cc|cc} 0 & 0 & 1 & 0 \\ 0 & 0 & -1 & -1 \\ \hline -1 & 1 & 0 & 0 \\ 0 & 1 & 0 & 0 \end{array} \right] \begin{bmatrix} f_1 \\ f_4 \\ e_2 \\ e_3 \end{bmatrix}
\tag{17-59}
$$

—an interesting result. We see that the effort outputs depend only upon the effort inputs, and the flows are similarly related. Also, we note that the only entries in the matrix are 0, $+1$, and -1, which is to be expected for simple JSs (except for some degenerate graphs). Finally, you may have observed that the matrices of both equations are skew-symmetric, a property we shall discuss in the next section.

Weighted Junction Structures

The addition of transformers to the element set lets us develop the properties of weighted JSs. Since a TF can have an arbitrary modulus, the resulting JS matrix can have arbitrary element values.

Figure 17-19a and b shows a mechanical system with a lever for linkage and its bond-graph model. Because the lever is assumed to act in an ideal manner, it is represented as a TF. The JS is identified in causal form as a subgraph in Fig. 17-19c. The causality was assigned at the system level previously. We have a weighted 6-port JS with three efforts and three flows as inputs.

The input-output equation for the weighted JS is

$$
\begin{bmatrix} e_1 \\ e_7 \\ e_{10} \\ f_2 \\ f_6 \\ f_9 \end{bmatrix} = \left[\begin{array}{ccc|ccc} & & & 1 & 0 & 0 \\ & \mathbf{0} & & m & -1 & 0 \\ & & & m & 0 & -1 \\ \hline 1 & -m & -m & & & \\ 0 & 1 & 0 & & \mathbf{0} & \\ 0 & 0 & 1 & & & \end{array} \right] \begin{bmatrix} f_1 \\ f_7 \\ f_{10} \\ e_2 \\ e_6 \\ e_9 \end{bmatrix}
\tag{17-60}
$$

where m denotes the lever ratio. Again we see that input efforts carry into output efforts only, and flows behave similarly. So the proper ordering of the port vectors can be very useful.

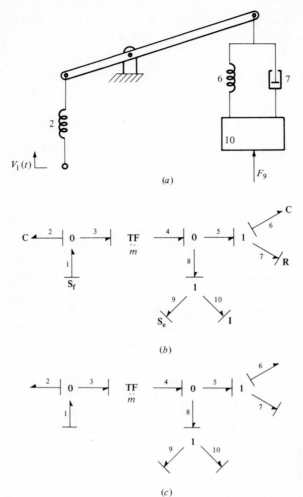

Figure 17-19 A weighted JS example: (a) mechanical schematic, (b) bond graph with causality, and (c) weighted JS with causality.

Careful inspection of the matrix shows that it is almost skew-symmetric. If the row 4 column 1 element were negative, for example, the array would be skew-symmetric. Consider the JS graph in Fig. 17-19c again. The 4,1 element relates f_1 to f_2, which occurs at the 0-junction with bonds 1, 2, and 3. If the power orientation on bond 1 only were to be reversed, this could be accomplished by redefining f_1 in the opposite sense. Then the graph would have all port powers directed the same way (outward, in this case) and the matrix would be skew-symmetric. You might reexamine the simple JS of Fig. 17-18 to verify that all port powers in that example are directed the same (inward) and the JS matrix is skew-symmetric. Generally, if external bonds to a JS have the sign half arrows pointing all toward or all away from the JS, the matrix will be skew-symmetric.

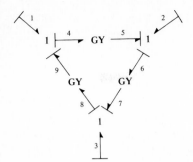

Figure 17-20 A general JS example.

General Junction Structures

General junction structures, which include 0, 1, TF, and GY elements, are not as common as simple and weighted JSs. They are power-conserving, but beyond that not much more can be said about the structure of their describing matrices. For example, a general JS is given in Fig. 17-20. This particular one is called a *circulator*. With the causality and power orientations as shown, we can derive the general JS matrix as

$$
\begin{bmatrix} e_1 \\ e_2 \\ e_3 \end{bmatrix} = \begin{bmatrix} 0 & r_{45} & -r_{89} \\ -r_{45} & 0 & r_{67} \\ r_{89} & -r_{67} & 0 \end{bmatrix} \begin{bmatrix} f_1 \\ f_2 \\ f_3 \end{bmatrix} \tag{17-61}
$$

As expected, it is skew-symmetric (since all powers are directed inward); but no longer are the effort variables at the ports decoupled from the flow variables. In fact, in this case they are fully coupled. Skew symmetry comes from power conservation, but gyrators always couple efforts and flows.

An example derived from a transducer problem is shown in Fig. 17-21 a. If we arrange effort outputs first in the output vector and then write the JS matrix equation, we find

$$
\begin{bmatrix} e_1 \\ e_3 \\ e_4 \\ f_2 \end{bmatrix} = \begin{bmatrix} 0 & -r & -r & -1 \\ r & 0 & 0 & 0 \\ r & 0 & 0 & 0 \\ 1 & 0 & 0 & 0 \end{bmatrix} \begin{bmatrix} f_1 \\ f_3 \\ f_4 \\ e_2 \end{bmatrix} \tag{17-62}
$$

Notice that the effort outputs are coupled to both effort and flow inputs in this case. No significant partitioning of the JS matrix can be achieved on a priori grounds, i.e.,

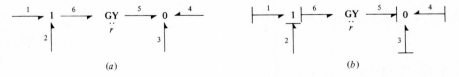

(a)   (b)

Figure 17-21 A second general JS example: (a) bond graph and (b) causality added.

before we derive the specific equations for the case at hand, when gyrators are present.

17-6 SOME PROPERTIES OF FIELDS

The examples of this chapter, which include storage fields, dissipative fields, and junction structures, display several interesting characteristics of matrices for the linear case. Here we present some useful general properties that can serve as checks on the accuracy of results in deriving field equations. We work in reverse, from junction structures through dissipative fields to storage fields.

Junction-Structure Properties

Suppose a given JS is characterized by an input-output equation

$$V_o = SV_i \tag{17-63}$$

where V_o = output vector
V_i = input vector
S = JS matrix of proper size

Assume that all port powers of the JS are directed the same (either inward or outward). Also assume that V_o and V_i are correlated by the ports; i.e., the same element of each vector refers to the same port. Then the following statement is true (T denotes the transpose):

$$P(t) = V_o^T V_i = 0 \tag{17-64}$$

That is, the net port power at every instant must be zero (since the JS is power-conserving), and the vector product shown will express the net sum of port powers.

In particular, consider a 2-port example. We have

$$P(t) = [v_{o1} \quad v_{o2}] \begin{bmatrix} v_{i1} \\ v_{i2} \end{bmatrix} = v_{o1}v_{i1} + v_{o2}v_{i2} = 0 \tag{17-65}$$

Furthermore

$$\begin{bmatrix} v_{o1} \\ v_{o2} \end{bmatrix} = \begin{bmatrix} s_{11} & s_{12} \\ s_{21} & s_{22} \end{bmatrix} \begin{bmatrix} v_{i1} \\ v_{i2} \end{bmatrix} \tag{17-66}$$

Now the power condition can be written as

$$P(t) = V_o^T V_i = V_i^T S^T V_i \tag{17-67}$$

Using Eq. (17-66) in (17-67) and expanding, we get

$$P(t) = s_{11}v_{i1}^2 + (s_{21} + s_{12})v_{i2}v_{i2} + s_{22}v_{i2}^2 = 0 \tag{17-68}$$

Since this must be true identically (under all possible port input conditions and for all values of s_{ij}), we are led to conclude that

$$s_{11} = 0 \qquad s_{22} = 0 \qquad \text{and} \qquad s_{21} = -s_{12}$$

In other words, *the matrix (or its transpose) must be skew-symmetric to ensure the power-conservation condition*. This important result is true for junction structures of any size provided the port powers are oriented the same way. It serves as a convenient check on the correctness of junction-structure matrices.

Dissipation-Field Properties

Now we turn our attention to R fields. Consider a linear R field in full resistance causality, so that

$$e = Rf \qquad (17\text{-}69)$$

where e = vector of effort outputs
f = vector of flow inputs
R = square (resistance) matrix of proper size

For an R field all powers at the ports are conveniently directed into the field, since the field dissipates power. Then the total power into the R field at any time is given by

$$P_d(t) = f^T e = f^T R f \qquad (17\text{-}70)$$

Although we have written the power as $f^T e$, it could as well have been written as $e^T f$. For a 2-port field the expression is

$$P_d = r_{11} f_1^2 + (r_{12} + r_{21}) f_1 f_2 + r_{22} f_i^2 > 0 \qquad (17\text{-}71)$$

What we need is to determine the conditions on R such that if f_1 or f_2 or both are different from zero, P_d is positive. This guarantees that power will be dissipated for any possible activity at the R-field ports. Clearly, from Eq. (17-71), if both f_1 and f_2 are zero, P_d equals zero, which is sensible.

Imagine that R is skew-symmetric. Then $r_{11} = r_{22} = 0$ and $r_{21} = -r_{12}$, so $P_d = 0$ independent of port activity. (We have already seen that this was true from the junction-structure properties.) Now we observe that if a given R matrix is partitioned into its symmetric and skew-symmetric parts, the skew-symmetric part contributes nothing to the power dissipated. Therefore, we can test only the symmetric part of the R matrix to see whether it obeys the property of Eq. (17-71), which is called *positive definiteness*. The generalization of Eq. (17-71) is that P_d, as given by Eq. (17-70), should be a positive definite quantity for any R field which we believe must dissipate power under all conditions.

To ensure this condition on power, we *test the R matrix* itself *for positive definiteness*. The easiest way is to use Sylvester's criterion, which applies to a symmetric matrix. Form the successive principal discriminants $\Delta_1, \Delta_2, \ldots \Delta_n$, where Δ_1 denotes

r_{11}, Δ_2 the determinant of the upper left 2×2 submatrix, Δ_3 the determinant of the upper left 3×3 submatrix, etc. Δ_n is, of course, the determinant of R. The test is simple. An n-port R field is positive definite if (and only if) each Δ_i is positive, $i = 1 \dots, n$.

For example, let us examine a proposed R field given by Eq. (17-72). First, we separate R

$$R = \begin{bmatrix} 2 & 0 & -2 \\ -1 & 3 & 0 \\ 0 & 0 & 4 \end{bmatrix} \tag{17-72}$$

into its symmetric and skew-symmetric parts. The symmetric part is $(R + R^T)/2$, and the skew-symmetric part is $(R - R^T)/2$

$$R = \begin{bmatrix} 2 & -\frac{1}{2} & -1 \\ -\frac{1}{2} & 3 & 0 \\ -1 & 0 & 4 \end{bmatrix} + \begin{bmatrix} 0 & \frac{1}{2} & -1 \\ -\frac{1}{2} & 0 & 0 \\ 1 & 0 & 0 \end{bmatrix} \tag{17-73}$$

Next, we evaluate Δ_1, Δ_2, and Δ_3 from the symmetric part

$$\Delta_1 = 2 \qquad \Delta_2 = {}^{23}\!/_4 \qquad \text{and} \qquad \Delta_3 = 20$$

Since all the Δ_i are positive, the R matrix is positive definite, and it corresponds to a purely dissipative field.

Storage-Field Properties

Multiport C and I fields occur in physical systems in many forms. When they do, the systems can no longer be treated by the same type of methods used for electric circuits or simple mass-spring-dashpot mechanical systems. Because multiport constitutive laws are by nature more complex than 1-port laws, it is helpful to have some checks on the correctness of laws used in a model. Reference 1 contains an extensive discussion of multiport fields, but here we discuss only a reciprocity condition which must hold if our models of C and I fields are to be energy-conservative. The conditions is called *Maxwell reciprocity* when it arises because of energy conservation.

Consider first a C field with n ports. The efforts e_i depend on the displacements q_i, which are the time integrals of the flow f_i. The stored energy E is assumed to depend only on the q's. It can be evaluated as the integral in time of all the power flows to the C. If all bonds have their power half arrows pointing toward C, then

$$E = \int^t (e_1 f_1 + e_2 f_2 + \cdots + e_n f_n)\, dt \tag{17-74}$$

but if we use $f_i\, dt = dq_i$, we can write

$$E(q_1, q_2, \dots, q_n) = \int_{0,0,\dots,0}^{q_1, q_2, \dots, q_n} e_1\, dq_1 + e_2\, dq_2 + \cdots + e_n\, dq_n \tag{17-75}$$

If we imagine varying only q_i, dE will be given by $e_i\,dq_i$ and we can write

$$\frac{\partial E}{\partial q_i} = e_i(q_1, q_2, \ldots, q_n) \tag{17-76}$$

Thus, if we can assume some expression for the stored energy, it will produce all the constitutive laws by differentiation. We have already used this idea for our solenoid model.

The reciprocity condition comes by rewriting Eq. (17-76) for the jth port

$$\frac{\partial E}{\partial q_j} = e_j(q_1, q_2, \ldots, q_n) \tag{17-77}$$

and differentiating Eq. (17-76) with respect to q_j and Eq. (17-77) with respect to q_i. The results are two versions of the mixed second partial derivative of E

$$\frac{\partial^2 E}{\partial q_i\,\partial q_j} = \frac{\partial e_i}{\partial q_j} = \frac{\partial e_j}{\partial q_i} \tag{17-78}$$

This means that $e_i(q_1, q_2, \ldots, q_n)$ and $e_j(q_1, q_2, \ldots, q_n)$ cannot be assumed as independent functions but must obey Eq. (17-78). Failure to recognize the reciprocal relations could result in a C field which would not conserve energy in a cyclic change in the q's. If such nonreciprocal C fields really did exist, one could extract any amount of energy from them.

For multiport storage-field modeling it is quite safe to assume energy functions and then to deduce the constitutive laws by differentiation, since then Maxwell reciprocity is automatically assured. If one assumes the constitutive laws directly, the reciprocity relations should be checked to make sure the laws imply conservation of energy.

For I fields, the constitutive laws relate flows f_i to momenta $p_1, p_2, \ldots, p_n$

$$f_i = f_i(p_1, p_2, \ldots, p_n) \tag{17-79}$$

The stored energy must be a function of the p's and by an argument similar to that given for C fields it can be shown that

$$\frac{\partial f_i}{\partial p_j} = \frac{\partial f_j}{\partial p_i} \tag{17-80}$$

which is another Maxwell reciprocal relation.

The reciprocal relations are valid for linear and nonlinear models. In the linear case, when the constitutive laws are expressed with matrices (a stiffness matrix for C and an inertia matrix for I), these matrices must be *symmetric*. Also, the inverses of these matrices are symmetric; as a result, compliance, capacitance, mass, and inductance matrices are all symmetric for any I or C field with all power signs on the ports oriented toward the multiport. These relations thus serve as a useful check on the correctness of the multiport constitutive laws.

As an example of a nonlinear C field, consider the 2-port C field described by Eqs. (17-31). The reciprocity condition can be applied to check them for consistency; i.e., we should find that

$$\frac{\partial F_x}{\partial y} = -\frac{1}{2}\frac{kxL_0}{(s^2 + y^2)^{3/2}}(2y) = \frac{-kL_0xy}{(x^2 + y^2)^{3/2}}$$

and

$$\frac{\partial F_y}{\partial x} = -\frac{1}{2}\frac{kyL_0}{(x^2 + y^2)^{3/2}}(2x) = \frac{-kL_0yx}{(x^2 + y^2)^{3/2}}$$

are equal. And they are.

Bond graphs, with their emphasis on power and energy, provide significant help in modeling complex physical systems using multiport elements.

PROBLEMS

17-1 Show that all multiport systems can be put into the form of Fig. 17-1. In particular, show how a system that includes the situation

$$S_f - C \rightarrow R$$

can be included. *Hint:* Consider all possible types of bond connections.

17-2 In the transition from Fig. 17-2 to 17-3, a set of field port variables was ignored. Where is that set? Why was it omitted?

17-3 Check the example result given by Eq. (17-11) by deriving state equations directly from the graph of Fig. 17-4b or c. Could the formal results in Eq. (17-6) still be used if b goes to 0? Explain.

17-4 For the hydraulic example of Fig. 17-5, described by Eqs. (17-17) to (17-20), assume that both the tanks and valves have constant coefficients, i.e., are linear. Modify the system equations as required and derive state equations.

17-5 For the nonlinear electrical example of Fig. 17-6, complete the reduction to state form as far as you can. Start with Eqs. (17-24) to (17-26). If the R elements have constant parameters, obtain the state equations. Note how the linear condition permits an explicit solution compared with the nonlinear case.

17-6 The stiffness matrix for a cantilevered beam is given by Eq. (17-28). Suppose we wish to input the loading (F and M) and compute the deflections (y and θ). Can this always be done? Explain. Derive the compliance (inverse stiffness) matrix for the beam.

17-7 Suppose a point mass is constrained to move in the x-y plane and is attached to a linear spring as shown. The spring is free to rotate about point P and has stiffness k and free length L_0. Equation (17-31) characterizes the spring in the x-y plane.

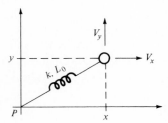

Figure P17-7

(*a*) Write state equations.

(*b*) Modify your model to include a gravity force in the negative y direction and repeat part (*a*).

(*c*) Show that your result in part (*b*) yields a suitable steady state if the system settles down due to parasitic damping (not shown).

17-8 For the rigid bar shown in Fig. 17-10*a* an inverse mass matrix was obtained; see Eq. (17-41). Does a characterization of the bar in derivative causality always exist? That is, is there always a mass matrix? Explain.

17-9 For the same rigid-bar problem referred to in Prob. 17-8, assume that the bar is not uniform. In particular, assume that point 1 is distance L_1 from P and point 2 is distance L_2 from P. Derive the inverse mass matrix and compare it with the result in Eq. (17-41).

17-10 A solenoid model was shown in Fig. 17-11*a* and its constitutive equations were described by Eqs. (17-43) and (17-45). Assume that the $L(x)$ inductance function looks as shown in the figure.

(*a*) Find an expression for $L(x)$ that can be used to fit the curve as shown.

(*b*) Evaluate the constitutive equations in terms of x and λ for the expression you chose.

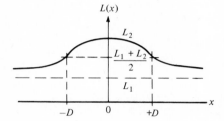

Figure P17-10   **Figure P17-11**

17-11 Consider the circuit shown, which is a 2-port R field. Assume that the R elements have constant parameters (R_1, R_2, R_3) and obtain the equivalent R matrix. *Hint:* Choose port causality carefully for best results.

17-12 Refer to the simple junction structure shown in Fig. 17-18*a*. Choose a causality different from that in Fig. 17-18*b* and derive the JS matrix. Show that your result agrees with that of Eq. (17-59).

17-13 Reconsider the general JS of Fig. 17-20. Assign efforts at the ports 1, 2, and 3 as inputs. Try to complete the causality on the graph. Describe the difficulties you may have found to the problem of obtaining the inverse solution for Eq. (17-61).

17-14 For the R matrix given by Eq. (17-72) convince yourself that only the symmetric part contributes to the power-dissipation expression. Do this by evaluating $P_d = f^T e$ directly from the R matrix given and show that the same P_d expression is obtained directly from the symmetric part of Eq. (17-73).

17-15 Show that the 2-port equivalent resistance matrix for the R field illustrated is symmetrical and is positive definite for all positive values of R_3 and R_4.

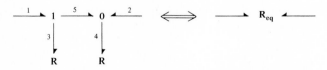

Figure P17-15

17-16 For what range of m could the following mass matrix correspond to an I field? *Hint:* Kinetic energy is inherently positive definite.

$$M = \begin{bmatrix} 4 & 2 & m \\ 2 & 4 & 3 \\ m & 3 & 1 \end{bmatrix}$$

REFERENCE

1. D. Karnopp and R. Rosenberg: *System Dynamics: A Unified Approach,* chap. 7, Wiley, New York, 1975.

EIGHTEEN

INTRODUCTION TO MECHANICAL VIBRATION

18-1 INTRODUCTION

The subject of mechanical vibration deals with oscillations of mechanical systems. The oscillations are often periodic or nearly so and, while the motion may be small in amplitude, the motion may cause annoyance and the resulting stresses may be large enough to cause breakage or a steady accumulation of fatigue damage. There are positive uses for vibration, but much of the effort in analysis and design of machines and structures is directed at eliminating unwanted and harmful vibration. No matter what turns technology takes in the future, you can expect vibration problems to continue to arise.

Because of the commonly small amplitudes of vibrations, geometric nonlinearities are usually not a problem, and natural frequencies calculated using linear models often are instrumental in explaining the problems in a device and in suggesting solutions. Nonlinearities are more common in damping effects, where coulomb friction and other nonlinear resistive laws may always be present. Unfortunately, it is very hard to justify specific loss models in complex systems, particularly at the design stage, so we often must get by with using linear dampers in our models. Such dampers should generally be regarded as equivalent energy absorbers and not as precise models of damping effects. Many of the worst problems in vibration have to do with lightly damped systems, which are easily excited by inputs with components near the system's resonant frequencies. In such cases, linear dampers may represent the system quite well even when nonlinear effects are clearly present.

A lot of skill is involved in reducing a real system to a vibration model, but the models themselves are not hard to represent or analyze. In most cases, the models are linear I and C elements connected by linear junction structures and possibly including linear resistance elements. From our point of view, such models are like the elementary mechanical systems we studied at the beginning of this book. You must remember, however, that real systems are not made up of little lumps of mass bouncing back and forth on springs. The schematic diagrams are just models, and only a skillful engineer can decide how many lumps are appropriate for the analysis and can estimate the mass and stiffness parameters. With this in mind, let us see how vibration studies fit in with our general approach to system dynamics.

Free Vibrations of Undamped Systems

The simplest case in vibrations occurs when an inertial element is connected to a compliant element. Figure 18-1 shows two schematic diagrams, one for translation and one for rotation. They will oscillate if disturbed, and the main use of such a simple model is to estimate a natural frequency. If, on the basis of an equivalent mass m and spring constant k or moment of inertia J and torsional spring constant k_t, a natural frequency is computed that correlates with the frequency at which a vibration problem is occurring, further analysis may be unnecessary. The solution may lie in a change in inertia or compliance big enough to shift the natural frequency away from the problem region.

The two bond graphs shown are very simple. We now demonstrate that the presence of the steady weight force mg for this linear system does not affect the natural-frequency calculation. The two equation sets for the examples are readily written.

$$\dot{p} = -kq + mg \qquad (18\text{-}1a)$$

$$\dot{q} = \frac{p}{m}$$

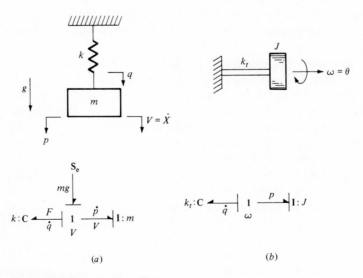

(a)   (b)

Figure 18-1 Two simple oscillators.

for Fig. 18-1 *a* and

$$\dot{p} = -k_t q \tag{18-2a}$$

$$\dot{q} = p/j$$

for fig. 18-1 *b*. The similarities between the equations are even more evident if they are expressed in matrix form

$$\begin{bmatrix} \dot{p} \\ \dot{q} \end{bmatrix} = \begin{bmatrix} 0 & -k \\ \dfrac{1}{m} & 0 \end{bmatrix} \begin{bmatrix} p \\ q \end{bmatrix} + \begin{bmatrix} mg \\ 0 \end{bmatrix} \tag{18-1b}$$

$$\begin{bmatrix} \dot{p} \\ \dot{q} \end{bmatrix} = \begin{bmatrix} 0 & -k_t \\ \dfrac{1}{J} & 0 \end{bmatrix} \begin{bmatrix} p \\ q \end{bmatrix} \tag{18-2b}$$

These 2×2 matrices determine the eigenvalues of the systems. (In these examples the eigenvalues are pure complex and have the form $s = \pm j\omega_n$, where ω_n is the radian natural frequency.) The *mg* term in Eq. (18-1) merely adds a constant to the solution and has no effect on its oscillatory part; therefore it is common to write equations in such a way that *mg* does not appear.

We usually express the compliance relation in the simple form

$$F = kq \tag{18-3}$$

where F is the tension force in the spring shown on the bond graph of Fig. 18-1 and in the free-body diagram of Fig. 18-2. This means that $F = 0$ when $q = 0$ or q is spring deflection measured from the point of no spring force. There is another way of expressing the spring force in terms of deflection away from the equilibrium under

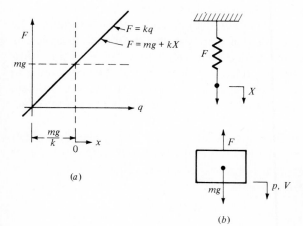

(a)

(b)

Figure 18-2 Compliance constitutive relation expressed in terms of (*a*) deflection q and (*b*) deviation from equilibrium position X.

the gravity force. In equilibrium, both p and $\dot{q}$ must vanish, which implies that $F = mg$ and $p = 0$, as can be seen from the bond graph or Eqs. (18-1).

We define X to be another deflection of the spring but measured from the point at which $F = mg$. The idea is shown in Fig. 18-2. Now, the law is

$$F = mg + kX \tag{18-4}$$

but

$$\dot{X} = \dot{q} = V = \frac{p}{m} \tag{18-5}$$

As you can see from the figure, X and q just differ by a constant [the constant of integration of Eq. (18-5)]

$$X = q - \frac{mg}{k} \tag{18-6}$$

Now we can write new equations of motion but using X instead of q for the displacement relation. The first of Eqs. (18-1 a) is now

$$\dot{p} = -F + mg = -(mg + kX) + mg \qquad \text{or} \qquad \dot{p} = -kX$$

and the second is just

$$\dot{X} = \frac{p}{m} \tag{18-5}$$

resulting in the matrix form we have

$$\begin{bmatrix} \dot{p} \\ \dot{X} \end{bmatrix} = \begin{bmatrix} 0 & -k \\ \dfrac{1}{m} & 0 \end{bmatrix} \begin{bmatrix} p \\ X \end{bmatrix} \tag{18-7}$$

Now you see that mg has been made to disappear by redefinition of the capacitor law.

Equation (18-7) is the same equation set we would get if we forgot about mg altogether. Here is a general rule. As far as calculations about natural frequencies or motion responses are concerned, linear vibratory systems are not affected by the presence of constant forces such as those of gravity if we consider all deflections q_i to be away from equilibrium positions and use the simple force law $F_i = k_i q_i$. Most of the time we simply neglect gravity altogether in the model. The only thing to remember in leaving out gravity forces is that kX in Eq. (18-7) does not represent all the force in the spring. The constant force mg has been canceled out of the equations, as we have just shown.

While there is no problem in finding system response or eigenvalues from first-order equation sets, as you have seen, there is a long tradition of using second-order equations in terms of displacements in vibrations. To convert Eq. (18-7) to this form we first note that the velocity $\dot{X}$ is just p/m, so

$$\dot{p} = m\ddot{X} \qquad \text{and} \qquad \dot{p} = m\ddot{X} = -kX$$

or

$$m\ddot{X} + kX = 0 \tag{18-8}$$

which is the result one would derive from the free-body diagram of Fig. 18-2 if mg were neglected.

There are any number of ways to see that sinusoidal solutions satisfy Eqs. (18-7) and (18-8) and to relate the frequency to m and k. You could use Laplace transforms or just assume a solution form to see if it will work. In this case, we try

$$X = A \sin(\omega_n t + \phi) \qquad (18\text{-}9)$$

where A and ϕ are constants, and look for an appropriate frequency ω_n. Two differentiations lead to

$$\ddot{X} = -A\omega_n^2 \sin(\omega_n t + \phi) \qquad (18\text{-}10)$$

which in Eq. (18-8) yields

$$(-m\omega_n^2 + k)A \sin(\omega_n t + \phi) = 0 \qquad (18\text{-}11)$$

The only way this can be satisfied without having $X = 0$ in Eq. (18-9) is for

$$\omega_n^2 = \frac{k}{m} \qquad (18\text{-}12)$$

a result you probably knew already.

In the first-order form of Eq. (18-7) it is easier to assume complex exponential solution forms, e.g.,

$$\begin{bmatrix} p \\ X \end{bmatrix} = \begin{bmatrix} A_1 \\ A_2 \end{bmatrix} e^{st}$$

where A_1 and A_2 are constants and we look for special values of s. Now

$$\begin{bmatrix} \dot{p} \\ \dot{X} \end{bmatrix} = s \begin{bmatrix} A_1 \\ A_2 \end{bmatrix} e^{st}$$

and thus

$$s \begin{bmatrix} A_1 \\ A_2 \end{bmatrix} e^{st} = \begin{bmatrix} 0 & -k \\ \frac{1}{m} & 0 \end{bmatrix} \begin{bmatrix} A_1 \\ A_2 \end{bmatrix} e^{st} \qquad (18\text{-}13)$$

which is the same as

$$\begin{bmatrix} s & 0 \\ 0 & s \end{bmatrix} \begin{bmatrix} A_1 \\ A_2 \end{bmatrix} e^{st} = \begin{bmatrix} 0 & -k \\ \frac{1}{m} & 0 \end{bmatrix} \begin{bmatrix} A_1 \\ A_2 \end{bmatrix} e^{st}$$

or

$$\begin{bmatrix} s & +k \\ -\frac{1}{m} & s \end{bmatrix} \begin{bmatrix} A_1 \\ A_2 \end{bmatrix} e^{st} = \begin{bmatrix} 0 \\ 0 \end{bmatrix} \qquad (18\text{-}14)$$

when the two matrices are combined.

The situation is now similar to that in Eq. (18-11). Since the term e^{st} will not be zero, we could divide it out. Then the equation could be satisfied if $A_1 = A_2 = 0$, but this is a trivial solution which merely means that both p and X can be zero. If we think of solving for A_1 and A_2 in Eq. (18-14), we see two equations in two unknowns which are *uniquely* solvable if the determinant of the matrix of coefficients does *not* vanish. Since we know one trivial solution, it is the only one unless this determinant does vanish. When s is an eigenvalue, this determinant *does* vanish. The *characteristic equation* results by forcing the determinant to vanish

$$s^2 + \frac{k}{m} = 0 \tag{18-15}$$

This requires

$$s = \pm \sqrt{\frac{-k}{m}} = \pm j \sqrt{\frac{k}{m}} = \pm j\omega_n \tag{18-16}$$

where

$$\omega_n = \sqrt{\frac{k}{m}} \tag{18-17}$$

The real sinusoidal form of Eq. (18-9) can be made up of complex exponentials if you remember the identities

$$\cos \omega_n t = \frac{e^{j\omega_n t} + e^{-j\omega_n t}}{2} \tag{18-18}$$

$$\sin \omega_n t = \frac{e^{j\omega_n t} - e^{-j\omega_n t}}{2j} \tag{18-19}$$

We shall not go into the details here, but just remember that eigenvalue *pairs* $s = \pm j\omega_n$ correspond to real sinusoids of the general form $A \sin(\omega_n t + \phi)$ or, for that matter, $B \cos(\omega_n t + \psi)$. You should be able to see by analogy that for the torsional oscillator of Fig. 18-1b

$$\omega_n = \sqrt{\frac{k_t}{J}} \tag{18-20}$$

Figure 18-3 shows a so-called two-degree-of-freedom system. The idea is that the complete equations of motion can be expressed in terms of two position variables such as the positions of the two masses. A system of this sort can be expressed by two second-order equations. This is a fourth-order system with two natural frequencies and four eigenvalues, $s = \pm j\omega_1$ and $\pm j\omega_2$, where ω_1 and ω_2 are the two natural frequencies. Expressed in first-order form, the state equations would normally number four, so we expect four state variables for a bond graph.

Zero-valued eigenvalues disturb this neat description somewhat. If one of the second-order-system eigenvalues happens to be zero, a third-order bond-graph equation set may suffice. Also, in some cases a fifth-order bond graph may have the four eigenvalues of the fourth-order system plus one extra zero value. The zero eigenval-

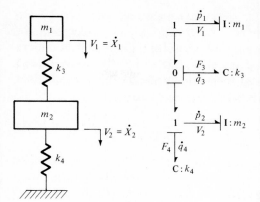

Figure 18-3 A two-degree-of-freedom system.

ues correspond to constants in the solution and often are less important than the nonzero eigenvalues.

The bond graph shown in Fig. 18-3 is very simple, particularly since we have neglected gravity forces and have chosen to measure all deflections away from equilibrium positions. At this point, you should be able to write the following equations with no problem

$$
\begin{bmatrix} \dot{p}_1 \\ \dot{p}_2 \\ \dot{q}_3 \\ \dot{q}_4 \end{bmatrix} = \begin{bmatrix} 0 & 0 & -k_3 & 0 \\ 0 & 0 & k_3 & -k_4 \\ 1/m_1 & -1/m_2 & 0 & 0 \\ 0 & 1/m_2 & 0 & 0 \end{bmatrix} \begin{bmatrix} p_1 p_2 q_3 q_4 \end{bmatrix} \quad (18\text{-}21)
$$

Converting this result into the one we would get by writing equations in terms of mass positions X_1 and X_2 is straightforward. First, we note that

$$
V_1 = \dot{X}_1 = \frac{p_1}{m_1} \tag{18-22}
$$

$$
V_2 = \dot{X}_2 = \frac{p_2}{m_2} \tag{18-23}
$$

so that

$$
\dot{p}_1 = m_1 \ddot{x}_1 \tag{18-24}
$$

$$
\dot{p}_2 = m_2 \ddot{x}_2 \tag{18-25}
$$

We are almost ready to substitute Eqs. (18-24) and (18-25) in the first two of Eqs. (18-21), but first we need to use the last two equations to relate q_3 and q_4 in terms of X_1 and X_2

$$
\dot{q}_3 = V_1 - V_2 = \dot{X}_1 - \dot{X}_2 \qquad \dot{q}_4 = V_2 = \dot{X}_2 \tag{18-26}
$$

Integrating them gives

$$
q_3 = X_1 - X_2 \quad \text{and} \quad q_4 = X_2 \tag{18-27}
$$

where we need no constants of integration since the q's are zero when the X's are zero, assuming no gravity forces. (There may be gravity forces, but they will not affect natural-frequency calculations in any case, and if X_1 and X_2 are deviations from equilibrium, they will cancel out of the final equations as explained above.) Now we use Eqs. (18-24) and (18-25) and substitute for the q's in the p equations using Eqs. (18-27). All lined up in matrix form, the results are

$$\begin{bmatrix} m_1 & 0 \\ 0 & m_2 \end{bmatrix} \begin{bmatrix} \ddot{X}_1 \\ \ddot{X}_2 \end{bmatrix} + \begin{bmatrix} k_3 & -k_3 \\ -k_3 & k_3 + k_4 \end{bmatrix} \begin{bmatrix} X_1 \\ X_2 \end{bmatrix} = \begin{bmatrix} 0 \\ 0 \end{bmatrix} \qquad (18\text{-}28)$$

The four first-order equations (18-21) are equivalent to the two second-order equations (18-28), which we could have derived using careful free-body diagrams.

The concept of Eqs. (18-12) can be used on Eqs. (18-28) to find the characteristic equation for the eigenvalues. If

$$\begin{bmatrix} X_1 \\ X_2 \end{bmatrix} = \begin{bmatrix} A_1 \\ A_2 \end{bmatrix} e^{st}$$

then

$$\begin{bmatrix} \ddot{X}_1 \\ \ddot{X}_2 \end{bmatrix} = s^2 \begin{bmatrix} A_1 \\ A_2 \end{bmatrix} e^{st} \qquad (18\text{-}29)$$

so Eqs. (18-28) become

$$\left\{ \begin{bmatrix} m_1 s^2 & 0 \\ 0 & m^2 s^2 \end{bmatrix} + \begin{bmatrix} k_3 & -k_3 \\ -k_3 & k_3 + k_4 \end{bmatrix} \right\} \begin{bmatrix} A_1 \\ A_2 \end{bmatrix} e^{st} = \begin{bmatrix} 0 \\ 0 \end{bmatrix} \qquad (18\text{-}30)$$

or

$$\begin{bmatrix} m_1 s^2 + k_3 & -k_3 \\ -k_3 & m_2 s^2 + k_3 + k_4 \end{bmatrix} \begin{bmatrix} A_1 \\ A_2 \end{bmatrix} e^{st} = \begin{bmatrix} 0 \\ 0 \end{bmatrix} \qquad (18\text{-}31)$$

We use the same argument again. We know that $A_1 = A_2 = 0$ is a (trivial) solution. It is the only solution unless

$$\det \begin{bmatrix} m_1 s^2 + k_3 & -k_3 \\ -k_3 & m_2 s^2 + k_3 + k_4 \end{bmatrix} = 0 \qquad (18\text{-}32)$$

which, symbolically at least, is the characteristic equation. Working out the determinant is a little messy, but the result is

$$m_1 m_2 s^4 + [m_1(k_3 + k_4) + m_2 k_3]\, s^2 + k_3(k_3 + k_4) + -k_3^2 = 0 \qquad (18\text{-}33)$$

Since this is a fourth-order equation in s, we shall find four eigenvalues. In fact, in this case, s^3 and s^1 terms are absent, which means that we could first solve for two s^2 values. These turn out to be negative, and thus the four roots come in two complex-conjugate imaginary pairs, $s = \pm j\omega_1$ and $\pm j\omega_2$, where ω_1 and ω_2 are the two natural frequencies. Generally, an n-degree-of-freedom system without damping will have n natural frequencies and $2n$ eigenvalues coming in complex-conjugate pairs. There are several important features of these natural frequencies.

An unforced, lightly damped vibratory system will vibrate approximately with a sum of sinusoids, each with one of the natural frequencies. When forced sinusoidally, the system will resonate or respond strongly when the forcing frequency coincides with, or is near to, one of the natural frequencies. Also, each natural frequency corresponds to a shape of vibratory motion called a *natural mode*. We do not intend to try to illustrate all these aspects of mechanical vibration, but perhaps you get the picture that natural frequencies are very important. Luckily, undamped systems yield fairly simple equation sets which can be analyzed to yield natural frequencies, and these frequencies retain some significance even when a fair amount of damping is present.

18-2 FORCED VIBRATION; RESONANCE

In Fig. 18-4 we return to the single-degree-of-freedom type of system but with sinusoidal forcing. Notice that we assume no constant forces such as a gravity force, so that the spring deflection q and the mass position measured from the equilibrium position are identical. The equations derived from the bond graph are

$$\begin{bmatrix} \dot{p} \\ \dot{q} \end{bmatrix} = \begin{bmatrix} 0 & -k \\ \dfrac{1}{m} & 0 \end{bmatrix} \begin{bmatrix} p \\ q \end{bmatrix} + \begin{bmatrix} F(t) \\ 0 \end{bmatrix} \tag{18-34}$$

where we shall assume

$$F(t) = F_0 \sin \omega_f t \tag{18-35}$$

Using $X = q$ for the mass position, and following the same steps as those used to derive Eq. (18-8), we can derive a single second-order equation from Eqs. (18-34)

$$m\ddot{X} + kX = F(t) = F_0 \sin \omega_f t \tag{18-36}$$

Since Eqs. (18-34) and (18-36) are linear, we know that superposition of solutions is possible. In this case, this means that the general solution will consist of a free vibration at frequency ω_n added to a forced vibration at ω_f. The free vibration will depend on the initial conditions we specify. The forced vibration will have to make the equations balance when $F(t)$ is given by Eq. (18-35). We want now to concentrate on the forced part of the solution.

Forced vibrations are quite important in practice since (as we shall see) there is a magnification of effects whenever a forcing frequency happens to be close to a

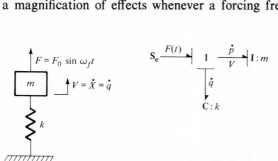

Figure 18-4 A single-degree-of-freedom oscillator under sinusoidal forcing.

natural frequency. Usually, the forced part of the vibration is thought of as the steady-state vibration, which applies after the forcing has been present for a long time and any free response has had time to die away. In our present model, the damping is neglected completely, so the free vibration will not die away. As we assumed in Eq. (18-9), the variables are sinusoidal without any exponential decay factors, but in fact every real system has some damping, however small, so that the free vibration would really disappear in time and we are justified in focusing our attention on the forced vibration.

It is easy to guess a solution form for Eq. (18-36). If

$$X(t) = A \sin \omega_f t \tag{18-37}$$

then
$$\ddot{X} = -A\omega_f^2 \sin \omega_f t \tag{18-38}$$

and, substituting into Eq. (18-36), the result is

$$(-m\omega_f^2 + k)A \sin \omega_f t = F_0 \sin \omega_f t \tag{18-39}$$

Then the amplitude of vibration is

$$A = \frac{F_0}{k - m\omega_f^2} = \frac{F_0/k}{1 - (\omega_f/\omega_n)^2} \tag{18-40}$$

after using Eq. (18-17). Equation (18-40) is plotted in Fig. 18-5.

There are several things to notice about this solution. First, near $\omega_f = 0$ the amplitude of motion A is just F_0/k, which is related to the idea of *static deflection*. If a constant force F_0 were applied to the mass in Fig. 18-4, the deflection would be F_0/k. Thus, at very low frequencies a sinusoidal force simply makes the system move sinusoidally as if it were always in equilibrium. At very low frequencies $F(t)$ is balanced by the spring force almost as if no mass were present. At high frequencies

$$A \rightarrow \frac{-F_0}{k\omega_f^2} \omega_n^2 = \frac{-F_0}{m\omega_f^2} \qquad \frac{\omega_f}{\omega_n} \gg 1 \tag{18-41}$$

in which two effects are interesting. The motion amplitude is negative, which means that when the force is positive, X is negative and the magnitude of the motion decreases as $1/\omega_f^2$. In the high-frequency region, the force is balanced off against the

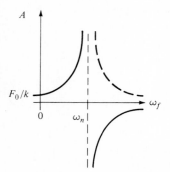

Figure 18-5 Forced-vibration amplitude response.

mass times the acceleration almost as if no spring were present. The $-m_f^2$ term in Eq. (18-39) is responsible for the high-frequency behavior in Eq. (18-41).

There is a minor problem with our solution, Eq. (18-40), when $\omega_f = \omega_n$. At this frequency our model predicts an infinite amplitude response. This is called *resonance,* and the phenomenon occurs for lightly damped systems whenever a forcing frequency and a natural frequency coincide. As you may have guessed, the prediction of an infinite amplitude of response occurs only because we have neglected all dissipation effects. With a small amount of damping, the response amplitude when $\omega_f \approx \omega_n$ would be large but not infinite. Since it is often difficult to estimate damping in the design phase, much vibration analysis is done on undamped models. The idea is to try to design the system so that natural frequencies do not coincide with forcing frequencies. If this can be achieved, the damping actually present usually will only help the situation.

For two or more degrees of freedom, a similar analysis can be carried out. Suppose in the system of Fig. 18-3 a force $F_0 \sin \omega_f t$ were applied to m_2. Then Eq. (18-29) would become

$$\begin{bmatrix} m_1 & 0 \\ 0 & m_2 \end{bmatrix} \begin{bmatrix} \ddot{X}_1 \\ \ddot{X}_2 \end{bmatrix} + \begin{bmatrix} k_3 & -k_3 \\ -k_3 & k_3 + k_4 \end{bmatrix} \begin{bmatrix} X_1 \\ X_2 \end{bmatrix} = \begin{bmatrix} 0 \\ F_0 \sin \omega_f t \end{bmatrix} \quad (18\text{-}42)$$

Since there is no damping, we can again assume that $\ddot{X}_1$, $\ddot{X}_2$, X_1, and X_2 will all be sinusoidal at frequency ω_f for our forced-vibration solution. Substituting into Eqs. (18-42) and combining terms, we find

$$\begin{bmatrix} k_3 - m_1\omega_f^2 & -k_3 \\ -k_3 & k_3 + k_4 - m_2\omega_f^2 \end{bmatrix} \begin{bmatrix} A_1 \sin \omega_f t \\ A_2 \sin \omega_f t \end{bmatrix} = \begin{bmatrix} 0 \\ F_0 \sin \omega_f t \end{bmatrix} \quad (18\text{-}43)$$

where we assume

$$X_1 = A_1 \sin \omega_f t \qquad \text{and} \qquad X_2 = A_2 \sin \omega_f t \quad (18\text{-}44)$$

Now we can divide out the $\sin \omega_f t$ terms and have two equations for the two amplitudes A_1 and A_2. If you recollect Cramer's rule for solving sets of algebraic equations in terms of ratios of determinants, the form of the solution is easy to write down

$$A_1 = \frac{\begin{vmatrix} 0 & -k_3 \\ F_0 & k_3 + k_4 - m_2\omega_f^2 \end{vmatrix}}{\begin{vmatrix} k_3 - m_1\omega_f^2 & -k_3 \\ -k_3 & k_3 + k_4 - m_2\omega_f^2 \end{vmatrix}} \quad (18\text{-}45)$$

$$A_2 = \frac{\begin{vmatrix} k_3 - m_1\omega_f^2 & 0 \\ -k_3 & F_0 \end{vmatrix}}{\begin{vmatrix} k_3 - m_1\omega_f^2 & -k_3 \\ -k_3 & k_3 + k_4 - m_2\omega_f^2 \end{vmatrix}} \quad (18\text{-}46)$$

Although we could work out these expressions, you can imagine that with several degrees of freedom this is a job for a computer. But we can notice an interesting fact by comparing the determinant in the denominators with the determinant form of the characteristic equation (18-32).

We argued that the eigenvalues were $s = \pm j\omega_1$ and $\pm j\omega_2$. If one of these values is substituted into Eq. (18-32) the determinant will vanish. But the s^2 terms would then be $-\omega_1^2$ or $-\omega_2^2$, and the determinant would look just like the denominators of Eqs. (18-45) and (18-46) with ω_1 or ω_2 instead of ω_f. Thus, the expressions for A_1 and A_2 will blow up if ω_f coincides with ω_1 or ω_2, the natural frequencies.

The amplitudes A_1 and A_2 are sketched in Fig. 18-6 as a function of the forcing frequency for a typical case. Note that, as before, when ω_f equals a natural frequency, the amplitudes of motion are unbounded and there is a change in sign of the A's as ω_f is varied from below to above a natural frequency.

The study of resonance diagrams like those in Fig. 18-6 is important because they not only show potential problems in structural design but also can be produced experimentally quite easily. The experimental resonant frequencies serve as a good check on the accuracy of mathematical models.

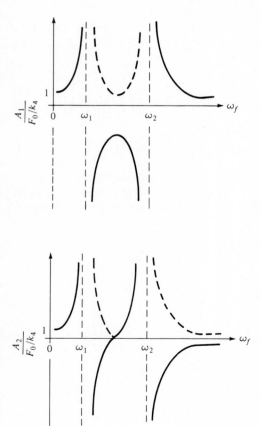

Figure 18-6 Forced-vibration amplitudes for a two-degree-of-freedom system.

18-3 DAMPED SYSTEMS

When we consider damped vibratory systems, and when we consider systems mounted on moving surfaces, as in Fig. 18-7, some of the techniques used in elementary vibration analysis are not so useful. The systems then are just general mechanical systems, and our previous state-equation methods still apply, but some of the simple assumed solutions we used for undamped systems do not apply. Free responses are *damped* sinusoids, and forced responses are sinusoidal but with phase angles depending in a continuous way on the forcing frequency.

Consider, for example, the oscillator of Fig. 18-7. It is forced by mg and a time-varying force as well as by the given velocity of its base $V_0(t)$. Although it seems natural to think of forces as being the cause of vibration, it is very common in practice to find vibrations caused by a connection to another system which is shaking.

Using the bond graph of Fig. 18-7, the state equations for p and q are readily written. As output variables we choose the mass velocity V and suspension force F

$$\begin{bmatrix} \dot{p} \\ \dot{q} \end{bmatrix} = \begin{bmatrix} \dfrac{-b}{m} & k \\ \dfrac{-1}{m} & 0 \end{bmatrix} \begin{bmatrix} p \\ q \end{bmatrix} + \begin{bmatrix} 1 & b \\ 0 & 1 \end{bmatrix} \begin{bmatrix} mg + F_0(t) \\ V_0(t) \end{bmatrix} \qquad (18\text{-}47)$$

$$\begin{bmatrix} F \\ V \end{bmatrix} = \begin{bmatrix} \dfrac{-b}{m} & k \\ \dfrac{+1}{m} & 0 \end{bmatrix} \begin{bmatrix} p \\ q \end{bmatrix} + \begin{bmatrix} 0 & b \\ 0 & 0 \end{bmatrix} \begin{bmatrix} mg + F_0(t) \\ V_0(t) \end{bmatrix} \qquad (18\text{-}48)$$

Standard computer routines can find the transient response or the steady-state frequency response for such equation sets.

The second-order forms for these equations are not quite as simple as for the undamped systems studied above. In terms of mass position X we have

$$p = mV = m\dot{X} \qquad \dot{p} = m\ddot{X}$$

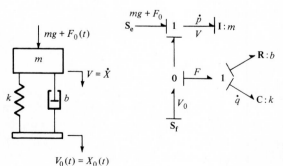

Figure 18-7 A damped system with force and motion inputs.

as before. So the p equation becomes

$$m\ddot{X} = mg + F_0(t) + kq + b[V_0(t) - \dot{X}]$$

Using the $\dot{q}$ equation, we have

$$\dot{q} = V_0(t) - V = \dot{X}_0(t) - \dot{X}$$

and so

$$q = X_0(t) - X + \text{constant} \tag{18-49}$$

where we need to use the given position of the mounting surface X_0 corresponding to its velocity V_0. The integration constant will vanish if we define X_0 and X so that $q = 0$ when $X_0 = X$. The final equation in terms of mass position is then

$$m\ddot{X} + b\dot{X} + kX = mg + F_0(t) + b\dot{X}_0(t) + kX_0(t) \tag{18-50}$$

Now we must remember that $V_0(t) = \dot{X}_0$ and $X_0(t)$ are not independent, although we need to use these two aspects of the input motion in Eq. (18-50). (If $b = 0$, this is not necessary.) For this second-order form we need an effort, a flow, and a position input, while for the first-order form we need only an effort and a flow.

Another second-order formulation can be made in terms of q. Now

$$V = -\dot{q} + V_0(t) \qquad p = mV = m[V_0(t) - \dot{q}]$$

so that

$$\dot{p} = m\dot{V} = m[\dot{V}_0(t) - \ddot{q}]$$

and the p equation becomes

$$m[\dot{V}_0(t) - \ddot{q}] = -\frac{b}{m}m[V_0(t) - \dot{q}] + kq + mg + F_0(t) + bV_0(t)$$

or

$$m\ddot{q} + b\dot{q} + kq = -mg - F_0(t) - m\dot{V}_0(t) \tag{18-51}$$

Now you see that the input quantities are forces and the acceleration of the moving surface.

What this shows is that the input signals for a system may change as the formulation is changed. Only for our standard first-order form can we predict that inputs will be forces and velocities. Naturally, all formulations must give identical results if the same inputs are specified and the same outputs are computed. We simply must be careful to use the correct input quantities, and for vibratory systems it is fairly complicated when there are damping forces and motion inputs.

18-4 CONCLUSIONS

We have shown that vibratory-system models are often fairly elementary systems since they are linear in the main. They are special if damping can be neglected, since then natural frequencies and certain shapes of vibration called normal modes can be found rather easily. Damped systems are often treated as if they were undamped, and at least preliminary design modifications can be made on the basis of the natural frequencies. When heavier damping is present, our first-order state equations may be more convenient than second-order formulations, which might require special forms for input variables.

PROBLEMS

18-1 The mass is supported by the rigid massless 2:1 lever by a spring of constant k. Using a bond graph incorporating a transformer, find state equations for the system and compute the natural frequency. Assume small motions about the horizontal position.

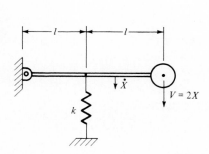

Figure P18-1

$$I_0 = \tfrac{1}{3} m l^2$$
$$\tau_0 = -mg\tfrac{l}{2} \sin \theta$$

Figure P18-2 $I_0 = \tfrac{1}{3} m l^2 \quad \tau_0 = -mg\tfrac{1}{2} \sin \theta$

18-2 The pendulum shown is a thin rod of length l and mass m. Represent the gravity torque τ by means of a C element. Write both nonlinear and linearized state equations and give the natural frequency for small oscillations near $\theta = 0$.

18-3 The bar is supported by two springs and pivoted at one end. Write three state equations, for p, q_1, and q_2. Show that they can be converted to a single second-order equation for the angle θ. What is the natural frequency?

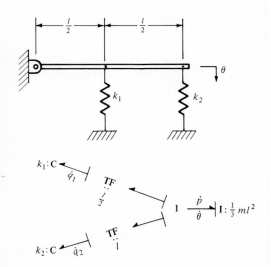

Figure P18-3

18-4 Show that five state equations, for p_1, p_2, q_3, q_4, and q_5, can be written. Combine them into equations for the mass positions X_1, X_2.

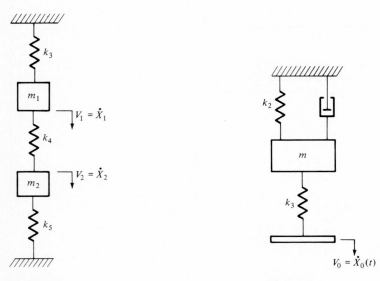

Figure P18-4   Figure P18-5

18-5 Write state equations for p_1, q_2, q_3 and convert them into a single equation for the mass position X_1. What is the *undamped* natural frequency, and what is the input quantity for the X_1 equation?

18-6 At equilibrium, both simple pendulums hang straight down, and the spring is unstretched. Represent the identical pendulums as I's with ml^2 as the inertia parameters and C's with mgl as the torsional spring constant. Show a bond graph and state equations.

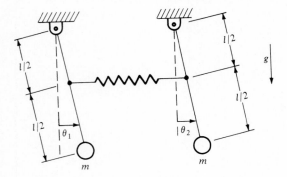

Figure P18-6

18-7 Consider the damped system

$$m\ddot{x} + b\dot{x} + kx = F_0 \sin \omega_f t$$

Show that if $\omega_f = \omega_n = \sqrt{k/m}$, the forced solution can be expressed as $x = A \cos \omega_f t$. Give an expression for the magnitude of the amplitude of motion $|A|$ exactly at $\omega_f = \omega_n$.

INTRODUCTION TO FEEDBACK CONTROL

19-1 INTRODUCTION

Automatic control involves coupling a low-power sensing and information-processing system with a high-power system. In most cases, there is a feedback of signals from instruments to actuators, which adjust the behavior of the energetic part of the system to optimize its performance. As the cost, size, and power requirements of electronic elements have decreased, the opportunities for applying automatic-control techniques have grown dramatically. Today, almost every area of engineering is impacted by automatic control.

Control-system engineering design makes major use of the dynamic-modeling techniques we have developed, because it is not possible to design a useful control system by considering only the steady-state behavior of the system to be controlled. Think of a high-performance aircraft, for example. Many classes of passive systems are known to be very stable; consequently, they approach steady state after a disturbance. But an active control element, intended to change the dynamic performance, can add energy to the controlled system, and an inappropriate control law can so alter the system that it becomes unstable and never reaches a steady state. In fact, since a badly designed or misadjusted controller can even destroy the system, a careful dynamic analysis is necessary before attempting to construct a controller.

The sets of first-order state equations we have been deriving throughout this text are the fundamental form used to describe systems in modern automatic control. State equations can be used to describe systems of any order, and they are convenient when computer-aided design techniques are to be applied. On the other hand, for low-order systems, block diagrams or transfer functions can be used, and these methods of description are commonly employed to introduce the basics of automatic control. As you know, bond graphs are particularly well adapted to the action-reaction finite-power interchanges between energetic components; but an augmented bond graph can be routinely converted into a block diagram or, if linear, into a transfer function, making it possible to represent simple systems with block diagrams or transfer functions rather than state equations if we wish. This will be our approach here.

The control system, consisting of sensors, signal processors, and actuators, is composed of components designed for one-way signal flow; therefore block diagrams and transfer functions are particularly useful means of description for this part of the overall system. Again, as the order of the control system increases, computer-aided analysis becomes a necessity, and equation sets typically are more useful than block diagrams or transfer functions. In this chapter we introduce some of the principles behind feedback control and show how dynamic modeling and analysis relate to control-system design.

19-2 THE FEEDBACK CONCEPT

Most automatic control is concerned with *closed-loop* systems, in which there is a feedback path from the output of the system through some signal-processing elements back to the inputs of the system. In the simplest case, the system's actual output variable is measured and compared with the desired output to form an error signal. This error signal is then used to drive the system so that the error tends to decrease. There are many variations on this scheme, but the feedback of information is central to most of them.

Although it was a great step forward to use feedback *consciously* in devices such as steam-engine governors and thermostats, almost all physical devices can be thought of as feedback systems. For example, the simple *RC* circuit of Fig. 19-1 can

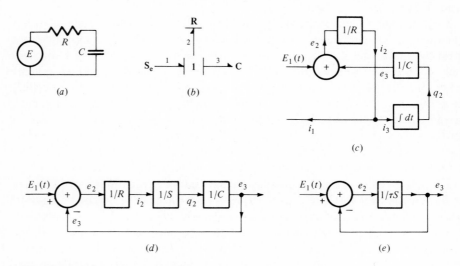

Figure 19-1 Feedback in a passive *RC* system.

readily be put into a feedback form which, at least mathematically, resembles a feedback control system.

The steps in Fig. 19-1a to c should be familiar to you by now. In the steps in Fig. 19-1d and e we simply rearrange the block diagram to look more like a control system with a single feedback loop. Note the emphasis on the output e_3 and the input E_1. We replace the time-integral block with the Laplace-transform equivalent $1/s$, and in Fig. 19-1d we combine the three blocks into one, defining the time constant τ by

$$\tau = RC \qquad (19\text{-}1)$$

The circuit compares the capacitor voltage e_3 with the source voltage $E_1(t)$ and generates by subtraction an error signal e_2. If e_3 is less than E_1, e_2 is greater than 0 and the $1/\tau s$ block will begin to increase e_3. If e_3 is greater than E_1, e_2 is less than 0 and e_3 will begin to decrease. Thus you could say that the circuit is controlling the capacitor voltage to follow the source voltage.

We might ask whether the system is stable (even though we know the answer). One way to find out would be to find the transfer function from E_1 to e_3 and then to check to see whether the characteristic equation has stable eigenvalues. A control engineer might use the identity illustrated in Fig. 19-2 and might think of $G(s)$ as a plant to be controlled and $H(s)$ as a feedback transfer function. The identity is easily proved in the Laplace domain. If $X(s)$ and $Y(s)$ are the Laplace transforms of the signals $x(t)$ and $y(t)$, the feedback block diagram yields

$$Y(s) = G(s)[X(s) - H(s)Y(s)]$$

A little algebra yields the transfer function

$$\frac{Y(s)}{X(s)} = \frac{G(s)}{1 + G(s)H(s)} \qquad (19\text{-}2)$$

as indicated in the figure.

For our example of Fig. 19-1 since $G(s) = 1/\tau s$ and $H(s)$ is just unity, by using Eq. (19-2) we have

$$\frac{e_3(s)}{E_1(s)} = \frac{1/\tau s}{1 + 1/\tau s} = \frac{1}{\tau s + 1} \qquad (19\text{-}3)$$

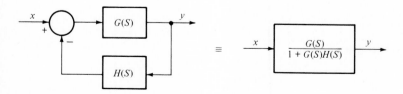

Figure 19-2 Reduction of a block diagram with feedback to a single transfer function.

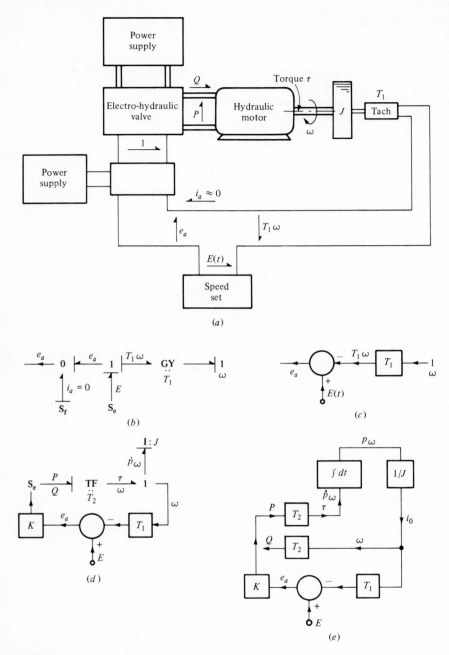

Figure 19-3 Speed-control system.

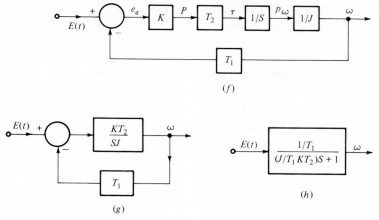

Figure 9-3 (*Continued*)

and we find a single eigenvalue from $s + 1 = 0$, or

$$\tau s = \frac{-1}{\tau} \tag{19-4}$$

so the system is stable. This means that if E_1 is constant and there are no external disturbances to the system, the error signal e_2 will go to zero eventually as e_3 approaches E_1.

Naturally, we could have treated this system in our normal way and come to the same conclusions. We have merely tried to show that energetic systems with their power-paired signal flows *do* exhibit a form of feedback and can be studied with techniques developed for control systems. Now let us construct a control system mathematically similar to our passive *RC* system in which *we* design in the feedback path.

Figure 19-3 shows a physical system intended to provide control of the speed of a rotating shaft. The shaft with its moment of inertia J may be part of a larger system (perhaps J represents the inertia of a papermill roll), but for now we do not consider any extra torques on the shaft from the rest of the system. As you can see from the schematic diagram in Fig. 19-3a, this control system seems quite complex at first glance. Mechanical, electrical, and hydraulic parts are involved, and we could make a detailed bond-graph model of every component. However, the feedback path is by design characterized by low-power one-way signal-transmission paths which allow drastic simplifications in the model. Furthermore, most control design is based on linearized models, a fact which also simplifies the modeling process. However, the ultimate performance limitations of control systems often are due to nonlinear effects. The linearized models provide a starting point for the design, which then is typically tuned up either on the actual device or on a more realistic computer model.

Figure 19-3b shows a section of the sensing and error circuit modeled as if it were a power system. The tachometer is basically a permanent-magnet dc machine with a gyrator character. The inertia and friction of the small tachometer have been neglected. The gyrator voltage $T_1\omega$ is combined with a variable speed-set voltage E to produce e_a, the voltage to the amplifier, in a 1-junction connection. However, the amplifier input impedance is so high that virtually no current flows, as indicated explicitly in Fig. 19-3b. This means that even if we include resistance in the tachometer windings there will be no voltage drop. Also, the back torque from the tachometer vanishes, shown by the active bond.

The result is that only part of the block diagram corresponding to Fig. 19-3b is really necessary for our control study. That part is shown in Fig. 19-3c. Now the gyrator constant T_1 appears as a gain relating angular speed ω to a voltage. This is the sensor part of the control system, and it can be very accurate since it operates at very low power.

The next element, the electric amplifier, needs a power supply, but the amount of power is still small. The amplifier needs only to have enough power to drive the electrohydraulic valve controlling the hydraulic power to the motor. We suppose that the amplifier and valve function to produce a pressure P proportional to e_a (Fig. 19-3d). Actually, the design of the amplifier, valve actuator, and valve is a story in itself. If you look at the curves for the four-way valve in Chap. 8, you can see that even when the spool position is moved proportionally with e_a, the pressure will normally depend upon Q to some extent. We have neglected this effect, and indeed it may be that the valve system has its own control system to make sure that the pressure is really proportional to e_a independent of Q. If so, our assumption must be that the valve response is so fast compared with the response of the whole system that we can neglect any valve dynamics.

The bond-graph–block-diagram mixture of Fig. 19-3d clearly separates the low-power feedback path from the high-power part of the system. But the block-diagram equivalent of the bond graph in Fig. 19-3e shows that we do not have to keep the bilateral signal paths if we do not want to. The hydraulic flow Q has no influence in the system since we have assumed a controlled pressure source. Thus we can eliminate some signals and can arrange the block diagram in the classical single-loop feedback form (Fig. 19-3f). This system is very similar to the one in Fig. 19-1d for the RC circuit. If we continue (Fig. 19-3g and h), we find the transfer function

$$\frac{\omega(s)}{E(s)} = \frac{1/T_1}{(J/T_1KT_2)s + 1} \tag{19-5}$$

The transfer function has a steady-state gain $1/T_1$ (from the tachometer gyrator parameter) and time constant J/T_1KT_2 (composed of the moment of inertia M, tachometer constant T_1, valve voltage to pressure gain K, and motor pressure to torque constant T_2).

If all this makes you think that designing a control system is as easy as soldering up an RC circuit, we'd better discuss this further. The RC circuit is stable for all values of R and C with its eigenvalue at $s = -1/RC$, but the speed-control system may not be stable. It is true that when $T_1\omega$ equals E the pressure and torque vanish, so that p_ω and ω stay constant; but this is true for any gain K whether it is positive or negative. The system eigenvalue is at $s = -T_1KT_2/J$, which is stable only for $K > 0$. Thus, even though the steady-state system behavior is all right no matter what, a crossed wire in the controller would make the system speed diverge. Most control engineers have learned to be very careful when turning on a control system for the first time, since one wrong sign in a feedback path can be very dangerous. A builder of RC filters has no such worries.

The stability of classes of passive systems can be traced to energy storage and dissipation. Control systems have controlled energy sources which sometimes can lead to unstable, rather than stable, system response.

19-3 HIGHER-ORDER SYSTEMS

The first-order system we just discussed showed some of the benefits of feedback automatic control. By means of an input signal E we were able to make the system change its output ω until the measured output was equal to the desired output. The sign of the gain is in principle easy to determine. If the output is too low, make sure the gain multiplied by the error signal e_a results in an increase in the output. For higher-order systems, such intuitive ideas alone will not suffice.

To see why, we extend the simple ideas of our speed-control example to position control. Figure 19-4a shows a different sensor element in place of a tachometer. It

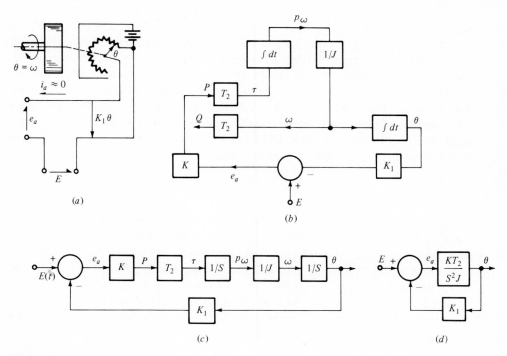

Figure 19-4 Shaft-position controller.

is a rotary potentiometer, which is attached to a constant-voltage supply. The wiper voltage is proportional to the angle θ because, again, almost no current flows into the amplifier. For this to be the case, the resistance of the potentiometer must be low compared with the input impedance of the amplifier. We assume that the potentiometer voltage is $K_1\theta$, where the gain K_1 includes the potentiometer parameters and the voltage supply.

Figure 19-4b is a modified version of Fig. 19-3e. An extra integrator is involved because we feed back a signal proportional to θ instead of one proportional to ω. In Fig. 19-4c and d we simplify and combine the block diagram as before. Using the feedback-reduction scheme of Fig. 19-2, we find the closed-loop transfer function to be

$$\frac{\theta(s)}{E(s)} = \frac{1/K_1}{(J/KK_1T_2)s^2 + 1} \tag{19-6}$$

This system has two eigenvalues, at

$$s = \pm \sqrt{\frac{-KK_1T_2}{J}} \tag{19-7}$$

There is no way to make this into a satisfactory system. If $KK_1 > 0$, we get two roots on the $j\omega$ axis, which means that the system will oscillate forever. If $KK_1 < 0$, we get one positive and one negative eigenvalue and the positive one will result in an unstable, divergent system. Clearly, the idea of simply feeding back a measured signal to form an error signal and using that to drive the system will not always work.

Although there are many solutions to the problems of higher-order systems, a general one is to consider feeding back signals from *all the state variables* in the system to be controlled. We can show how this would work in the example at hand by combining two feedback signals, from ω and θ. In Fig. 19-5 the ω feedback from Fig. 19-3f has been combined with the θ feedback from Fig. 19-4c. Figure 19-5a shows the two signals independently, and Fig. 19-5b shows a single equivalent feedback. The idea behind this equivalent feedback is as follows. First we observe from the block diagram that in the Laplace domain

$$\theta(s) = \frac{1}{s}\omega(s) \quad \text{or} \quad \omega(s) = s\theta(s)$$

Then, instead of feeding the ω signal through the T_1 block, we could imagine feeding the θ signal first through an s block and then a T_1 block. Now two feedback signals stem from θ, one going through an ST_1 block and the other going through a K_1 block, and both are then subtracted from E to get e_a. Figure 19-5b shows a single block of $K_1 + sT_1$, which first combines the two signals and then subtracts the combination so that e_a is determined just as before. The advantage of this single equivalent feedback is that since now the system is in the basic feedback pattern of Fig. 19-2, we can get the complete transfer function quite easily

$$\frac{\theta(s)}{E(s)} = \frac{KT_2/s^2J}{1 + (KT_2/s^2J)(K_1 + sT_1)} = \frac{1/K_1}{(J/KK_1T_2)s^2 + (T_1/K_1)s + 1} \tag{19-8}$$

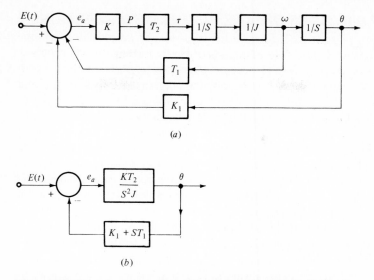

(a)

(b)

Figure 19-5 Example system with position and angular-velocity feedback.

If you compare this transfer function with Eq. (19-6) for angle feedback alone, you will see that although the characteristic equation is still second order, there is now a coefficient for s which was absent previously. In fact, given the parameters J, T_1, and T_2, we can still adjust the s^2 coefficient and the s coefficient independently by choosing K and K_1. This gives us the freedom to pick *any* eigenvalues for the closed-loop system that may seem desirable. We can make the system respond as fast as we like and with whatever damping we like. This is complete control. Simple output feedback usually does not work for high-order systems because there are not enough gains available to adjust coefficients in the closed-loop characteristic equation; as a consequence some eigenvalues may remain in unfavorable locations in the s plane. In the example of Eq. (19-6) the characteristic equation always is missing the s^1 term, so that a truly stable system is not possible.

A general way to start the practical design of a control system is to study *state-variable feedback*. First, imagine that each state variable in an nth-order system is measurable and feed it back to the system input through its own gain. Then if the system can be controlled using the input, the closed-loop eigenvalues can be placed arbitrarily by choosing the feedback gains. The large literature in modern automatic control treats the next series of questions which then arise naturally: Where should the eigenvalues be placed? How can one deduce the values of state variables which cannot be measured from signals which are measurable? What can one do about unknown disturbances on the system? How powerful do the actuators need to be? What if the system's parameters change during its useful life? The list goes on, but you get the idea. For most modern control techniques, the starting point is a set of state equations for the system to be controlled of exactly the sort we have been deriving. As you can imagine, state variables closely related to the physics of the system are particularly useful in implementing a control-system design.

19-4 SOME LIMITATIONS OF LINEARIZED SYSTEM DESIGNS

Although no actual systems are truly linear, control-system design leans heavily on linear models. This is understandable, since only for linear systems is there a generally complete set of available tools for design. However, we must keep in mind that what linear models tell us is possible may be possible only to a limited extent. Physical limitations, or nonlinearities, often provide constraints not obvious in the mathematical model.

For example, if in our speed-control example of Fig. 19-3 J represents the inertia of a high-tonnage set of rollers in a steel rolling mill, you should not expect that by making the gain K large enough you can accelerate the roller with a nanosecond time constant. But this is exactly what the transfer function of Eq. (19-5) would predict. (Where might the limitations come from?)

Similarly, when we said that the coefficients in the denominator of Eq. (19-7) could be set arbitrarily by adjusting gains, this statement should be taken with a grain of salt. The problem is that although linear systems always yield an output proportional to the input, real systems always are limited in some way. When we try to use large gains, the output can follow a linear law only for small input signals. For larger inputs, the output will *saturate* or become some nonlinear function of the output, and all real systems have limitations on the outputs they can create. For our speed-control system, large gains will merely mean that the hydraulic valve will open completely, the pressure will approach the supply pressure, and the torque will approach its maximum until the error signal becomes small. Beyond a certain value, increases in gain will have almost no effect, even though the linear model predicts a steady decline in time constant as the gain is increased. Similarly, in the position-control system of Eq. (19-7) if we attempt to achieve eigenvalues too far from the origin in the s plane, the linearized equations will no longer apply for finite changes in $E(t)$, due to saturation.

If the only effect of nonlinearity were to slow down the response of a controlled system compared with the response predicted by its linearized model, automatic control would not be the challenging field that it is. In fact, it is quite possible for a system to move to a new region of its state space in which the linearized parameters have changed so much that a control system designed for small variations around the original state is no longer even stable. For example, an autopilot for a space shuttle must operate at subsonic, transonic, supersonic, and hypersonic speeds. This is not an easy design problem. Gains appropriate for one regime may not be appropriate for another. All dedicated control engineers occasionally have nightmares in which one of their systems, under extreme conditions, reaches a state where the control system is unstable and destroys the system it is supposed to control.

There is a theory of *adaptive control* which allows control parameters to change as the system to be controlled changes parameters. The most common approach to nonlinear problems is to design a controller based upon a linearized model and then to test the controller either on the real system or on a more complex computer model that includes significant nonlinearities. For large, expensive, and potentially dangerous systems, such as nuclear power plants and chemical refineries, computer simulation is the only reasonable alternative. Perhaps this will help you to see why mathematical modeling of dynamic systems is so important.

19-5 CONCLUSIONS

In this brief introduction to automatic control we have emphasized the crucial role that modeling plays in designing control systems. A faulty model of a system very likely will lead to an unsatisfactory control system. Also, we have shown that it is possible to blur the distinction between an energetic system to be controlled and an information-feedback system which does the controlling. This can lead to a uniform signal-oriented representation of the system. A passive system characterized by bilaterally oriented signal flows can be represented in the classical feedback-control form. Similarly, a block diagram for a controlled system can be shown in such a way that the high-power and the zero-power parts are treated similarly. We suggest, however, that when you study control-system analysis and design, you try to maintain the important distinction between the parts of the system which operate at high power and those which transmit information.

PROBLEMS

19-1 Make a bond graph for this system, convert it into a single-feedback-loop configuration with $V(t)$ as an input and v as an output, and use Fig. 19-2 to find the transfer function $v(s)/V(s)$.

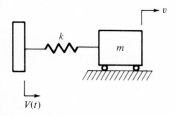

Figure P19-1

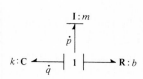

Figure P19-2

19-2 Prove that the mass-spring-damper oscillator is stable when m, b, and k are positive. Use the following steps:

(a) Write the state equations for p and q.

(b) Show that the only equilibrium point in the state space is $p = q = 0$.

(c) Show that the stored energy function $p^2/2m + kq^2/2$ is always positive except at $p = q = 0$, where the energy assumes its minimum value of zero.

(d) Show that the derivative of the energy function is always negative except at equilibrium, so the system will always tend to approach equilibrium. (A generalization of this idea underlies Liapunov stability theory, which can be used for nonlinear control systems.)

19-3 Suppose in Fig. 19-3*d* we consider that the pressure supplied by the valve system drops as the flow increases. This can be modeled by putting a resistance *R* in series with the source (Fig. P19-3). Show how the addition of this element would change the diagrams in Fig. 19-3*e* and *f*. Since this *R* element acts rather like the controller itself, this sort of effect is sometimes called *self-regulation*. In the original system, in the absence of feedback, *P* would cause a continuous increase in ω, but with *R* included the motor pressure drops as ω increases, so the speed self-limits.

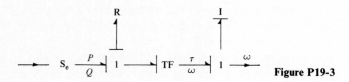

Figure P19-3

19-4 Show that the proportional, i.e., direct-output, feedback of this three-integrator plant is always unstable no matter what values of *K* and K_1 you pick.

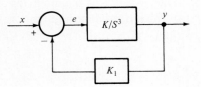

Figure P19-4

19-5 Show that if the pressure-drop resistor of Prob. 19-3 is included in the position-feedback scheme of Fig. 19-4, the system will at least be stable, even without ω feedback as in Fig. 19-5. (This is a benefit of self-regulation.)

19-6 The three-integrator system of Prob. 19-4 can be controlled if each state or integrator output is fed back through its own gain. Show this by reducing the scheme to a single equivalent feedback from *y* using the trick of Fig. 19-5 and then using Fig. 19-2.

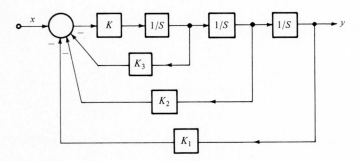

Figure P19-6

19-7 A servomechanism is to be constructed from an electrohydraulic valve, which supplies a pressure proportional to the control voltage. The hydraulic pressure *P* is applied to a ram of area *S*, which moves

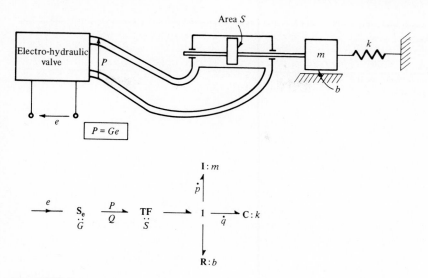

Figure P19-7

a mass m against a load represented by a spring of constant k and a damper of constant b. Find matrices A and B such that the state equations assume the form

$$\begin{bmatrix} \dot{p} \\ \dot{q} \end{bmatrix} = [A]\begin{bmatrix} p \\ q \end{bmatrix} + [B][e]$$

Using matrix methods, show that if $e = -k_q[q - X(t)] - k_p p$, where $X(t)$ is a command position, the eigenvalues of the closed-loop system can be arbitrarily chosen by adjustments of the feedback gains k_q and k_p.

APPENDIX
A

SI UNITS

Table A-1 SI units used in this book

Unit	Abbreviation	Equivalent	Unit	Abbreviation	Equivalent
ampere	A		newton	N	$kg \cdot m/s^2$
coulomb	C	$A \cdot s$	pascal	Pa	N/m^2
farad	F	C/V	radian‡	rad	
henry	H	Wb/m	second	s	
hertz	Hz	cycles/s	tesla	T	Wb/m^2
joule	J	$W \cdot s$, $N \cdot m$, $A \cdot Wb$, $V \cdot C$	volt	V	W/A
Kelvin	K	†	watt	W	$N \cdot m/s$, J/s, $V \cdot A$
kilogram	kg		weber	Wb	$V \cdot s$
meter	m		ohm	Ω	V/A

   † The intervals on the Kelvin and Celsius scales are identical (1 K = 1 Celsius degree), but K is used in writing units.
   ‡ The radian is not an SI unit but is included for completeness.

Table A-2 Some SI prefixes

Multiple	SI prefix	Symbol	Multiple	SI prefix	Symbol
10^{18}	exa	E	10^{-1}	deci	d
10^{15}	peta	P	10^{-2}	centi	c
10^{12}	tera	T	10^{-3}	milli	m
10^{9}	giga	G	10^{-6}	micro	μ
10^{6}	mega	M	10^{-9}	nano	n
10^{3}	kilo	k	10^{-12}	pico	p

MATHEMATICAL BACKGROUND

B-1 DERIVATION OF THE INITIAL- AND FINAL-VALUE THEOREMS

Initial-Value Theorem

Start with the first-derivative relation

$$sX(s) - x(0) = \mathcal{L}\left\{\frac{dx(t)}{dt}\right\} = \int_0^\infty e^{-st}\frac{dx}{dt}\,dt$$

Take the limit of both sides as

$$\lim_{s\to\infty}[sX(s)] - x(0) = \lim_{s\to\infty}\int_0^\infty e^{-st}\frac{dx}{dt}\,dt$$

Interchange the limit and the integration on the right to get

$$\lim_{s\to\infty}[sX(s)] - x(0) = \int_0^\infty \lim_{s\to\infty} e^{-st}\frac{dx}{dt}\,dt$$

or $$\lim_{s\to\infty} sX(s) = x(0) \qquad \text{initial-value theorem}$$

since the integral goes to zero.

Final-Value Theorem

Start with the first-derivative relation

$$sX(s) - x(0) = \int_0^\infty e^{-st} \frac{dx}{dt} \, dt$$

Then take the limit of both sides as

$$\lim_{s \to 0} [sX(s) - x(0)] = \lim_{s \to 0} \int_0^\infty e^{-st} \frac{dx}{dt} \, dt$$

or

$$\lim_{s \to 0} [sX(s)] - x(0) = \int_0^\infty \lim_{s \to 0} e^{-st} \frac{dx}{dt} \, dt$$

This becomes

$$\lim_{s \to 0} [sX(s)] - x(0) = x(t \to \infty) - x(0)$$

or

$$\lim_{s \to 0} sX(s) = x(t \to \infty) \qquad \text{final-value theorem}$$

B-2 EVALUATION OF DETERMINANTS

The most practical procedure for evaluating symbolic determinants (involving s) is to use the expansion rule about a row or column. The procedure is summarized here and illustrated by example.

1. Select a row or column to use as the basis for expansion.
2. Assign to each element in the row or column a *position sign* according to the following rule: the sign is positive if $i + j$ is even and the sign is negative if $i + j$ is odd, where i is the row index and j is the column index.
3. Associate with each element e_{ij} in the basis row or column a subdeterminant D_{ij}, evaluated from the initial array with row i and column j deleted.
4. The determinant D is the sum of $\{(\text{sign}) \, e_{ij} D_{ij}\}$ for the basis row or column.

Example B-1 Evaluate a 3×3 matrix using the expansion procedure

$$A = \begin{bmatrix} a_{11} & a_{12} & a_{13} \\ a_{21} & a_{22} & a_{23} \\ a_{31} & a_{32} & a_{33} \end{bmatrix}$$

$$D = (+)a_{31} \begin{vmatrix} a_{12} & a_{13} \\ a_{22} & a_{23} \end{vmatrix} (-)a_{32} \begin{vmatrix} a_{11} & a_{13} \\ a_{21} & a_{23} \end{vmatrix} (+)a_{33} \begin{vmatrix} a_{11} & a_{12} \\ a_{21} & a_{22} \end{vmatrix}$$

$$= a_{31}(a_{12}a_{23} - a_{22}a_{13}) - a_{32}(a_{11}a_{23} - a_{21}a_{13}) + a_{33}(a_{11}a_{22} - a_{12}a_{21})$$

$$= a_{12}a_{23}a_{31} + a_{32}a_{21}a_{13} + a_{33}a_{11}a_{22} - a_{22}a_{13}a_{31} - a_{11}a_{32}a_{23} - a_{33}a_{12}a_{21}$$

This is a well-known result which is worth memorizing. It involves the combination of all elements, one from each row and column, with signs adjusted for relative position. Notice that the expansion procedure about any row or column yields the same end result.

Example B-2

$$A = \begin{bmatrix} 2 & 3 & 0 & 4 & 0 \\ 0 & 6 & 2 & 0 & 1 \\ 0 & 0 & 4 & 1 & -2 \\ 4 & -1 & 7 & 0 & 1 \\ 0 & 1 & 0 & -1 & 0 \end{bmatrix}$$

Since column 1 and row 5 each have three zero elements, we can easily use either one as a basis. Let us use row 5. Then

$$D = (-)1 \begin{vmatrix} 2 & 0 & 4 & 0 \\ 0 & 2 & 0 & 1 \\ 0 & 4 & 1 & -2 \\ 4 & 7 & 0 & 1 \end{vmatrix} \quad (-)-1 \begin{vmatrix} 2 & 3 & 0 & 0 \\ 0 & 6 & 2 & 1 \\ 0 & 0 & 4 & -2 \\ 4 & -1 & 7 & 1 \end{vmatrix}$$

Now we can expand each of the 4×4 arrays one step further, using column 1 in each case

$$D = (-1) \left\{ 2 \begin{vmatrix} 2 & 0 & 1 \\ 4 & 1 & -2 \\ 7 & 0 & 1 \end{vmatrix} \quad (-)4 \begin{vmatrix} 0 & 4 & 0 \\ 2 & 0 & 1 \\ 4 & 1 & -2 \end{vmatrix} \right\}$$

$$(+1) \left\{ 2 \begin{vmatrix} 6 & 2 & 1 \\ 0 & 4 & -2 \\ -1 & 7 & 1 \end{vmatrix} \quad (-)4 \begin{vmatrix} 3 & 0 & 0 \\ 6 & 2 & 1 \\ 0 & 4 & -2 \end{vmatrix} \right\}$$

Finally,

$$D = (-1)\{2(2-7) - 4(16+16)\}$$
$$(+1)\{2(24+4+4+84) - 4(-12-12)\}$$

or

$$D = (-1)(-10-128) + 1(232+96) = 466$$

Example B-3

$$M = \begin{vmatrix} s & 0 & 2 & 0 \\ 1 & s+1 & 0 & - \\ -1 & 0 & s & 1 \\ 0 & 0 & -2 & s+2 \end{vmatrix}$$

Expand about row 1.

$$D = (+)s \begin{vmatrix} s+1 & 0 & 4 \\ 0 & s & 1 \\ 0 & -2 & s+2 \end{vmatrix} \quad (+)2 \begin{vmatrix} 1 & s+1 & -4 \\ -1 & 0 & 1 \\ 0 & 0 & s+2 \end{vmatrix}$$

$$= s\{(s + 1)[s(s + 2) + 2]\} + 2\{-(s + 1)[-1(s + 2)]\}$$
$$= s(s^3 + 3s^2 + 4s + 2) + 2(s^2 + 3s + 2)$$
$$= s^4 + 3s^3 + 6s^2 + 8s + 4$$

B-3 SUMMARY OF MATRIX TERMS

Assume a matrix A having m rows and n columns. The element a_{ij} occupies the ith row and jth column position.

The Identity Matrix

The identity matrix I is a square matrix with the following elements:

$$[I_{ij}] = \begin{cases} 1 & i = j \\ 0 & i \neq j \end{cases}$$

Transpose

The transpose of A, denoted A^T, is defined by $[a_{ij}]^T = [a_{ji}]$.

Adjoint

The adjoint of square matrix A, denoted Adj A, can be found by the following sequence of operations on A:

1. Find the minor of A, denoted M. $[m_{ij}]$ is found by evaluating det A with row i and column j deleted.
2. Find the cofactor of A, denoted C. $[c_{ij}]$ is found from

$$[c_{ij}] = (-1)^{i+j}[m_{ij}]$$

3. Find the adjoint from the cofactor transposed

$$\text{Adj } A = C^T$$

Inverse

The inverse of square matrix A, denoted A^{-1}, is given by

$$A^{-1} = \frac{\text{Adj } A}{\det A}$$

where det A denotes the determinant of A. A matrix whose determinant is zero is said to be *singular;* its inverse does not exist.

Skew-Symmetry

A square matrix A is skew-symmetric if $[a_{ij}] = -[a_{ji}]$.

Susan Allen Ackman